Introducing
FOOD
SCIENCE

Introducing

FOOD

SCIENCE

ROBERT L. SHEWFELT

CRC Press
Taylor & Francis Group
Boca Raton   London   New York

CRC Press is an imprint of the
Taylor & Francis Group, an **informa** business

Photo of students with M&M'S® taken by Sharon Dowdy. Photo of decayed orange provided by Dr. Larry R. Beuchat.

CRC Press
Taylor & Francis Group
6000 Broken Sound Parkway NW, Suite 300
Boca Raton, FL 33487-2742

© 2009 by Taylor & Francis Group, LLC
CRC Press is an imprint of Taylor & Francis Group, an Informa business

No claim to original U.S. Government works
Printed in Great Britain
10 9 8 7 6 5 4 3 2 1

International Standard Book Number-13: 978-1-58716-028-8 (Hardcover)

Library of Congress Cataloging-in-Publication Data

Shewfelt, Robert L.
   Introducing food science / Robert L. Shewfelt.
      p. cm.
   Includes bibliographical references and index.
   ISBN 978-1-58716-028-8 (hardcover : alk. paper)
   1. Food industry and trade. 2. Food--Microbiology. 3. Food--Quality control. 4. Food handling. 5. Nutrition. I. Title.

TP370.S424 2009
664--dc22   2008054116

Visit the Taylor & Francis Web site at
http://www.taylorandfrancis.com

and the CRC Press Web site at
http://www.crcpress.com

Dedication

This book is dedicated to my parents, the two people who introduced me to food science.

In memory of my father, Lorne, a food scientist who encouraged me to apply for an IFT (Institute of Food Technologists) scholarship, recruited me into the major, served as my academic advisor during my freshman and sophomore years, and serves as a source of inspiration today.

In memory of my mother, Doris, a home economist who taught me my basic food groups, fixed many nutritious meals for me, baked the best pies in the world, and pointed out the foods that were good for me on my weekly visits with her.

Contents

Section II: Commercial food products

Section III: Functions of food scientists

Preface

This book is designed to provide an overview of the field of food science for the major and nonmajor alike. It proceeds from a nontechnical discussion of food issues that concern today's student to an in-depth technical overview of the basic principles as they relate to the study of food. Sandwiched in between are descriptions of the types of commercial products and processes, with particular reference to nutritional consequences and primary functions and roles of food scientists. Each section builds on the previous section, providing for greater technical depth.

Food science is unique among scientific disciplines because of its heavy commercial orientation. Food science serves as the technical arm of the food industry, where most food science graduates are employed. Although academicians acknowledge the importance of the commercial aspects of the field, they are also dedicated to being sources of unbiased knowledge for consumers and regulatory agencies. This book attempts to convey both commercial and scientific perspectives of the field in the interest of providing a true flavor of food science.

Food issues in the news comprise the first section of this book. From disease outbreaks to health benefits and detriments to international trade, foods and beverages provide a steady source of stories in every type of news medium. As consumers of foods and beverages daily, everybody has interests, concerns, opinions, and questions about what they eat. Unfortunately, most issues in the news pose nontechnical questions that defy technical answers. The purpose of this section is to enter the student's world and reframe food issues presented from a journalistic viewpoint into a more scientific perspective. Stories are selected as illustrations for Chapters 1, 2, and 3—"Food safety," "Healthiness of foods," and "Choosing the food we eat." While widely held misconceptions are presented and corrected, the text avoids pat answers. Rather, emphasis is placed on separating what we know about such issues from what we do not know, and how they can be reformulated into testable hypotheses. For greater depth on specific issues, the student is referred to specific passages later in the book.

The second section introduces students to commercial food products. Chapters 4 through 6—"Processed foods," "Formulated foods," and "Chilled and prepared foods"—provide the basic principles of food manufacture and food preservation with a strong emphasis on unit operations. Both traditional (canning, freezing, fermenting, and drying) and more modern (aseptic, irradiation, and high pressure) processes are presented. The importance of individual ingredients and how they are displayed on food labels are vehicles used to introduce formulated foods, a major component of the student diet. Food service operations, distribution systems, and packaging technology highlight the discussion on chilled and prepared foods. Material in these three chapters is sprinkled with flowcharts, diagrams, ingredient statements (can you guess what this product is?) and pictures to illustrate the major points. The importance of food preservation and sanitation in preventing food-associated outbreaks raised in Chapter 1 provides a major theme.

The profession of food science provides the basis for Section III. Career opportunities, necessary background, and professional perspective permeate these chapters. Chapter 7, "Quality assurance," introduces the topic of quality (nutritional, microbial, and sensory), its measurement, process control, HACCP (Hazard Analysis and Critical Control Point), shelf life, and commercial sanitation. Chapter 8, "Product and process development," describes the creative process and what is required to provide the wealth of options available to the modern consumer with sufficient quality and shelf life. Chapter 9, "Government regulation and basic research," presents the importance of food regulations and a fundamental understanding of foods to ensure a healthy and safe food supply. Knowledge derived in Section II provides the basis for the topics covered in Section III.

The book culminates with an introduction to the basic principles of the primary subdisciplines within food science. This section represents the in-depth technical detail needed to answer questions posed earlier in the book. A chemical basis for safety, nutrition, preservation, ingredient functionality, and quality is provided in Chapter 10. The basic principles of nutrition and the contribution of foods to our health are presented in Chapter 11. An understanding of the factors that affect growth and inhibition of fermentative, spoilage, and pathogenic microorganisms is presented in Chapter 12. Critical engineering concepts needed to design and monitor food processes are emphasized in Chapter 13. Fundamentals of sensory perception of foods and systematic means of measuring sensory attributes provide the basis of the final chapter. The section is designed to whet the student's appetite for a deeper taste of food science in more detailed books and courses.

While no single volume will satisfy needs for all introductory courses, this book offers a range of material that should be useful for many of those courses. Introductory courses for nonmajors would focus on the first three sections and use the last section primarily as a reference source for scientific principles. In contrast, courses for food science majors would emphasize the final three sections and use the initial section, Food Issues in the News, as a basis for outside readings and assignments. Courses that include both majors and nonmajors would require more careful tailoring of course content to meet the specific needs of the students.

Acknowledgments

This book did not write itself, and it is hard to recognize everyone who contributed to the effort.

First I would like to thank Eleanor Riemer, who approached me to write this book and became my first editor, and Stephen Zollo, who pressured me to finish the manuscript. I appreciate the help of Carlos Margaria, Anne Morrison, Karen Simmons, and Kathryn Acosta with many things, particularly the illustrations, and Therese Bartlett, who helped me adjust my writing style. Also, I am grateful to all of my students who have critiqued early versions of the chapters and have taught me as much about food and culture as I have taught them about food science.

Finally, I must acknowledge all of the support I have had from Betty and Katie, who gave up precious time together with me so that I could complete my task.

About the Author

Dr. Robert Shewfelt is the Josiah Meigs Distinguished Teaching Professor of Food Science and Technology at the University of Georgia. He advises more than 50 students and has taught 11 different courses in the past 2 years, ranging from freshman seminars in chocolate science and coffee technology to graduate-level courses in flavor chemistry and evaluation, and food research and the scientific method.

He started his career in Georgia at the University of Georgia Griffin campus as part of an interdisciplinary research team on postharvest handling of fresh fruits and vegetables. The team developed a unique management style that took advantage of situational leadership, and its efforts have changed the way postharvest technologists here and abroad view fresh fruits and vegetables in the handling and distribution system. The team was awarded the USDA (U.S. Department of Agriculture) Superior Service Award for Group Research in 1988. In 1996, Dr. Shewfelt relocated to the main campus in Athens to teach and become the undergraduate coordinator of the food science program.

Known for his extensive shirt collection with food-related themes and an unconventional style in the classroom, Dr. Shewfelt has been recognized with teaching and advising awards at the departmental, college, university, and national levels. He teaches courses in food processing, where the laboratory is run as a virtual food company and the class selects, develops, manufactures, and tests at least four food products. *Introducing Food Science* has been designed as a textbook for his introductory course, Food Issues and Choices.

Dr. Shewfelt received his B.S. from Clemson University, M.S. from the University of Florida, and Ph.D. from the University of Massachusetts.

section one

Food issues in the news

chapter one

Food safety

It is 3:00 a.m. Amy's intimate companion for the last ninety minutes has been her toilet. She has been intermittently spewing from both ends and feels horrible. The intense stomach cramps have just let up, but she is so nauseous she's afraid to venture out of the bathroom. Her condition did not give much warning. What has she done to deserve this? She hasn't eaten for hours, and she only had two glasses of wine! She tries to think what she ate that made her so sick! It couldn't have been the grilled chicken salad. Salads are healthy. She'd baked up fresh cookies for dessert, but not as many as she had planned. She and her roommates had consumed about half of the raw dough that afternoon. It was tasty. Oh no, here comes another vicious cramp!

Brian and Carlos were on their way back to school from spring break. They had decided to spend the last night in Daytona and leave early the next morning. They were out by 9:30 a.m. and tooling down the road trying to make good time. By late afternoon, both were starving and decided to stop at an all-you-can-eat place. Although there weren't many people in line, this was an awesome food bar. Although some of the food was cold, it was all great. Brian filled his plate four times, Carlos three. They left full and satisfied, ready to make the next ten hours. About 150 miles down the road, Brian started getting kind of queasy. He pulled off the road and got out and clear of the SUV before puking his guts out. As soon as Brian got back in the vehicle, Carlos needed a restroom, quick. Both were very sick, but they finally arrived back on campus many stops later.

Dana was in the hospital on a kidney dialysis machine in a coma. It wasn't clear whether she would live. The doctor tried to be reassuring as he talked with her mom. Her dad was out of state and wouldn't be able to get there anytime soon. The doctor wanted to know if she'd eaten hamburger recently. Hardly! Dana is a strict vegetarian. She had been one for years. The only burgers that touched her lips were made from soy. Except for some problems with anemia, she had been in good health before all of this happened. She loved whole wheat bread, veggies, fresh fruits, and sprouts. "Sprouts? What kind of sprouts?" asked the doctor.

Dana, Carlos, Brian, and Amy represent an estimated 76 million Americans who come down with food poisoning each year. This figure seems unreasonably high for a country that is supposedly so wealthy and sophisticated. What is wrong with the American food supply? Why isn't the government doing anything to prevent this problem? Why do some foods go bad so quickly and others stay safe forever? What can we do to lower our risk? This chapter seeks to answer these and other questions about the safety of the food supply. Some answers are easy. Others are clear, but may be unexpected. Still others require either a deeper understanding of science than a simple answer can provide, or are the basis of ongoing research at universities and governmental laboratories. Food science is at the forefront of research and education programs on the safety of foods.

Food in the news

Newspapers, magazines, television, and radio are filled with news stories about food-associated outbreaks. On the one hand we hear that the American food supply is the safest ever. On the other hand it seems like more people are getting sick every year. Somebody must be hiding something. Salmonella, E. coli, botulism, and listeria have become household words. We know they make us sick, but it can be hard to establish how serious the risks are and how likely we are to get sick. Our response to the media bombardment varies from being afraid of everything we eat to ignoring any report on food.

While there is a great deal of information on food-associated illness available, not all of it is accurate or useful. It is often difficult to separate fact from fiction. Certain widely held beliefs about food poisoning are completely wrong. Other simple guidelines that could dramatically reduce food-associated illness are widely ignored. Consumers point fingers at the food industry and governmental agencies that point fingers back at the consumer. Blaming somebody else may make the pointer feel good, but it doesn't help solve food safety problems.

The news media are very good at raising awareness, but they are not as effective at providing solutions. It is easy to blame the media for this situation, but that ignores simple problems in transmitting information. Primary limitations of the news media in communicating food issues are:

- It is easier to grab attention and present the problem than it is to convey factual information and provide the solution.
- Up-close-and-personal stories are more likely to grab attention than more general ones.
- Obscure and unusual hazards tend to become amplified while more common and widespread ones tend to be diminished.
- Journalists are not typically trained in science while scientists are not good at communicating in simple language.

Many food issues are complex and require an understanding of science. Journalists looking for a killer story become frustrated with scientists, who use big words, provide long answers to simple questions, and are reluctant to identify a simple cause and solution. Scientists become frustrated with the quick deadlines, the desire to oversimplify, and the lack of appreciation for basic scientific principles on the part of their interviewer. Part of the problem is that many of the questions we ask are not answerable by science, while many of the answers science can provide do not fit into a sound bite a journalist can communicate. See Insert 1.1 for some celebrities who have been in the news in the past few years.

The remainder of the chapter will introduce us to

- how food makes us sick,
- some common misconceptions about the safety of our foods,
- some of the causes for the apparent rise in food-associated illnesses in the United States in recent years,
- how food scientists work to decrease our chances of getting sick from the food we eat, and
- some simple rules we can follow to reduce the risks associated with our foods.

This chapter provides a food scientist's perspective on the subject. More scientific details will follow in Chapters 4, 7, and 12. For a journalistic perspective, read *Spoiled* (1997) by Nicols Fox. Or if you prefer a novel written by someone trained as a medical doctor, try *Toxin* (1998) by Robin Cook.

Unsafe foods

Simply put, an unsafe food is one that makes us sick when we eat it. There are at least five ways a food can make us sick: (1) presence of harmful

Insert 1.1 These celebrities have something in common. Do you know what it is? The answer is at the end of the chapter.

Muse
Mariah Carey
Allen Iverson
John McCain
Jane Seymour
Roger Federer
Kelly Osbourne
Elizabeth Hurley
Black-Eyed Peas
My Chemical Romance

microbes (including viruses and parasites), (2) presence of natural toxins, (3) presence of environmental contaminants (such as pesticides), (4) presence of harmful additives (such as preservatives), or (5) presence of an allergen. Food scientists are much more concerned about illness caused by microbes than the other sources, as it is estimated that more than 95% of food-associated illnesses are due to microbes. In addition, food scientists believe natural toxins present a greater risk to consumers than either pesticides or preservatives.

There is a general belief that foods are inherently safe in their natural state and that they become unsafe when exposed to technology. Food scientists believe that many natural foods, like those derived from animals, are inherently unsafe, and that a scientific understanding of what causes safety problems leads to development of technology that provides safer foods. An important goal of this book is providing a better understanding of our foods and what we can do to keep foods safe.

Microbial hazards

It comes as a surprise to most consumers to learn that fresh foods are more likely to contain harmful microbes than processed products (see Insert 1.2. to rank the safety of several products). The primary sources of harmful microbes in foods can be traced back to feces (human or animal waste). In fact, when food microbiologists start to look for a microbe of interest they frequently collect soil or feces samples to isolate their microbe of choice. The reason why we wash raw foods is to remove soil or feces from their surface. Unfortunately, even if we remove all visible evidence of the soil or feces, the associated microbes may cling to the surface of the food.

Insert 1.2 Rank the dangers. Listed below are ten different foods. Rank them in order from most likely to cause food-associated illness (1) to least likely (10). Answers appear at the end of the chapter.

____	alfalfa sprouts
____	rare hamburger
____	chocolate éclair
____	canned tuna fish
____	pasteurized milk
____	raw cookie dough
____	refrigerated yogurt
____	salad from a food bar
____	undercooked chicken
____	unpasteurized orange juice

Fresh fruits, vegetables, and grains may become contaminated from the soil; from insect, rodent or bird droppings; or from organic fertilizers.

As disgusting as it may seem, when we eat meat we are eating the muscle tissue of formerly live animals. Theoretically, the inside of the muscle is sterile and does not pose a threat to human health. Unfortunately, all these animals produce feces as well as muscle. During livestock production and during the slaughter process, the surface of meat is likely to become contaminated with feces. Some advocates for safer foods believe it is careless handling of animal carcasses in fresh meat and poultry packing plants that causes contamination with visible feces, but this is not the case. It is doubtful that we could provide sterile steaks, pork chops, or chicken thighs (or even less-contaminated ones than commercial packing operations) if we slaughtered our own steers, hogs, or chickens and then dressed and cut up the meat. See Insert 1.3 for some common microbes responsible for food-associated illness.

Food products are processed primarily to either partially or completely kill microbes or at least slow their growth in or on foods. Proper handling and storage of perishable products, such as fresh fruits and vegetables and raw meats, slows the growth of microbes. The most important weapons used to control food-associated illnesses are proper sanitation and storage during handling and distribution from the farm all the way to the consumer. Proper sanitation involves doing everything to reduce contamination at each step along the way. Proper storage involves maintaining conditions that lower the chance of a food becoming unsafe. Packaging of foods helps prevent microbes from touching the food, but even packaged foods can become dangerous if mishandled. Do you know why food scientists wipe off the top of their soda cans before popping off the top, or why they rarely eat unpackaged mints available at the cash register of steak restaurants? (Answers are at the end of the chapter.)

Insert 1.3 Disease incidence. Rank the following microorganisms causing food diseases with respect to the percentage of hospitalizations and percentage of deaths due to food poisoning in the United States (1 = most, 8 = least). Answers can be found at the end of the chapter.

_____	*C. botulinum* (botulism)
_____	*Campylobacter* (campylobacteriosis)
_____	*E. coli* O157:H7
_____	*Listeria* (listeriosis)
_____	Noroviruses
_____	*Salmonella* (salmonellosis)
_____	*Staphylococcus* (staphylococcal intoxication)
_____	*Toxoplasma gondii* (toxoplasmosis)

Insert 1.4 Spoiled bananas. (Photo provided by Dr. Larry R. Beuchat, Center for Food Safety, University of Georgia.)

Spoiled: When good food goes bad

Another misconception most consumers have about food is that it is easy to tell if a food is safe by looking at it, smelling it, or taking a small bite. If it looks, smells, or tastes nasty, then it must be dangerous, right? Not necessarily! When we say a food has gone bad, we are probably talking about spoilage (see picture in Insert 1.4). Spoiled food is no longer pleasant to eat. While it is never a good idea to eat spoiled food, a spoiled food is not necessarily an unsafe food. Even more important, an unsafe food is not necessarily spoiled. If all unsafe foods were spoiled, there would be much less food poisoning. If we remember the examples of Carlos, Brian, and Amy at the beginning of the chapter, none of them reported anything tasting strange that led to their illness. Dana was in a coma and unavailable for comment.

What spoiled and unsafe foods have in common is that microbes probably caused the problem. The difference is that the types of microbes that cause spoilage are not usually the same ones that make foods unsafe. By now it should be clear that spoilage is not a good indicator of a safety risk because many of the microbes that make food unsafe do not change the color, taste, or smell of the food. However, the handling and storage conditions that lead to growth of spoilage microbes tend to be similar to those that lead to the growth of microbes that make the food unsafe.

Food poisoning

Before discussing food poisoning, we need to know what it is and what it is not. Food poisoning is a sickness caused by consuming a contaminated food. There are many possible causes, but the most likely cause is microbial. When we talk about food poisoning, we are not talking about the stomach turning we might get when we smell spoiled food. We are talking about symptoms that send us to bed or the bathroom for about twenty-four hours or to something that can lead to a coma or death. Most cases result in severe gastric distress with accompanying diarrhea or vomiting or both. What people refer to as a stomach virus is usually the result of food poisoning.

Microbial food poisoning occurs as an infection or intoxication. An infection happens when a microbe is present in a food or beverage, we eat the contaminated item, and the microbe grows in us like a typical infectious disease. An intoxication happens when the microbe grows in the food or beverage and produces a toxin. After the toxin has been produced, the item might be heated to the point where the microbe is killed, but the toxin is still present. We then consume it and the toxin makes us sick.

There are many things about food poisoning that most people do not know. For example, the last meal consumed is not usually the meal responsible for the outbreak. When people get sick and suspect food poisoning, they blame the last thing they ate. One type of food poisoning, a staph intoxication, can happen within one to six hours after the offending food is eaten. A staph intoxication is a likely diagnosis for the illness Brian and Carlos experienced on the way back from spring break at the beginning of the chapter. Most other food poisoning outbreaks take at least twelve hours after the offending food is eaten to cause distress, like those encountered by Amy (probably due to Salmonella) or Dana (probably E. coli). Some outbreaks take as much as a week to develop. At that point it can be hard to figure out the cause. Symptoms of common food-associated illnesses are shown in Insert 1.5.

Tracking down the culprits

Scientists who try to track down the cause of an outbreak are called *epidemiologists*. Epidemiologists go to the scene of an outbreak and compile a list of those who became sick to get a clue as to the responsible organism. They also list foods and beverages consumed by those who became sick and those closely associated who did not get sick to narrow down the number of suspected foods. Some patients may have more severe symptoms than others based on the amount of food consumed, different levels of contamination in different parts of the food, and personal sensitivities to the microbe. Food handling and preparation practices are studied to identify potential causes while potential offending foods are collected and tested. Samples from patients' feces are also collected and tested to see if microbes in the food match microbes found in the stool. In addition, the distribution patterns of the offending foods are traced back to their sources and studied to find the main point of contamination. Most

Insert 1.5 Symptoms of food-associated illnesses. Match the symptoms and the time they take to develop with the responsible microbe. Answers can be found at the end of the chapter.

Microbes:		Symptoms (times for development in parentheses):	
Campylobacter jejuni	1	A	Slight fever, stomach cramps, diarrhea, abortion, meningitis, etc. (1 day–many weeks)
Clostridium botulinum (botulism)	2	B	Vomiting, watery turning to bloody diarrhea, kidney failure (3–9 days)
Escherichia coli O157:H7	3	C	Nausea, vomiting, stomach cramps, intense diarrhea
Listeria monocytogenes	4	D	Chills, fever, nausea, vomiting, stomach cramps, diarrhea (24–36 hours)
Salmonella enterididis	5	E	Blurred vision, breathing and swallowing difficulties, dry mouth, respiratory failure (12–36 hours)
Staphylococcal enterotoxin	6	F	Sweating, chills, nausea, vomiting, stomach cramps, diarrhea (1–6 hours)

outbreaks are caused by a combination of several food handling errors, permitting contamination and growth of one of a small number of harmful microbes. The longer the time lapse between food consumption and illness, the more difficult it is to find the cause. A picture of a notorious microbe is shown in Insert 1.6.

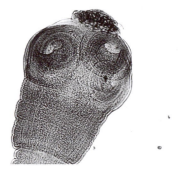

Insert 1.6 Scolex of the pork tapeworm cestode *Tenia solium*. (Photo provided by Dr. Ynes Ortega, Center for Food Safety, University of Georgia.)

Expiration dates

Many packaged foods, particularly the more perishable ones, come with an expiration date. The expiration date represents the food scientist's best guess on how long a food will last before it spoils. A food scientist calculates the expiration date by first determining the product's shelf life (how long it will last under typical storage). The expiration date is usually set before the end of shelf life, but it is not an exact day. Too many things can affect spoilage of food products. The food scientist does not want to cut the day too close as consumers get very upset when they find something has spoiled before the date on the package. On the other hand, if the date is set long before it actually spoils then consumers are unlikely to buy it, and perfectly good food will be thrown out before it has spoiled, leading to increased costs for the food company and increased prices for the consumer. Expiration dates are useless if products are not stored at the right temperature. For example, no one should expect fresh milk to last until the expiration date if it is kept at room temperature rather than in the refrigerator. Since expiration dates are mostly related to spoilage, they are not much better indicators of safety than smelling a food. It is still not a good idea to eat a food that is well past its expiration date, but it is important to understand that expiration dates are not guarantees of safety.

Food preservation

Fresh foods are more likely to contain harmful microbes than commercially processed products. Food preservation involves reducing the chances that food will spoil or be unsafe. One way to preserve food is to kill microbes by heating the food or beverage. Canned foods are essentially sterile. *Canning* refers to the process and not necessarily the package. Many products in plastic or glass jars have also been canned and are thus essentially sterile. Some products in cans have not been heated at all. Other methods of heating that lower the number of microbes include cooking (in either a microwave or regular oven) and pasteurization. Pasteurization kills harmful microbes but not spoilage microbes. Another way of preserving foods involves slowing or preventing the growth of microbes. A refrigerator slows the growth of microbes in highly perishable foods. Freezing and drying are much more effective in slowing microbial growth, but it is important to realize that microbes are still present. Once we thaw a frozen food or add water to a dried food they are just as likely to spoil or become unsafe as any other unprotected food. Even canned food can become contaminated after it has been opened. Another method of food preservation actually encourages the growth of microbes. In fermented products (yogurt, alcoholic beverages, most pickles, some sausages, and sauerkraut) beneficial microbes are added to produce that particular product, and they grow so fast that spoilage and harmful microbes don't have a chance to grow.

Processed foods are handled, prepared, treated, and packaged under carefully controlled conditions. These processes must be approved by the appropriate governmental agency. An important safeguard in approving these processes is that a processed food should be designed to spoil before it becomes unsafe, to make it less likely that someone will eat the unsafe food and become sick. An important part of any food is its package. The most important purpose of any package is to protect the food from contamination by microbes. Once a food has been removed from the container it can become contaminated quickly by kitchen utensils, the air, or by touching contaminated food or water. Improper storage can cause harmful microbes to grow leading to spoilage or unsafe foods.

Preservatives

The term *preservative* is considered a dirty word by many consumers. Many look for the words "no preservatives" on food labels. Actually, preservatives are just food ingredients that slow spoilage and prevent food-associated illnesses. They achieve this effect by preventing or slowing the growth of microbes. The two most common preservatives are sugar and salt. Preservatives are food additives that extend shelf life and lower costs, which can lead to less waste, more profit for the food company, and lower prices for the consumer. Food additives must not be added to conceal diseased, decayed, or other defective products. These additives must serve a useful purpose and must be effective for the intended use.

One reason consumers object to food additives is that they are chemicals. All the components in our foods and in our bodies are chemicals, including water, sugar, salt, vitamins, minerals, proteins, and many, many more. Food scientists are not concerned about the actual presence of chemicals in our food, but they are concerned about reducing the ones that are harmful and increasing the ones that are beneficial. Food scientists believe that consumers are overly concerned about the safety of preservatives. Food additives are required by law to undergo rigorous safety testing and be effective for their intended use.

Safety of the American food supply

If we know so much about food poisoning, then why do we have so much of it? In fact it seems like food is getting more dangerous, not safer. The American food supply has undergone revolutionary changes in the last thirty years. Several possible reasons for this situation have been advanced, but the truth is we really don't know.

Some of the possible reasons for more food-associated illness today than thirty to forty years ago include

- new and newly recognized harmful microbes have come into our food supply;
- harmless microbes have mutated into harmful ones;
- greater consumption of imported foods such as fruits, vegetables, and seafood, from countries in which good hygienic practices are not used;
- consumption of more raw and undercooked foods;
- factory farms lead to concentration of microbes in raw foods;
- less knowledge by food preparers on proper food handling techniques;
- more centralization of food processing and distribution;
- increased use of antibiotics in animals leading to greater resistance of microbes to antibiotics;
- our food supply has fewer microbes so we are less likely to build up resistance to them when we are young;
- modern medicine is prolonging life and making individuals more susceptible to harmful microbes in food through treatments that weaken their immunity; and
- more people are exposed to other diseases (such as AIDS), which compromise the body's ability to resist harmful microbes.

Many items on this list have led to food poisoning outbreaks that were unheard of just a few years back. For some items it is very difficult to prove the cause. Advanced knowledge has led to more effective control of microbes in food processing plants and restaurants. However, greater centralization has led to more people becoming sick when a mistake occurs.

Many food scientists do not believe that there has been a great increase in cases of food poisoning. They believe that we are just becoming more aware of the dangers that lurk in our foods. Epidemiologists have become more effective in identifying the cause of food-associated illnesses that might have been blamed on some other cause. The long times between eating the food and the start of the illness are not as big a problem in identifying the cause as they once were. Physicians have also become better at diagnosing food-associated illnesses. Research in food microbiology has lead to the discovery of microbes that we did not know were harmful. These microbes may have been causing illnesses for years, but we did not know it. One very important contributing factor to detection of modern-day, food-associated disease outbreaks that would have previously gone undetected is a national and international network (PulseNet) that collects molecular subtyping results of human isolates of food-associated pathogens. Finally, sophisticated news-gathering organizations have the ability to tell us about small outbreaks in faraway places that previously would only be known in the local town or city in which they occurred.

Safety in the home

If we are going to decrease the number of cases of food poisoning, then everyone who interacts with the food in any way from the farm to the table must take responsibility for the safety of his or her food. Food scientists believe that one of the least controlled steps from farm to table is in the home or wherever consumers handle and prepare the food after they buy it. This belief goes back to the idea that foods are inherently unsafe in their natural state and that a scientific understanding of what causes safety problems leads to development of technology that provides safer foods. Food companies that produce the packaged foods we buy and restaurants that prepare the meals we eat have access to this science and technology, but many consumers do not.

When food inspectors went into homes using the same forms they use in restaurants, they found that more than 99% of home kitchens failed! See what the inspectors were looking for in Insert 1.7. There are some simple rules we can follow in our apartments, dorm rooms, or homes that can reduce our chances of getting sick from the food we eat. Our efforts

Insert 1.7 Would the health department shut you down? If the area in which you prepare your meals was investigated by your local health department it would fail you if you have one critical or four major violations. (Adapted from R. W. Daniels, "Home Food Safety," *Food Technology* 52, no. 2 (1998): 54.)

Critical violation examples:
- ☐ Cross-contamination of foods that will not be cooked
- ☐ Failure to wash hands
- ☐ Failure to keep hot foods hot
- ☐ Failure to properly cool leftovers
- ☐ Use of swollen or rusty cans
- ☐ Preparation of foods by someone who is sick

Major violation examples:
- ☐ Evidence of roaches, other insects, or mice
- ☐ Eating or drinking by person preparing food
- ☐ Failure to have separate towels for drying hands and for wiping counters, etc.
- ☐ Failure to properly store leftovers
- ☐ Failure to use thermometers to monitor cooking of meats
- ☐ Use of foods beyond the stated expiration date
- ☐ Refrigerated storage temperatures too high

should start in the supermarket. We want to make sure that no juices from raw meats (which will eventually be cooked) drip on fresh fruits and vegetables that we plan to eat raw. If we worry about raw meat contaminating other foods, we can always buy precooked meats as long as we are willing to pay extra for them. In the supermarket it is also a good idea to look at expiration dates to get as much life out of a perishable product as we can. Damaged packages can lead to contamination of a packaged food, either deliberately or accidentally. Foods that should be cold but are not provide evidence that the product has been mishandled by the supermarket. After we buy our food and take it to our vehicle, we want to make sure that we don't become guilty of food abuse. Perishable foods (anything normally stored in a refrigerator of freezer) begin to spoil rapidly in a hot environment. It's a good idea to make the supermarket the last stop on the way home. Highly perishable foods such as raw meats should never be left out of refrigerated conditions for more than one hour. The longer perishable foods are left at room temperature the more quickly they will spoil and the more likely they are to become a safety hazard. When we get home, it is important to put away the perishables as quickly as possible. Again, we need to make sure juices from raw meats do not come in contact with any food that will not be fully cooked. We also should make sure the food is out of reach of any companion animal that might not be as concerned about personal hygiene as we are.

The most frequent factors in food poisoning outbreaks are improper storage temperatures, poor personal hygiene, cooking temperatures that are too low, cross-contamination, and food from unsafe sources. Personal hygiene should be practiced before and during food preparation. That means we should wash our hands thoroughly before handling any foods and make sure that all surfaces that will touch the food are clean. Frequent hand washing during meal preparation is a good idea, particularly after touching raw meat or uncooked eggs. The towel we use for hands and counters should never be used to dry dishes. Tasting while fixing a meal is not a good idea, either, particularly if a heated food has not been cooked enough to kill the harmful microbes. Care should be taken not to use the same cutting boards or surfaces for raw meats and salad materials unless they have been thoroughly cleaned. Adequate cooking is needed to kill harmful microbes in raw foods. When the meal is over, food should be put away, but not in the container in which it was prepared. Cooking followed by contamination and improper storage can accelerate growth of harmful microbes. Proper refrigeration slows growth of harmful microbes in leftovers. Food in hot pans takes a long time to cool allowing for growth of microbes. Hot foods should be transferred to shallow containers and refrigerated immediately. Two simple rules for handling food are: (1) keep hot food hot and cold food cold and (2) when in doubt, throw it out. Despite concerns about increasing incidents of food poisoning, the recommendations described above have not changed much in the last thirty years.

Pesticides and other contaminants

It would be nice to live in a world where we did not need pesticides, but worldwide, pests represent the greatest threat to availability of food. Many poor people in countries with adequate food supplies are unable to obtain sufficient food, partly because insects, rodents, and plant diseases get to the food first. Insects devour crops before or after harvest in the fields and damage food in storage. Rodents can spread disease and consume food meant for humans. Plant diseases lower yield in the field and spoil harvested crops. Stress to plants in the field from flooding, drought, and low or high temperatures also lower yields and increase susceptibility to pests. Pesticides are used to help prevent crop loss due to insects, rodents, and diseases. Contaminated food and water cause illness and death.

Pesticides are applied to crops to kill pests. They are carefully monitored to ensure the safety of the food supply. Pesticides are very toxic chemicals when applied to the plant. These types of chemicals are also likely to be toxic to us. Pesticides are designed to break down to less-harmful levels before they reach the consumer. Federal and state governmental agencies carefully monitor residue levels of pesticides in raw and processed products. Many techniques such as Integrated Pest Management (IPM) are practiced to reduce pesticides on food crops and their residues in our foods. IPM involves the use of scouts who travel throughout the fields checking insect traps. When the number of insects in a trap reaches a certain level, the fields are sprayed. This practice contrasts with periodic spraying of a crop whether it needs it or not. When pests are particularly numerous, IPM can result in increased pesticide use. In most cases, however, IPM reduces the number of sprayings, thus reducing the amount of pesticide applied. Steps in food processing tend to decrease pesticide levels in foods but do not eliminate them. While pesticides poison and kill agricultural workers who don't take proper precautions when applying them, there have been few cases of consumers becoming ill due to pesticides in the foods they ate.

Organic food production attempts to reduce the use of fossil fuels and synthetic chemicals. Animal manure and composted plant materials serve as fertilizers for organic crops while synthetic nitrogen products are used in modern agriculture, which produces most of the food we eat. Organic fertilizers that come in contact with food present a hazard, and contaminated irrigation wastes are a primary source of harmful microbes. Proper composting of animal waste is supposed to kill all harmful microbes, but it is not clear that current practices are completely effective. Some of the most serious E. coli outbreaks have occurred by contamination of a water supply by animal manure. Most crop production in the United States relies on pesticides, while organic production emphasizes biological and cultural alternatives. These alternatives can be effective, but they require much more work, as anyone who has tried to garden without pesticides can tell us. Biological control of pests on a large scale can lead to creating new pest problems, with the control agent becoming a bigger pest problem than the original pest. The cane toad in Australia and the mongoose in Hawaii are bigger problems than the original pest they were imported to control. Organically grown products are desired by many consumers, but these products usually cost more, have a shorter shelf life, may not look as nice as those grown with pesticides, and may be less safe than those grown with pesticides.

Natural toxins

Despite the widespread belief that natural is better than artificial, food scientists see little scientific evidence to support this claim. In fact, food scientists claim that natural chemicals and products are not necessarily superior to artificial chemicals and processed products. The idea that natural is good and artificial is bad is at best oversimplistic and at worst dangerous. That doesn't stop food companies and makers of dietary supplements from taking advantage of our admiration of all things natural. Organic chemists can synthesize artificial chemicals (sometimes called *nature-identical*) that are chemically indistinguishable from natural compounds and can perform the same functions. Some of the deadliest chemicals are natural, like the botulinal or paralytic shellfish toxins. We avoid some plants such as poison ivy and hemlock because we know they are poisonous even though we are unable to tell the difference by looking at them. Some foods that are normally safe become poisonous under special conditions. Documented cases of deaths from poisonous shellfish and green potatoes provide examples. Chemical synthesis can produce toxic impurities, but there is generally more control over the end products. Artificial ingredients tend to be more pure and chemically defined than their natural counterparts.

Whole foods generally provide a complementary package of nutrients with minimal toxic consequences and maximum bioavailability, but when natural compounds are separated (extracted) from a natural source and concentrated, safety risks increase. First, organic solvents may be needed to separate the natural chemical from the plant, animal, or insect (cochineal is an approved natural red food coloring extracted from beetles). Many of these solvents are toxic and remain in the final product at trace levels. Natural extracts are usually less pure than chemical synthetics, and some of the impurities are toxic. We should not be worried about the safety hazard when present at the low levels in nature. When natural compounds are used as ingredients, they must be concentrated. During concentration and purification, natural toxins can be concentrated and purified in the process. What poses no risks in whole foods can be dangerous at the higher levels in natural ingredients. Also, there is no such thing as natural cooking, just as there is no natural food processing or natural synthesis of artificial compounds. *Homo sapiens* is the only species that has developed the sophistication to modify its environment to overcome natural barriers for good or for ill. Food scientists think that it makes more sense to judge a chemical by understanding its properties than by whether it is natural or artificial.

Allergies and food sensitivities

Individual sensitivities to foods are controversial. Skin or blood tests can determine true allergies to specific foods. More general reactions to foods may occur, but these sensitivities are more difficult to diagnose. Many physicians and scientists who study allergens are skeptical about these sensitivities (called atopic responses), suggesting that such sensitivities reside only in the mind of the victim. Suffering victims are likely to seek help outside the medical or scientific community when told they are making up their illness.

True food allergens induce an abnormal immune response in susceptible individuals. Celery, eggs, fish (including shellfish), grains (corn, wheat, and rice), legumes (soybeans, peanuts), milk, and yeasts are common sources of food allergens. Although we link an allergy to a complete food, usually only a few proteins in a food product are responsible for the allergic reaction. Symptoms of allergic reactions are triggered when specific protein molecules in the food are linked on the surface of special cells in the intestine. These cells then send out signals to other cells leading to an allergic reaction. Many of these reactions result in gastroenteritis leading to stomachaches, diarrhea, and other intestinal difficulties. These reactions can also occur in the food processing workplace through inhaling of the allergen from the air. Some allergies, particularly an allergy to peanut proteins, can be very serious, leading to anaphylactic shock and death in less than an hour after exposure.

Some consumers have atopic reactions to foods or additives that are not true allergens. Sulfites can induce severe reactions including abdominal pain, diarrhea, cramping, nausea, difficulty in swallowing, headache, chest pain, faintness, and loss of consciousness in a few very susceptible consumers. Consumption of large amounts of MSG (monosodium glutamate) on an empty stomach has been referred to as Chinese Restaurant Syndrome leading to headaches, burning neck, chest tightness, nausea, and profuse sweating. Despite widespread publicity, the Feingold Hypothesis, which claims associations among artificial colors, other additives, and sugar to hyperactivity in children, has not been proven in controlled tests. These atopic responses are much more difficult to study than true allergies, which produce true immunological responses that can be measured in the blood. That is why many scientists and physicians are skeptical about them. One clearly documented sensitivity is lactose intolerance. Many Africans, Asians, Native Americans, and their descendants lose the ability to break down lactose.

Consuming a diet that eliminates any food or beverage that might cause a reaction is the surest way of preventing allergic reactions. Skin tests narrow down the number of possible true allergens but do not always work for atopic responses. All foods implicated by the skin tests are eliminated from the diet, with a single food added back in a week, with more tried each time. In addition, shots are given to lessen the chances of reactions. A combination of shots and dietary restrictions works for many individuals, while others must quit eating the suspected food. Atopic responses are usually tested by placing a small amount of the suspected food in the mouth and then waiting for symptoms to develop. This type of test is less likely to provide results than those looking for true allergic reactions.

Government regulation

Many governmental agencies are responsible for monitoring the safety of our foods. The U.S. Food and Drug Administration (FDA) determines which food additives are allowed in foods and matches the ingredient statement on the label with the ingredients in the product. The FDA approves processing methods to make sure that the products are safe with the idea that any process should be designed for spoilage before it becomes unsafe. It has developed a list of Good Manufacturing Practices that help food processors produce safer foods. It also sets microbial standards for foods to ensure that they do not contain levels of microbes that could make us sick, and prohibits the distribution of any food that has ingredients that are unsafe at levels they would normally be consumed. The FDA inspects food processing plants, analyzes food off the supermarket shelf, and inspects incoming overseas shipments to see if its regulations are being followed. It has the power to shut down a processing plant that fails to meet its guidelines and the power to issue product recalls of unsafe foods.

The United States Department of Agriculture (USDA) regulates all meat plants that distribute meat across state lines. It has developed strict guidelines on methods of slaughter and other aspects of handling fresh products. USDA inspectors look at every carcass of fresh meat slaughtered in the country to ensure that diseased animals are not distributed to the American public. The USDA works with food companies to develop

Hazard Analysis and Critical Control Point (HACCP) plans, which are more sophisticated in detecting microbial problems than visual inspection. Some of the USDA and FDA guidelines overlap, such that many food companies must comply with slightly different regulations. The USDA also inspects food-processing plants and can shut down plants that do not comply.

The U.S. Environmental Protection Agency (EPA) regulates pesticides in foods. All pesticides must be registered for their intended use. Growers are required to follow the instructions on the labels. The EPA sets tolerances for the levels of a pesticide that can be present in a food. It measures pesticide residues on these products to make sure that the level in the product is lower than the accepted tolerance. If a pesticide is concentrated in a processed product, it becomes considered a food additive and is then regulated by the FDA.

Restaurants are inspected by the local or state heath department. These inspections involve conditions in a restaurant that could result directly in a health hazard or those that are indications of poor sanitation practices. Restaurants can be shut down until adequate corrections are made if serious infractions are found. Poor sanitation practices result in lower health rating scores. Most restaurants are required to post their most recent inspection results.

Another governmental agency that is an important contributor to food safety is the Centers for Disease Control and Prevention (CDC). While the CDC does not make or enforce regulations, it is the CDC that sends epidemiologists to the scene of food-associated outbreaks (both microbial and nonmicrobial) to help identify the cause of the outbreak and provide guidelines on how future outbreaks can be prevented. It also conducts research on the behavior of harmful microbes to provide a greater understanding.

These governmental agencies are not perfect, but they provide a strong basis for a safe food supply. Agencies like those mentioned above can provide guidelines, inspect processing plants or restaurants, and investigate companies that are violating the law or regulations. They do not have the budgets to constantly inspect every food operation every day, week, or even month. They can't place agents in every public or commercial restroom to make sure we all wash our hands (and most of us wouldn't want them to be there). They can't be in every field, backyard, or kitchen to make sure proper procedures are being followed. Every company and every consumer needs to assume the responsibility of improving the safety of our food. For more information on food safety, check out the Web sites in Insert 1.8.

Insert 1.8 Useful Web sites.

The following Web sites offer more information on food diseases.

Institute of Food Technologists: http://www.ift.org

U.S. Food and Drug and Drug Administration: http://www.fda.gov

Centers for Disease Control and Prevention: http://www.cdc.gov

U.S. Department of Agriculture: http://www.usda.gov

World Health Organization: http://www.who.int/en

Gateway to Food Safety Information: http://www.foodsafety.gov

www.topix.net/wire/health/food-poisoning

Remember this!

This chapter is filled with much information, and it is not easy to remember it all. If you have limited brain space for an understanding of food science, please try to remember the key points listed below:

- Many governmental agencies are responsible for monitoring the safety of our foods.
- Natural chemicals and products are not necessarily superior to artificial chemicals and processed products.
- Worldwide, pests represent the greatest threat to availability of food.
- Preservatives are food additives that prevent or retard spoilage.
- Processed foods should be designed to spoil before they become unsafe.
- A key to preventing future outbreaks is to understand how each occurs.
- The last meal consumed is not usually the meal responsible for an outbreak of food poisoning.
- The expiration date represents the food scientist's best guess about how long a food will last before it spoils.
- Spoilage is not a good indicator of a safety risk.
- The most important weapons in control of food-associated illnesses are proper sanitation and storage during handling and distribution from the farm to the consumer.
- Fresh foods are more likely to contain harmful microbes than processed foods.

Looking ahead

This chapter was designed to introduce you to issues in food safety and entice you to read further. Chapter 2 introduces health issues associated with food, while Chapter 3 explores reasons why we choose the foods we eat. For more information on food preservation, flip to Chapter 4. For more on quality testing procedures to reduce the chances of spoilage and safety hazards, check out Chapter 7. You will find more on preservatives, natural toxins, and chemical reactions that occur in foods in Chapter 10. Chapter 12 discusses in much greater detail the types of microbes in foods and how the food and its environment affect their growth.

Answers to chapter questions

Insert 1.1

They all reported becoming sick from food poisoning.

Insert 1.2

Most dangerous to least dangerous, according to food microbiologist Dr. Mike Doyle, director of the Center for Food Safety at the University of Georgia:

alfalfa sprouts
undercooked chicken
raw cookie dough
rare hamburger
salad from a food bar
unpasteurized orange juice
chocolate éclair
pasteurized milk
refrigerated yogurt
canned tuna fish

Insert 1.3

Most hospitalizations to least:

1. Noroviruses (33%)
2. *Salmonella* (26%)
3. *Campylobacter* (17%)
4/5. (tie) *Listeria* (4%); *Toxoplasma gondii* (4%)
6/7. (tie) *Staphylococcus* (3%); *E. coli* O157:H7 (3%)
8. *C. botulinum* (0.1%)

Most deaths to least:

1. *Salmonella* (31%)
2. *Listeria* (28%)
3. *Toxoplasma gondii* (21%)
4. Noroviruses (7%)

5. *Campylobacter* (6%)
6. *E. coli* O157:H7 (3%)
7. Botulism (0.2%)
8. *Staphylococcus* (0.1%)

For more information, see the CDC (Centers for Disease Control) study by Mead et al. (1999).

Microbial hazards: A food scientist wipes off the top of a soda can because it may have become contaminated with urine or droppings from rodents or insects. Likewise, unpackaged mints in a restaurant may have become contaminated by customers who did not wash their hands after using the restroom.

Insert 1.5:
Campylobacter jejuni (C)
Clostridium botulinum (E)
Escherichia coli O157:H7 (B)
Listeria monocytogenes (A)
Salmonella enteritidis (D)
Staphylococcus aureus (F)

References

Cook, R. 1998. *Toxin*. New York: G. P. Putnam Sons.

Daniels, R. W. 1998. Home food safety. *Food Technology* 52(2): 54.

Fox, N. 1997. *Spoiled: The dangerous truth about a foodchain gone haywire*. New York: Basic Books.

Mead, P. S., L. Slutsker, V. Dietz, L. F. McCaig, J. S. Bresee, C. Shapiro, P. M. Griffin, and R. V. Tauxe. 1999. Food-related illness and death in the United States. *Emerging Infectious Diseases* 5(5): 607.

Further reading

Doyle, M. P., and L. R. Beuchat. 2007. *Food microbiology: Fundamentals and frontiers*, 3rd ed. Washington, DC: ASM Press.

Jay, J. M., M. J. Loessner, and D. A. Golden. 2005. *Modern food microbiology*, 7th ed. New York: Springer Science.

Maleki, S. J., A. W. Burks, and R. M. Helm. 2006. *Food allergy*. Washington, DC: ASM Press.

Montville, T. J., and K. R. Matthews. 2005. *Food microbiology: An introduction*. Washington, DC: ASM Press.

Ortega, Y. R. 2006. *Foodborne parasites*. New York: Springer Science.

Ray, B. 2004. *Fundamental food microbiology*, 3rd ed. Boca Raton, FL: CRC Press, Taylor & Francis Group.

Riemann, H. P., and D. O. Cliver. 2006. *Foodborne infections and intoxications*. San Diego, CA: Academic Press.

Wheeler, W. B. 2002. *Pesticides in agriculture and the environment*. New York: Marcel Dekker.

chapter two

Healthiness of foods

Emily decided to weigh herself on the scale so thoughtfully provided by the cafeteria. She knew she had been gaining weight (or all her jeans were shrinking), but wasn't sure how much. She looked down and couldn't believe it. She was only halfway through her freshman year, and she had already gained thirteen pounds of the dreaded freshman fifteen! It was time for some serious weight loss. What could she do? She'd read that carbs were the problem and decided to try the latest diet, which replaces all those nasty carbs with protein. It was time to start eating healthy and get away from all that pizza and other junk food served up by the cafeteria.

Frank is into working out. He is on his university baseball team, and he needed to bulk up. He worked the weight machines every day, but he realized that exercise was only part of a healthy lifestyle and that nutrition was the other part. He learned that most foods were unhealthy for one reason or another and that the best way to bulk up was to get rid of carbs and get his energy from protein powders. He could get all his other nutrients from dietary supplements and high-protein bars.

Greg had grown up on a farm. He likes good food, and lots of it. When he was young, his mom fixed him lots of vegetables, but he never cared much for them. That was rabbit food, and he wanted man food. When he went off to college he became a meat-and-potatoes guy. Of course he liked farm-fresh eggs (fixed a different way each morning) and grits for breakfast. He washed his food down with a large glass of whole milk at every meal. He seems to be gaining weight since he started school. He didn't seem to be eating more, but he wasn't doing the physical labor he had been doing at home.

Hannah is a committed vegetarian. She has become disgusted with the American fast-food culture, which features fatty burgers, greasy fries, and sugar-laden drinks. She knows these foods are unhealthy. At first she just eliminated red meats from her diet. A little later, after hearing about the gross things that happen to chickens in a slaughterhouse, she did away with all meats. Then, in talking with some of her newfound friends who were also shunning meats, she decided to eliminate all dairy products—eggs and anything else that came from animals. When she stopped hanging out with Greg because he insisted on keeping his carnivorous habits, she became even more serious. One of the things she likes best is that her diet is so simple. All she needs to do is to eat anything that is not from animals. She was surprised at the diversity of products available. There are lots of grains, fruits, and vegetables, and they are actually very tasty. At first she was hungry all the time, but after a while the hunger pains went away. It has now been a month, and she feels healthier. She also notices that she has lost some weight.

Emily, Frank, Greg, and Hannah think their diets are reasonably healthy, but they all have misconceptions. This chapter will explore what healthiness means. We'll introduce the basic nutrients and describe what we need to consider in designing a healthy diet.

Looking back

The last chapter focused on issues concerning the safety of foods. Some key points in that chapter help prepare us for understanding processed foods.

- Fresh foods are more likely to contain harmful microbes than processed foods.
- Spoilage is not an indicator of a safety risk.
- The expiration date represents the food scientist's best guess about how long a food will last before it spoils.
- Processed foods should be designed to spoil before they become unsafe.
- Preservatives are food additives that prevent or retard spoilage.

Healthy and unhealthy foods

Many of us are interested in eating healthier, but sometimes it is difficult to know what is healthy and what is not healthy. Foods like carrots, apples, lettuce, sprouts, whole wheat bread, and yogurt have a healthy image, whereas colas, beer, chocolate bars, pizza, hamburgers, fries, and most appetizers in our favorite restaurants do not. It seems that healthy foods are often unappetizing or tasteless, but unhealthy foods are temptingly flavorful. Is it possible to eat healthy foods and enjoy them too?

When we talk about eating healthy, we are talking about nutrition. Good nutrition involves getting the proper nutrients without consuming too many calories. Obviously, any food that has a good balance of nutrients but can cause food poisoning is not very healthy. Thus a food must be safe as well as nutritious to be healthy. Food scientists and nutritionists tend to discuss healthy and unhealthy diets rather than healthy and unhealthy foods. Healthy diets require a balance of nutrients and yet allow for a few foods that might not qualify as healthy foods. Unhealthy diets generally do not provide adequate amounts of some nutrients and usually provide other nutrients in excess. Dieticians and food scientists disagree about how to get consumers to eat a healthy diet. Dieticians stress the development of good eating habits, such as eating a wide variety of foods. Food scientists prefer designing fun foods that are filled with added nutrients, even if they have a reputation of being unhealthy. For instance, vitamin- and mineral-filled chocolate bars are a food scientist's idea of a convenient component of a healthy diet, but many dieticians think these bars detract from a healthy diet.

Weight loss without pain

Whether it is fighting the "freshman fifteen" or trying to maintain one's appearance for the next social outing, weight control is of great interest on any college campus. Although most of us realize controlling our weight is about diet and exercise, we still hope there is an easier way to lose weight. Many advertisements for supplements, foods, or weight-loss programs offer remedies without pain. Unfortunately there is no simple way to lose weight. Fad diets, fat burners, and other miracle weight-loss regimes make empty promises. Counting calories is still the most effective way of monitoring food intake. A knowledge of food composition, a sensible exercise plan, and willpower are the three most successful elements of a good weight loss program.

Maintaining a healthy body weight is important in remaining fit and in feeling good about ourselves. We can determine if we have a healthy body weight by locating our body mass index in Insert 2.1. As an athlete, Frank works out frequently (more than two vigorous sessions of thirty minutes or longer per week) and appears to be overweight by the chart, but he realizes that the chart does not allow for extra muscle mass. Since Emily is becoming a little overweight, cutting down on calories and increasing exercise is probably a good idea. Hannah, however, is not heavy enough. She is probably not eating enough food and could be malnourished. Being underweight may be as much of a health risk as being overweight. Another serious health problem is weight cycling: the repeated loss and regaining of body weight. Genetics and early eating habits play a role in our cravings and future eating habits, which influences our body weight. One of my favorite diets is the one shown in Insert 2.2.

Cutting down on carbs and stocking up on protein

Although the Appalachian Trail Diet is effective, it is not popular because there are many competing diets that promise quick weight loss. Emily adopted one of the most popular diets by cutting down on carbohydrates (carbs) and increasing proteins. The idea is that consumption of carbs stimulates insulin production, which leads the body to convert the carbs to stored fat, whereas the consumption of proteins stimulates glucagon production, which leads the body to use proteins and fats for energy. This

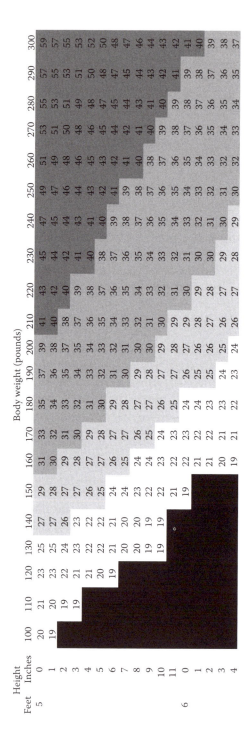

Insert 2.1 Body mass index chart. (From "Body Mass Index and Health," Nutrition Insights 13, no. 2 (2001), U.S. Department of Agriculture. http://www.cnpp.usda.gov/FENR/FENRv13n2/fenrv13n2.pdf.) Body mass index in this table was calculated as: BMI = 705 × [Weight (lb)] / [Height 2 (inches)].

Toll of the trail

How do eight weeks of hiking affect the human body? Prior to setting out on the Appalachian Trail, reporter Scott Huler, 35, dropped by the Raleigh Athletic Club to have some vital statistics recorded. Then, within a week after completing his 410-mile stint, Huler, who is 6'3", was re-examined. Some key findings:

Stat	Before (3/20)	After (5/10)
Chest	36$\frac{1}{2}$"	34$\frac{1}{2}$"
Arm (flexed)	12$\frac{3}{4}$"	12"
Waist	34$\frac{1}{2}$"	30"
Hips	37$\frac{3}{4}$"	36$\frac{1}{4}$"
Thigh (flexed)	22$\frac{1}{4}$"	21$\frac{1}{8}$"
Calf	15"	14$\frac{1}{2}$"
Weight	172 lbs.	162 lbs.
Blood pressure (resting)	106/76	110/72
Heart rate (resting) (In beats per minute)	62	66
Body-fat analysis	8.6%	6.2%
Bench press (max.)	160 lbs.	175 lbs.
Leg extension (max.)	135 lbs.	155 lbs.

Trail diet
An average day of food consumption for a hiker looks something like this:

Food	Calories	Calories from fat	Protein (grams)	Carbohydrates (grams)	Fat (grams)
BREAKFAST					
Oatmeal	160	16	4	33	2
Tangerine	65	0	1	16	0
Pop-Tart	210	70	3	34	7
LUNCH					
Cheese	300	210	21	3	24
Bread	140	5	6	31	0.5
Peanut butter	380	260	16	14	32
Jelly	50	0	0	0	0
Dried-meat stick	280	110	7	0	12
SNACKS					
Snickers bars (2)	560	240	8	72	28
Gorp*	1,240	320	30	174	49
DINNER					
Freeze-dried Lipton dinner	500	60	20	80	8
Total	**3,885**	**1,291**	**116**	**457**	**162.5**

* Includes raisins, peanuts, M&Ms and granola

The U.S. Department of Health and Human Services recommends a normal daily diet of about 2,000 calories.

A person Huler's size, carrying his backpack, walking 3 mph, would burn about 354 calories per hour. By comparison, he might expect to burn 432 calories per hour doing calisthenics or 630 calories per hour swimming.

Insert 2.2 Appalachian Trail Diet. (From S. Huler, "Appalachian Adventures," *Charlotte News and Observer*, 1995. With permission.)

theory permits unlimited consumption of fat and protein, because it is only carbohydrates that lead to weight gain. Although this theory sounds good, it ignores the energy needs of the body and the calorie levels of food. Most food scientists, nutritionists, and dieticians are skeptical about the effectiveness of low-carbohydrate/high-protein diets. They are concerned that these diets can cause long-term damage to strict followers. Unfortunately, these diets are very popular with many physicians, who tend to have limited training in nutrition, yet people trust them and consult them about their dietary needs.

In other parts of the world, the average person takes in 70% or more of their calories from carbohydrates, whereas the average American tends to take in less than 50% of daily calories from carbohydrates; still, Americans are more likely to be overweight than people from other countries. If these low-carbohydrate/high-protein diets don't work, why do so many people try them? Possibly, it is the difficulty of not really knowing how effective a diet is. Some dieters don't give a diet an opportunity to work; they cheat on it and defeat the purpose. For others, when a diet doesn't work, they assume they are not following it closely enough. For the dieters who notice weight loss, it may not be because of the diet but because they are actually cutting calories. Sugars and starches are the most widely consumed carbohydrates. Many times the sugars and starches contained in tempting treats have generous helpings of fats; therefore, when we reduce our carbohydrate consumption, we also reduce our fat intake. Furthermore, high-protein foods tend to be ones that are less tempting to eat in large quantities.

If low-carbohydrate diets can trick people into losing weight, why do nutritionists consider them harmful? Low-carbohydrate diets lead to imbalances in our bodies. When we don't consume enough carbohydrates and consume too much protein, our bodies begin to use the protein as fuel and not as material to make new proteins. Too much protein in our diet also leads to excretion of excessive urea in urine. Too much excretion of urea can tax the kidneys and lead to severe kidney damage. At least two high-profile athletes have experienced kidney problems. It is not clear whether the cause of the problem was an overconsumption of proteins, but with increased reliance on protein powders in training regimes, we may see more athletes experiencing kidney difficulties. High-protein diets can also lead to a loss of calcium, and they also frequently contain high levels of fat. In addition, the body cannot store protein, so it converts the excess protein to fat and then stores it. Furthermore, low-carbohydrate diets result in an imbalanced state called *ketosis*. In ketosis the body cannot maintain enough blood sugar to meet the needs of the brain, with extended periods of ketosis possibly resulting in brain damage. Proteins and carbohydrates are important nutrients for the body. We will learn more about these nutrients in this and upcoming chapters.

Fake fats

If low-carbohydrate/high-protein diets don't work, then how about lowering fat consumption? Since an ounce of fat contains twice the calories in an ounce of protein or carbohydrate, we can cut back our calorie consumption twice as fast if we cut out fats. There are many ways we can reduce calories in our diet. One way is to just do without them. For example, we can eat bread, biscuits, or potatoes without butter or margarine. The problem is that we are not willing to cut back on using the spread because it helps lubricate the chewing action and makes it easier to swallow. Eating baked rather than fried potato chips is another way to reduce fat intake, since baked chips have not picked up fat. Unfortunately, many consumers don't like their chips without the oily coating. Another way to avoid fats is to eat foods typically low in fats (see Insert 2.3 for some helpful guidelines in designing your diet). Many fruits and vegetables combined with small portions of meat and cottage cheese are satisfying and low in fat. The problem with low-fat diets is that most people get easily discouraged and give up because they miss the fat flavor and end up cheating. They may also be eating fatty foods they do not realize contain fat.

The technological solutions designed by food scientists are low-fat versions of typically fatty foods, which would allow Hannah to enjoy her favorite foods without consuming large amounts of fat. It is not as simple as just removing the fat. For example, if you remove all the fat and sugar in a large cup of frozen yogurt, you just get a very small cup of yogurt. Still, there are several ways to lower the fat content in foods. One way is to replace the fat with carbohydrates, proteins, or a combination of both. Unfortunately this approach does not always result in caloric reduction. When Emily looked at the label of some reduced-fat peanut butter, she was surprised that it is actually higher in calories than the full-fat version! Another way of lowering the fat is by replacing it with something that adds bulk to the product but does not provide calories. Many low-fat ice cream and yogurt products use this trick. However, there is a small price to pay for consuming too much of these delicious treats—bloating and digestive distress.

Reducing fat consumption can also be achieved by using the so-called fake fats like Olestra®. These fake fats have all the taste sensations of true fats in the mouth, but the body is not able to absorb them, so they have no calories. Olestra is in many products, but people are probably most familiar with its use in fat-free potato chips. Unfortunately, Olestra® interferes with the body's ability to use vitamins A and D properly. This problem is remedied by replacing these vitamins. Another problem with Olestra® is that its lubricating properties in the mouth are duplicated in the lower intestines, resulting in some rather loose stools. This problem is

Insert 2.3 Making smart choices from MyPyramid's food groups. (From International Food Information Council Foundation, "Your Personal Path to Health: Steps to a Healthier You," available at http://www.ific.org/publications/brochures/pyramidbroch.cfm. Reprinted with permission.)

Grains: Make Half Your Grains Whole

What's in the Grains group: Any food made from wheat, rice, oats, cornmeal, barley or another cereal grain. "Whole grains" include whole-wheat flour, bulgur (cracked wheat), oatmeal, whole cornmeal, and brown rice.

- Get a whole grain head start with oatmeal or whole grain cereal.
- Use whole grains in mixed dishes such as barley in vegetable soup or stews, bulgur in casseroles, or brown rice in stir fries.
- Change it up. Make your sandwich on 100% whole-wheat or oatmeal bread. Snack on popcorn or whole grain crackers.

Vegetables: Vary Your Veggies

What's in the Vegetables group: Any vegetable or 100% vegetable juice. Vegetables may be raw or cooked; fresh, frozen, canned, or dried/dehydrated.

- It's easy going dark green. Add frozen chopped spinach, collard greens, or turnip greens into a pot of soup.
- Swap your usual sandwich side for crunchy broccoli florettes or red pepper strips.
- Microwave a sweet potato for a delicious side dish.

Fruits: Focus on Fruits

What's in the Fruits group: Any fruit or 100% fruit juice. Fruits may be fresh, canned, frozen, or dried; and may be whole, cut-up, pureed, raw, or cooked.

- Bag some fruit for your morning commute. Toss in an apple to munch with lunch and some raisins to satisfy you at snacktime.
- Buy fresh fruits in season when they taste best and cost less.
- Never be fruitless! Stock up on peaches, pears, and apricots canned in fruit juice or frozen so they're always on hand.

Milk: Get Your Calcium-Rich Foods

What's in the Milk group: All fluid milk products and many foods made from milk. Examples include cheese and yogurt. Make your Milk group choices fat-free or low-fat.

- Use fat-free or low-fat milk instead of water when you make oatmeal, hot cereals, or condensed cream soups such as cream of tomato.
- Snack on low-fat or fat-free yogurt. Try it as a dip for fruits and veggies and a topper for baked potatoes.
- Order your latte or hot chocolate with fat-free (skim) milk.

Special Tip: Although cream cheese, cream, and butter are made from milk, they don't count in the Milk group because they contain little or no calcium. Instead, if you eat these foods, count them as "extra" calories from solid fats.

Insert 2.3 Making smart choices from MyPyramid's food groups. (From International Food Information Council Foundation, "Your Personal Path to Health: Steps to a Healthier You," available at http://www.ific.org/publications/brochures/pyramidbroch.cfm. Reprinted with permission.) (Continued)

Meat & Beans: Go Lean with Protein

What's in the Meat & Beans group: All foods made from beef, pork, poultry, fish, dry beans or peas, eggs, nuts and seeds. Make your meat and poultry choices lean or low-fat.

- Trim visible fat from meat and remove skin from poultry.
- Broil, grill, roast, or poach meat, poultry, or fish instead of frying.
- Enjoy pinto or kidney beans on a salad or a hearty split pea or lentil soup for extra protein.

Oils—Know Your Fats: Oils are fats that are liquid at room temperature such as canola, corn, and olive oils. Mayonnaise and certain salad dressings are made with oils. Nuts, olives, avocados, and some fish such as salmon are naturally rich in oils.

- Use some vegetable oil instead of butter for cooking and baking.
- Toss salad with salad oil and flavored vinegar.
- Try thin slices of avocado on a sandwich or sprinkle some nuts on a salad.

Special Tip: Solid fats are different from oils because they are higher in saturated and/or trans fats, so they are considered extras. Solid fats are found in whole milk, cheese, higher-fat meats, and other foods such as butter, lard, chicken skin, and shortening. Some oils such as palm, palm kernel, and coconut are also higher in saturated fats.

---

not usually a major one for people who refrain from eating large amounts at a single sitting.

Although fat reduction can be an important part of a disciplined weight-loss plan, there are some typical mistakes dieters make with foods. First, the more they lower fats in their diet, the harder they find it is to stick with the diet because fats contribute to the enjoyment of foods. Generally a reduced-fat food is more likely to satisfy than a fat-free food. Regardless of how fats are reduced in the diet, weight loss will not occur unless there is a reduction in calories. In short, if we cut half the fat, yet eat twice as much, we will not lose weight.

Natural, organic, and whole foods

Many consumers are concerned about the foods produced by modern food technology. They long to return to simpler, more "pure" foods. These consumers reject many of the commercially packaged foods, believing that natural, organic, or whole foods provide more nutritious and safe alternatives. As you might expect, food scientists do not believe such alternatives are necessarily more nutritious or safe. Although natural, organic, and whole foods are similar, they also have important differences. Sometimes it is difficult to distinguish the differences among these foods. To confuse matters further, some consumers use different definitions to identify certain foods. Can you classify the foods in Insert 2.4?

Natural foods can be found in nature, unlike colas, potato chips, snack cakes, breakfast beverages, nutritional bars, and frozen entrees. What "natural" means, however, is not always clear. Few would argue that fresh garden vegetables, meat from animals hunted in the wild, or mushrooms and herbs gathered in the woods would qualify as natural foods. Most, but not all, consumers would consider fresh vegetables grown in greenhouses, beef steaks from a supermarket, or dried spices in a bottle also to be natural. Items such as canned corn, breads, pastas, and juices are more difficult to classify. Then there are unnatural products that are made from natural ingredients, such as stone-ground wheat flour, cochineal (a natural red colorant that is made from an extract of dried beetles) or WONF (with other natural flavors). Are natural products more nutritious or safer than "unnatural" ones? Food scientists believe we need more information to make that determination. As we will learn in Chapter 4, food processing can decrease vitamins and minerals. A freshly picked ear of corn is more nutritious if it is cooked properly shortly after harvest than if it is canned. Overcooking the ear of corn, however, can be just as damaging

SILK and SILK LIVE! are registered
trademarks of WhiteWave Foods Company;
use of photo is authorized

Courtesy Melissa's Organics, Melissa's
World Variety Produce, www.melissas.com

Courtesy The Beef Checkoff

Courtesy of Moringa Nutritional
Foods, Inc. www.morinu.com

Courtesy of Earthbound Farm
http://www.ebfarm.com/

Courtesy of McCall Farms and Margaret Holmes Foods
www.margaretholmes.com

Insert 2.4 Which of the following products are natural, organic, or whole foods? Answers are at the end of the chapter.

to the nutrients as the canning process. If we quickly blanch and freeze the ear of corn, it will retain its nutrients much better than if it is stored for several days in the refrigerator. Most food scientists would take their chances with a fresh beef steak from the supermarket, which has been inspected by the United States Department of Agriculture (USDA), over a fresh venison steak from a weekend hunter who has slaughtered and cut up the deer in a neighborhood deer cooler.

Organic food production attempts to minimize inputs of fossil fuels and synthetic chemicals. Animal manure and composted plant materials serve as fertilizers for organic crops, whereas nitrogen products are used in conventional agriculture. Conventional crop production relies on pesticides, but organic production emphasizes biological and cultural alternatives. Organic products have a marketing advantage, but they tend to cost more and have a shorter shelf life. As described above, a freshly harvested product is likely to be more nutritious than one that has been shipped and stored for a longer period of time. However, there is little, if any, scientific evidence to support the claim that not using pesticides or fossil fuels makes a food more nutritious. The use of animal manure as a fertilizer certainly does not make a crop safer, particularly a root crop like potatoes or carrots.

The term *organic* is used to describe meat products that come from animals that have not been treated with antibiotics or synthetic growth hormones and have consumed feeds that were not treated with pesticides. There are legitimate concerns about the overuse of antibiotics because it has been shown that antibiotic overuse can produce drug-resistant strains of bacteria, which threaten human health. Meats that claim to be "hormone free" are misleading because hormones are natural components of any plant or animal. Thus, the only truly hormone-free foods are distinctly "unnatural." To make a claim that meat is free of synthetic hormones, an affidavit is needed from the producer because there are no scientific tests that can be done to distinguish between synthetic hormones and natural hormones. Again, there is little or no evidence that organic meat is more nutritious or safe than conventional meat.

Whole foods are generally those foods that are readily identifiable by their original components. Milk, fresh and processed fruits and vegetables, as well as whole and ground meats would be considered whole foods. Many foods, such as the "unnatural" types listed at the beginning of this section, are clearly not whole foods. Low-fat and diet foods are generally not considered whole foods. Generally speaking, whole foods have more nutrients with less sugar and fat than many snack foods. There are exceptions, such as ripe bananas, which have very high levels of sugar, and avocados (or guacamole) and butter, which are very high in fat. Also, fortified breakfast cereals could hardly be considered whole foods, but they contain extensive vitamins and minerals. Food scientists consider synthetic nutrients to be just as effective as natural ones. Once again, a name such as "whole food" does not guarantee nutrition. We need more information about the product before we can determine how nutritious or how safe it is.

Food and disease

One thing many of us are concerned about is how the food we eat will affect our health. We learned in Chapter 1 that microbes can cause food poisoning and how we can reduce our chances of becoming one of its victims. Generally, though, when we talk about healthy foods we are talking about foods that will help us avoid the diseases of civilization (cancer, diabetes, heart disease, and obesity). Hunger, nutritional deficiency diseases, and infection are the major killers in less-industrialized societies. More industrialized countries tend to have higher standards of sanitation and better health care. As a result, the citizens tend to live longer and are more likely to develop cancer, diabetes, or heart disease. During the last one hundred years there have been major increases in life expectancy (see Insert 2.5). Although living one hundred years is not as rare as it used to be, few of us would welcome it if it meant suffering in extreme pain or if it

Insert 2.5 Life expectancy (at birth in years) and infant mortality rate (infant deaths per 1,000 births) for selected countries. (Adapted from http://www.infoplease.com as derived from the U.S. Census Bureau, International Database. Data is based on year 2006.)

Country	Life expectancy	Infant mortality
Angola	37.6	184.4
South Africa	42.5	59.4
Nigeria	47.4	95.5
Kenya	55.3	57.4
Bangladesh	62.8	59.1
Russia	65.9	11.1
India	68.6	34.6
Iran	70.6	38.1
China	72.9	22.1
Mexico	75.6	19.6
United States	78.0	6.4
United Kingdom	78.7	5.0
France	79.9	4.2
Canada	80.3	4.6
Australia	80.6	4.6
Japan	81.4	3.2

required the use of extensive respiratory equipment. Quality of life in our aging years is being seen as a more important goal as the aging population grows.

Most of the health problems associated with food are caused by either not getting enough food or getting too much food. First, our bodies must get enough energy to supply our needs. The unit of energy is the calorie (more appropriately, the kilocalorie). The energy provided by food comes from carbs, fats, and proteins, not caffeine or other stimulants. If our body does not get enough calories, then it starts to break down body stores. The first thing the body breaks down is carbohydrates, then fats, and finally protein. If we combine a low-calorie diet with a low-activity life, it becomes very difficult to lose weight because the body begins to shut down normal energy-using functions. There comes a point when our bodies start cannibalizing themselves because they do not have enough calories to support themselves. When a person reaches that point because there is not enough to eat, we call it starvation. When a person reaches that point out of a personal choice, we call it fasting. Either way will lead to death if more calories are not supplied. Not having enough calories and protein is called *marasmus*. Enough calories, but not enough protein is called *kwashiorkor*. A person who receives enough calories, but most of it from protein can starve to death as well. A certain amount of carbohydrate is needed in the diet to provide adequate levels of glucose in the blood to maintain proper brain function.

Most people get enough protein and calories, but they can still be malnourished. Not getting enough specific vitamins or minerals can result in a deficiency disease. For example, not getting enough vitamin C can result in scurvy, which is characterized by bleeding gums, blotchy bruises, and failure of wounds to heal. Lack of vitamin A can prevent cells from developing properly and lead to night blindness. Anemia can result from not enough iron or a lack of vitamin B_{12}. Osteoporosis is the result of a calcium insufficiency. Such deficiency diseases are worldwide problems. Deficiency diseases can be devastating, but the cause, preventive measures, and treatments are clear.

Cancer is the most dreaded disease of civilized societies. It has been estimated that about 20% of the deaths in the United States are attributed to cancer. Cancer is not a single disease; rather it is a term that groups many similar diseases characterized by abnormal cellular growth and malignancy. The good news is that many cancers are attributed to environmental causes and are presumed to be preventable. The bad news is that the causes, treatment, and prevention of cancer are not as simple as those for deficiency diseases. Each type of cancer has a different mechanism (or set of events that result in the disease) of development. Cancers affect forty-six sites in the body. Lung cancer accounts for 25% of cancer deaths and is directly attributable to tobacco smoke. Colon cancer and breast cancer are the next leading killers and are related to dietary causes. Cancer is not a disease related to deficiencies of a vitamin or mineral. It is not usually caused by the ingestion of a single carcinogen or prevented by a single nutrient or drug. Although diet plays a role in either increasing or decreasing the risks associated with a type of cancer, rarely is it the specific cause. Generally speaking, excessive fat intake can increase the chances of breast cancer and promote other types of cancer; heavily smoked foods, grilled foods, and pickled meats can increase the chances of stomach cancer; and excessive alcohol consumption can contribute to mouth, esophagus, and liver cancer. In contrast, increased fruit and vegetable consumption provides dietary fiber and antioxidant vitamins, which appear to lower cancer risk.

Heart disease is the leading killer in the world. It must be understood that in the classification of causes of death, no one dies of old age. If Frank lives until the age of ninety-three when his heart stops beating and nothing else is seriously wrong with him, he will be classified as dying of heart disease. Generally speaking, overconsumption of food, particularly fatty foods, increases the chances of developing heart disease. Smoking and excessive consumption of alcohol are also risk factors for heart disease. Obese smokers and drinkers, particularly those with a family history of heart disease, are at risk for premature strokes or heart attacks. Similar factors increase the risk of developing diabetes. Diabetes is a disease with several levels of treatment: diet, medication, and insulin injection. Although a diabetic must be very careful with dietary intake, it is not clear that overconsumption of sugar in a peripatetic state is the cause of the onset of the disease. Overconsumption of sugar has been linked with many conditions, from hyperactivity to obesity to cavities. In many of these conditions the linkage is not clear. Obviously, excessive consumption of anything with carbohydrates, and therefore calories, will result in weight gain. Sugar has been shown to be a cause of tooth decay, but it has not been demonstrated as a cause of hyperactivity. As with cancer, the cause, treatment, and prevention of heart disease and diabetes are not as clear as for malnutrition.

A word here about cause and effect might be appropriate. An article from the Internet proclaims, "Tea Consumption Doubles the Chance of Conception." The article points out that tea drinkers (presumably women) are twice as likely to become pregnant as those who do not drink tea. Now most of us are sophisticated enough to know that it takes more than tea for a woman to conceive. Is there something in the tea that makes a woman more fertile? Is tea really an aphrodisiac? The lack of a cause-and-effect relationship in this example is clear, but in many things we see in the media it is more difficult to separate the cause from the effect.

Dietary supplements, nutraceuticals, and functional foods

There are dietary supplements and super-healthy foods available today. The first dietary supplements were vitamin pills (many contained minerals as well). Now we can buy protein powders, medicinal herbs, super-nutritious chocolate bars, and many other foods that are supposed to be good for us.

Dietary supplements are substances that are taken to supply a need or reinforce one's diet. Until recently, the term was used to describe pills that were consumed to make up for an essential nutrient that was not supplied by the diet. Parents provided their children with vitamins and minerals in a single pill a day (sometimes in the form of a favorite cartoon character) to make sure they were not deficient in any essential nutrients. Most dieticians prefer that we obtain our vitamins and minerals from whole foods. Sometimes, however, supplements of specific nutrients are needed in special circumstances. For example, women of childbearing age generally need additional iron and folic acid, vegetarians need more iron and vitamin B_{12}, and lactose-intolerant individuals need additional calcium. Dietary supplements have now gone beyond the supplying of essential nutrients. We are now consuming garlic for infections and prevention of atherosclerosis, echinacea for the treatment of sore throats and colds, and St. John's wort for depression. We are beginning to think we can treat our symptoms at home for most diseases without the need of a physician.

Nutraceuticals are foods specifically designed to act as drugs. Nutraceuticals are the food industry's answer to the health-food market. Many of these ingredients are added to improve mood. Although there appears to be a relationship between foods and mood, the linkage has not been clearly established. With the potential profitability of nutraceuticals, food companies have rushed in to gain their share of the market. One problem that food scientists face in designing nutraceuticals is producing a truly healthy product that also has good sensory quality because many of the ingredients used to formulate the product have an objectionable taste. Food scientists need to find ways to build in flavor without losing the nutritional value or health benefits of the product.

Functional foods are marketed to perform a specific function. Two of the hottest trends in food products today are nutrition and convenience. Vitamin- and mineral-filled chocolate bars provide quick, easy nutrition with the extra fat and sugar providing a pleasant way to get all the nutrients one needs. Likewise, sugared and fortified breakfast cereals perform a similar function. Sports drinks provide vitamins and minerals, as well as electrolytes to replenish those lost in sweat during a workout. Calcium-fortified orange juice provides an alternative to dairy products for those who dislike or have sensitivities to milk. With the discovery of additional benefits of certain components, some traditional foods are now being promoted as functional foods. For example, ketchup is high in lycopene, which is a powerful antioxidant, and mustard is now being touted as being healthful due to the presence of antimicrobial compounds and antioxidants.

With all the information available, how is the consumer supposed to sort it all out when even professionals have trouble doing so. There are some guidelines we should consider the next time we buy a supplement, nutraceutical, or functional food:

- It is a fallacy that if a little is good, a lot is better (it might even be toxic—just like drugs, you can get an overdose of a supplement).
- Many of these substances have not been thoroughly studied, and when studied have generally been tested in isolation and not in the presence of the wide variety of substances that make up a typical diet.
- A substance may be present but not at a high enough level to be effective (you may need a lot more to cure your depression, acne, or warts than you are getting in that herbal tea).
- If you are getting an effectively high dose, it may be doing other damage (some of these supplements cause serious interactions with prescription drugs).
- There may be other components present that could have adverse effects when consumed at too high a level.

As medical and nutrition science advances, we will find substances and combinations of substances that are truly beneficial to health. Presently, however, there are more products available that consumers trust and are willing to buy than there are products with proven health benefits.

Enhancing athletic performance

There is one group of consumers who is particularly interested in nutrition and in dietary supplements––athletes. Like Frank at the beginning of the chapter, they are looking for performance enhancement through dietary supplements. The most important nutritional need of an athlete is additional calories. Insufficient calories to replace the ones burned off during exertion and exercise will result in weight loss, and excessive weight loss results in loss of strength. The increased calories of an athlete's diet require increased levels of vitamins, but there is little information to support megadosing with vitamins. Excess exercise can lead to amenorrhea in women, which can lead to other health problems. Women athletes generally can benefit from extra calcium and iron. It appears that athletes have a greater need for dietary iron than nonathletes. Remember, many athletes turn into couch potatoes during the off-season or when they hang it up for good. During those times, drastic reductions in calorie consumption may be necessary to keep from gaining weight.

Increased protein consumption may lead to increases in muscle nitrogen balance in active athletes but not in couch potatoes. Protein supplements are available as whole proteins or amino acids in the form of pills, powders, and potions. Protein powders are likely to contain nonfat dried milk, soy products, and other ingredients from nonanimal sources. Consumption of large amounts of animal proteins, like a big juicy steak, is likely to add excess fat and cholesterol to the diet. The building blocks of proteins are amino acids. Our digestive system breaks down the proteins we consume into amino acids. The consequences of consuming large doses of amino acids have not been widely studied, although some, like tryptophan, can have toxic impurities that can lead to illness or death. Consumption of individual amino acids is not recommended, as they may lead to metabolic imbalances. Branched-chain amino acids are particularly popular, as they are thought to be more likely to serve as energy sources and are thought to enhance serotonin production. Scratonin may enhance performance and raise the threshold for pain, but the short-term gain could have long-term consequences.

Optimal diets for athletes in training are 60 to 70% calories from carbohydrates, 25 to 30% calories from fat, and 10 to 15% calories from protein—not the low-carb, high-protein diets many are trying. Whether we are participating in intercollegiate athletics or just working out on our own, it is important that our diet meets both our long-term and short-term health needs. There are many good books on sports nutrition that are based on solid nutrition science. Some excellent ones are listed in the reference section of this chapter. There are many not-so-good books, articles, and individuals who have strange ideas about sports, training, exercise, and nutrition. We shouldn't let them experiment on us with their untested ideas, as we could be suffering the consequences many years after we are finished with competitive sports.

Fasting

People of many religious faiths practice fasting as a way to purify the mind and the body. Fasts generally involve the elimination, for a period of time, of all solid foods. Juice fasts permit the consumption of nutritious beverages, but strict fasts do not even permit the consumption of water. Ramadan is a month of strict fasting from sunup to sundown practiced by Muslims. Every Jewish person over the age of thirteen is expected to fast for twenty-four hours at Yom Kippur. Drinking of water is permitted. Fasting is an integral part of the Hindu religion. It was an important part of the early Christian religion, but is not widely practiced today.

Even for a short time, fasting is an effective expression of self-sacrifice. Spiritualists who fast recommend starting with small fasts before progressing to extended ones. They also caution against eating a large meal right before starting a fast. Hunger pangs can become intense early in the fast, followed by general weakness and difficulty concentrating. The tongue becomes coated and breath becomes bad, which spiritualists attribute to the removal of toxins normally consumed in our diets. Fasting can be combined with a spiritual retreat, either in solitude or with a like-minded group, as it is difficult to function normally while fasting. Fasting, combined with meditation, tends to help us focus on the things that are really important in life, but it can have nutritional consequences.

Fasting is self-starvation and can result in severe nutritional deficiencies. When we consume more energy than we burn, we store excess calories as animal starch (glycogen) and fat. During a fast, the body breaks down glycogen first, followed by fat. As stated previously, the brain needs glucose to function properly. When blood glucose is low, it sends signals out to break down fat and protein. In an extended fast, proteins are broken down producing ketone bodies leading to ketosis. It is these ketone bodies that are responsible for the faster's coated tongue and bad breath. During a total fast, up to a half-pound of fat can be burned, but at the potential cost of loss of brain function as the ketone bodies are not adequate substitutes for glucose. Strange as it might seem, an overweight person can starve to death before all of the fat has disappeared!

Eating disorders

Eating disorders develop when peer pressure clashes with biological needs. The most common eating disorder is anorexia. It involves a self-induced fast to lose weight. Advanced stages of anorexia can interfere with normal sleep, induce depression, and lead to malnutrition. A related disorder, bulimia, results in anorexia through self-induced vomiting or excessive use of laxatives, enemas, or diuretics. Since the food is not in the body long enough for most of the nutrients to be absorbed, weight loss and other symptoms of anorexia result. Binge eating is a frequent response to stress (exams, grades, dysfunctional personal relationships, peer pressure) resulting in weight gain. Binge-and-starve diets result in nutritional imbalances and interfere with the body's normal metabolic processes. People with this disorder are particularly susceptible to zinc deficiencies.

Studies have shown that women, particularly those of college age, tend to be more susceptible to eating disorders than men due to hormonal secretions and greater social pressure to be thin. Although society values a slender figure, food scientists continue to produce tempting sweets and fatty foods that are readily available (as close as the nearest vending machine) and convenient (can be eaten with our hands and choked down while in motion between classes). In the second stage of anorexia, anorexics become desensitized to hunger pangs and develop an aversion for most foods, eating just enough to survive but not enough to thrive. The body shuts down vital processes to conserve energy. The final stage of anorexia results in wasting, withering, and death, similar to starvation associated with famines. Bulimia involves the consumption of normal to excessive amounts of food but preventing digestion by inducing vomiting or by using laxatives, enemas, or diuretics. Bulimia leads to the development of anorexic symptoms. If we have eating problems or know a close friend who does, it is important to catch it before it gets out of hand. Treatment is painful, and serious digestive problems, such as spastic colon, can persist after a normal diet has been resumed. Most college campus health centers have programs that will provide counseling. For more insight into eating disorders see Insert 2.6.

Food fads and their consequences

The concept of a diverse and balanced diet appears to be giving way to diets tailored to meet unique nutritional needs. Although there may be a few individuals with special dietary needs, in general our nutritional requirements tend to be very similar. An overexaggeration of the different dietary requirements leads to food fads and can result in malnutrition. Some current fads are brightly colored food products and foods that "can make us smarter."

Brightly colored foods are not new. Even our parents can remember Kool-Aid®, Froot Loops®, and colored Easter eggs from their childhood. New food trends today have a wider range of products: green ketchup, sports drinks, Cheetos® that turn our tongues a weird color, including the currently popular color, blue. Although some consumer groups have denounced artificial food colors, regarding them as harmful and particularly guilty of inducing hyperactivity in children, there is little or no solid scientific evidence to support this assertion. A greater threat to human health is probably the products that are high in sugar and calories and low in vitamins and minerals. Green ketchup, which contains artificial food color, still has the red tomato pigment lycopene, which is one of the most sought-after compounds in nutraceuticals due to its antioxidant activity.

Although there are no proven links between food and intelligence, eating a balanced diet does help us make the most of what we have. Claims that food or supplements can make us smarter or help us concentrate are exaggerated. Stimulants such as caffeine can help us stay awake and increase our attention level; however, too much caffeine can make us irritable and lead to anxiety attacks. If you are a heavy caffeine user and don't believe this statement, ask someone close to you for an honest evaluation. Lack of enough calories interferes with our ability to stay alert and attentive, and

Insert 2.6 Signs and symptoms of anorexia and bulimia and how you can help. (Excerpted in part from A. S. Litt, *The College Student's Guide to Eating Well on Campus*, 2nd ed., Bethesda, MD: Tulip Hill Press, 2005, chap. 7. With permission.)

Signs and Symptoms of Anorexia Nervosa
- Weight loss leading to a body weight of 85% of what's considered acceptable
- Intense fear of being "fat" or gaining weight
- Frequent weighing
- Develops ritualistic eating habits, such as cutting food into tiny pieces, eating alone, and dragging out meals
- Loss of menstrual period
- Excessive exercise
- Increased sensitivity to cold
- Refuses to admit eating patterns are abnormal
- Withdraws socially

Signs and Symptoms of Bulimia
- Preoccupation with food, weight, and appearance
- Eats large volumes of food and then "gets rid" of it by vomiting, fasting, exercising, or taking laxatives
- Experiences mood swings and depression
- Dental problems
- Stomach and digestive problems—such as bloating, constipation, diarrhea
- Scratched or scarred knuckles from scraping against teeth to induce vomiting
- Irritation of the esophagus and throat
- Realization that eating pattern is abnormal
- Irregular menstrual periods

How to Help a Friend with an Eating Disorder
- Choose a time and place to talk away from distractions and other interruptions.
- Don't be judgmental.
- Be a good listener, but don't promise to keep serious information confidential.
- Don't assume the role of a therapist or nutritionist.
- Don't oversimplify the problem by saying, "All you have to do is eat."
- Don't engage in a battle, but don't ignore the problem.

we can develop a could-care-less attitude. On the other hand, too many calories can lead to sleepiness and laziness. Inadequate consumption of vitamins and minerals can also affect our ability to concentrate. There is some
evidence that certain foods (or their components) can affect our mood, but
the linkage has not been clearly established. There are few if any magical
foods, but a balanced diet contributes to health and well-being, whereas a
poor diet can lead to laziness, inattentiveness, and a bad attitude.

The sad part about many food fads is that people who intend to eat
healthy do not know the nutritional benefits or detriments of a particular
food, supplement, ingredient, or component in the context of an overall
diet. Thus, by trying to improve their health, they may actually be harm-
ing it. When food fads emerge, there is always a business willing to produce the product and an author willing to exploit it. It remains to be seen
if the growth of dietary supplements and functional foods is a fad, a fundamental change, or a springboard for new products that will be perma-
nently incorporated into our diets. It's hard to believe that the original
Kellogg's® and Post® breakfast cereals, which are so much a part of our
culture today, were part of the health-food craze that affected Americans
of the late 1800s.

Up to this point we have discussed the health issues of foods that we
hear about in the media. Although there is much good information about
health in the news media, there is also much misinformation. Before we
can sort it all out, we need to know some of the basic principles of nutrition. For some useful Web sites on nutrition see Insert 2.7.

Six glasses a day

Water is the major component of most foods. Next to oxygen, it is the most
important nutrient for our body. We can only live a few minutes in the
absence of oxygen and a few days without water. Water (dihydrogen oxide)
is a liquid chemical that is in virtually all foods and beverages. Although
we don't usually think of water as a chemical, it is the one chemical for

---

Insert 2.7 Some Web sites you might wish to consult to get more information
on nutritional quality of foods and developing reasonable diet plans.

Center for Nutrition Policy and Promotion, http://www.cnpp.usda.gov
American Dietetic Association, http://www.eatright.org
Elizabeth Somer, M.A., R.D., http://www.elizabethsomer.com/
FITDAY, http://www.fitday.com
Nutrition.gov, http://www.nutrition.gov
MyPyramid.gov, http://www.mypyramid.gov
Weight Watchers, http://www.weightwatchers.com

---

which everyone knows the formula—H_2O. It is recommended that we drink one and one-half quarts (approximately six glasses) of water per day. The idea that we can only get water from the bottle or the tap is misguided; we can obtain sufficient water from foods and beverages. Meats contain 50 to 70% moisture (another term for water), fruits and vegetables contain 70 to 95%, and even dry flour contains 10 to 20%.

Although tap or bottled water is not the only way we can get our liquids, we mustn't think we are getting sufficient water when we drink caffeinated or alcoholic beverages. Caffeine and alcohol are diuretics, which means they increase the frequency of urination. Thus, some of the water goes down the toilet rather than meeting our daily needs. That's why our trips down the hall or stops at the rest areas are more frequent when we drink too much coffee, cola, sweet tea, beer, or similar refreshments. Another problem with consuming too much sugared or alcoholic beverages is that we may be adding excess calories and thus adding to a potential weight problem. In dry climates we need more water because we lose more water through the pores in the skin in dry climates than in moist climates. Although water collects on our skin as beads of sweat in humid climates, it evaporates immediately into the dry air in dry climates. Electrolyte (sodium and potassium) balance is essential in maintaining proper water balance in our tissues. Although rare, there are cases where individuals have consumed too much water (water intoxication), such that the body is unable to function.

Energy from foods

In a time when it seems that everyone is either overweight, on a low-calorie diet, or both, it is easy to forget the main reason for eating—to get energy. We get energy from food in the form of calories. There are three main sources of calories—carbohydrates, proteins, and fats. We will discuss another source of calories later. Carbohydrates and proteins provide approximately four calories of energy per gram (114 calories per ounce). Fats provide approximately nine calories per gram (255 calories per ounce), more than twice as much as carbohydrates or proteins. That's why we are more likely to lose weight if we cut back on fat instead of proteins or carbohydrates. We need a balance in the sources of our calories. Although most of us could stand to cut back on the fat we consume, we do need some fat in our diet. Despite its deficiencies, fat slows stomach-emptying time. Meals with a modest amount of fat will slow the digestive processes, making the time between periods of hunger longer. We also need protein to perform other functions in addition to providing calories. As we learned earlier, when protein becomes the major source of energy, we may not produce enough glucose in the blood to maintain proper brain functioning.

Another danger many of us face is the snack attack that comes between meals and can lead to extra calories and unwanted pounds.

The other main nutritional reason for eating food is to get vitamins and minerals. Vitamins and minerals are needed to support bodily processes. Essential vitamins and minerals serve as enzyme cofactors or components of these cofactors. A special class of proteins called *enzymes* speeds up chemical reactions in the body. These enzymes need special substances (cofactors) to help them perform properly. The enzymes and other molecular components in our bodies are constantly under construction.

A third nutritional component is dietary fiber. Fresh fruits and vegetables are high in dietary fiber as are products containing whole grains. Dietary fiber is chemically a carbohydrate that comes from the cell walls of plants; much of it is indigestible by the body so it contributes fewer calories than other carbohydrates like sugars and starch. Dietary fiber adds bulk to the diet to help keep bowel movements regular. It also helps remove toxins from the intestines, and may help in the prevention of many of the diseases associated with overnutrition. Too much dietary fiber, however, can bind vitamins and minerals, thus making a malnourished condition worse.

Basic nutrition is more simple and straightforward than it is portrayed in the media and in popular books. On the other hand, human physiology and metabolism (what happens to those nutrients inside our bodies) is more complex than the oversimplified explanations in the same media and books. It is also important to include an exercise plan as part of an overall diet plan. The following sections in this chapter and the information covered in Chapter 11 should help us design a healthier diet.

Reading the label

Nutrition labels can help us with the nutritional quality of our diets. Although whole foods do not have nutrition labels, there are many resources we can refer to for these values. The Nutrition *Facts* part of the label indicates the serving size, how many servings per container, calories per serving, and the fat calories per serving in the left column. Then there is a chart that provides the amount per serving of the macronutrients (total fat, total carbohydrate, and protein). Fat is further broken down into saturated fat, polyunsaturated fat, monounsaturated fat, and trans fat. It is generally recommended that we limit our intake of trans fats. The hazards of trans fats represent a controversial issue in nutrition. At best they count as saturated fatty acids; at worst they could be toxic. The two categories listed under carbohydrates are dietary fiber and sugars. The amount per serving of sodium must also be listed. In addition, the percent daily value is included for all the nutrients except sugar and sodium. The percent daily values for vitamin A, vitamin C, calcium, and iron are also given as well as for the micronutrients that contribute significantly to our diet from this product.

It is important to keep in mind that all values are stated in terms of the amount per serving. A bag of chocolate chip cookies may contain 75 cookies, but the serving size may be only 5 cookies. The label indicates that the product provides 4% of the daily value of iron, 11% of the daily value of fat, and 150 calories. Five cookies can be a nice snack, but if we eat the whole bag, we will get 60% of our iron, 165% of our fat, and 2,250 calories! We probably should look elsewhere to get our iron.

Look at the Nutrition Facts statements of the six items in Insert 2.8. From them, Emily designed a simple diet for a day. She had a big bowl of item A containing 3 servings and a single serving of item B. That started her day off with 340 calories with 45 of them from fat (5 grams). She drank some water but was hungry by mid-morning, but she resisted the temptation to hit the vending machine. Lunch was rushed between classes, so she heated item C for 90 seconds in the microwave and chased it with the entire can of item D, which is two servings. She, of course, had another two glasses of water with items C and D. In the afternoon, she ignored her hunger pangs and looked forward to her evening meal. For supper she went to her favorite fast-food restaurant to consume a healthy item E topped with item F along with more water. They gave her two packs (one serving each) of item F, but she only used one. In Insert 2.9 she calculated her total intake for the day. There are several points to note. First, she did not get enough calories. She ended up tired and hungry at the end of the day. For her desired weight and age she should be getting about twice as many calories. Next, 37.5% of her calories came from fat. Most guidelines recommend getting the total down to under 30%. (It is good she didn't add the extra pack of Item F!) She consumed close to her %DV (daily value) for sodium, protein, vitamin C, and iron but was low on fiber, calcium, and vitamin D. If she ate twice as much of everything, she would get the right amount of calories, but she'd be eating too much fat and sodium, although most of her micronutrients would be provided. It is difficult to balance nutrients if we eat just a few food items. In addition, many commercial foods have extra fat and sodium.

There are other important items on a food label. One is the ingredient statement as shown in Insert 2.10. The ingredients are listed by quantity––from the most to the least. Items B and D are very simple processed products, whereas items A, C, and E are complex formulated foods with many ingredients. Note that some ingredients like the lasagna product in Item C have ingredients of their own, which are listed in parentheses. Parentheses are also used to help explain the common names of some strange-sounding chemicals like the vitamins listed in items A and F. Each label must also include the company address of the distributor. This address is not necessarily where the product was manufactured.

Product A

Nutrition Facts
Serving Size (21g)
Servings Per Container

Amount Per Serving

Calories 70	Calories from Fat 0

% Daily Value*

Total Fat 0g	0%
Saturated Fat g	
Trans Fat g	
Cholesterol 0mg	0%
Sodium 170mg	7%
Total Carbohydrate 15g	5%
Dietary Fiber 1g	4%
Sugars 2g	
Protein 4g	

| Vitamin A 10% | • | Vitamin C 15% |
| | | Iron 30% |

*Percent Daily Values are based on a 2,000 calorie diet. Your daily values may be higher or lower depending on your calorie needs:

	Calories:	2,000	2,500
Total Fat	Less than	65g	80g
Saturated Fat	Less than	20g	25g
Cholesterol	Less than	300mg	300mg
Sodium	Less than	2,400mg	2,400mg
Total Carbohydrate		300g	375g
Dietary Fiber		25g	30g

Calories per gram:
Fat 9 • Carbohydrate 4 • Protein 4

Product B

Nutrition Facts
Serving Size (236g)
Servings Per Container

Amount Per Serving

Calories 130	Calories from Fat 45

% Daily Value*

Total Fat 5g	8%
Saturated Fat 3g	15%
Trans Fat g	
Cholesterol 20mg	7%
Sodium 130mg	5%
Total Carbohydrate 13g	4%
Dietary Fiber g	
Sugars 12g	
Protein 8g	

| Vitamin A 15% | • | Vitamin C 4% |
| Calcium 30% | • | |

*Percent Daily Values are based on a 2,000 calorie diet. Your daily values may be higher or lower depending on your calorie needs:

	Calories:	2,000	2,500
Total Fat	Less than	65g	80g
Saturated Fat	Less than	20g	25g
Cholesterol	Less than	300mg	300mg
Sodium	Less than	2,400mg	2,400mg
Total Carbohydrate		300g	375g
Dietary Fiber		25g	30g

Calories per gram:
Fat 9 • Carbohydrate 4 • Protein 4

Product C

Nutrition Facts
Serving Size (213g)
Servings Per Container

Amount Per Serving

Calories 210	Calories from Fat 50

% Daily Value*

Total Fat 6g	8%
Saturated Fat 2g	10%
Trans Fat g	7%
Cholesterol 20mg	35%
Sodium 840mg	10%
Total Carbohydrate 31g	8%
Dietary Fiber 2g	
Sugars 14g	
Protein 9g	

| Vitamin A 6% | • | Vitamin C 6% |
| Calcium 6% | • | Iron 8% |

*Percent Daily Values are based on a 2,000 calorie diet. Your daily values may be higher or lower depending on your calorie needs:

	Calories:	2,000	2,500
Total Fat	Less than	65g	80g
Saturated Fat	Less than	20g	25g
Cholesterol	Less than	300mg	300mg
Sodium	Less than	2,400mg	2,400mg
Total Carbohydrate		300g	375g
Dietary Fiber		25g	30g

Calories per gram:
Fat 9 • Carbohydrate 4 • Protein 4

Product D

Nutrition Facts
Serving Size (121g)
Servings Per Container

Amount Per Serving

Calories 20	Calories from Fat

% Daily Value*

Total Fat g	
Saturated Fat g	
Trans Fat g	
Cholesterol mg	
Sodium 390mg	16%
Total Carbohydrate 4g	1%
Dietary Fiber 2g	8%
Sugars 2g	
Protein 1g	

| Vitamin A 6% | • | Vitamin C 4% |
| Calcium 2% | • | Iron 4% |

*Percent Daily Values are based on a 2,000 calorie diet. Your daily values may be higher or lower depending on your calorie needs:

	Calories:	2,000	2,500
Total Fat	Less than	65g	80g
Saturated Fat	Less than	20g	25g
Cholesterol	Less than	300mg	300mg
Sodium	Less than	2,400mg	2,400mg
Total Carbohydrate		300g	375g
Dietary Fiber		25g	30g

Calories per gram:
Fat 9 • Carbohydrate 4 • Protein 4

Product E

Nutrition Facts
Serving Size (302g)
Servings Per Container

Amount Per Serving

Calories 200	Calories from Fat 90

% Daily Value*

Total Fat 10g	15%
Saturated Fat 2g	10%
Trans Fat g	
Cholesterol 40mg	13%
Sodium 110mg	5%
Total Carbohydrate 7g	2%
Dietary Fiber 3g	12%
Sugars g	
Protein 21g	

| • |
| • |

*Percent Daily Values are based on a 2,000 calorie diet. Your daily values may be higher or lower depending on your calorie needs:

	Calories:	2,000	2,500
Total Fat	Less than	65g	80g
Saturated Fat	Less than	20g	25g
Cholesterol	Less than	300mg	300mg
Sodium	Less than	2,400mg	2,400mg
Total Carbohydrate		300g	375g
Dietary Fiber		25g	30g

Calories per gram:
Fat 9 • Carbohydrate 4 • Protein 4

Product F

Nutrition Facts
Serving Size (35g)
Servings Per Container

Amount Per Serving

Calories 210	Calories from Fat 190

% Daily Value*

Total Fat 21g	32%
Saturated Fat 3g	15%
Trans Fat 0g	
Cholesterol 0mg	0%
Sodium 160mg	7%
Total Carbohydrate 4g	1%
Dietary Fiber 0g	0%
Sugars 4g	
Protein 0g	

| Vitamin A 0% | • | Vitamin C 0% |
| | • | |

*Percent Daily Values are based on a 2,000 calorie diet. Your daily values may be higher or lower depending on your calorie needs:

	Calories:	2,000	2,500
Total Fat	Less than	65g	80g
Saturated Fat	Less than	20g	25g
Cholesterol	Less than	300mg	300mg
Sodium	Less than	2,400mg	2,400mg
Total Carbohydrate		300g	375g
Dietary Fiber		25g	30g

Calories per gram:
Fat 9 • Carbohydrate 4 • Protein 4

Insert 2.8 Nutritional labels for products A through F are described in more detail in Inserts 2.9 and 2.10.

Insert 2.9 Daily diet composed of just the items shown in Insert 2.8.

Item	Servings	Calories Total	From fat	Total Fat	Satur. Fat	Choles-terol	Sodium	Total Carb	Fiber	Protein*	Vitamins A	C	D	Calcium	Iron
							% Daily Value (DV)								
A	3	210	0	0	0	0	21	15	12	24	30	45	24	0	90
B	1	130	45	8	16	7	5	4	0	16	15	4	25	30	0
C	1	210	50	9	10	7	35	10	8	18	6	6	0	6	9
D	2	40	0	0	0	0	32	2	16	2	12	8	0	4	8
E	1	200	90	16	10	15	5	2	12	42	100	25	0	15	20
F	1	210	190	33	15	0	6	1	0	0	0	0	0	0	0
Total	9	1000	375	66	51	29	104	34	48	102	163	88	49	55	127

* % DV for protein is not given on the label but 50g is the recommended value.

Insert 2.10 Ingredient statements for items in Insert 2.8.

Item A

Rice, wheat gluten, sugar, defatted wheat germ, salt, high fructose corn syrup, dried whey, malt flavoring, calcium caseinate, ascorbic acid (vitamin C), reduced iron, niacinamide, zinc oxide, pyridoxine hydrochloride (vitamin B_6), riboflavin (vitamin B_2), thiamin hydrochloride (vitamin B_1), vitamin A palmitate, folic acid, and vitamin D. Quality protected with BHT.

Item B

Reduced fat milk, vitamin A palmitate, and vitamin D_3 added.

Item C

Tomatoes (water, tomato paste), lasagna macaroni product (semolina, water, egg whites, glyceryl monstearate), cooked beef, sugar, corn oil flavoring, salt, parmesan cheese (pasteurized cultured milk, salt, enzymes), citric acid, spices, autolyzed yeast, hydrolyzed corn, soy, and wheat protein, olive oil, mushroom flavor (maltodextrin and mushroom juice powder).

Item D

Green beans, water, salt (for flavor).

Item E

Lettuce, chicken tenders, tomatoes, carrots, celery, diced onions.

Item F

Soybean oil, balsamic vinegar (preserved with sulfites), olive oil, basil, water, high fructose corn syrup, sugar, salt, dehydrated garlic, spice, xanthan gum, iron, niacinamide, zinc oxide, pyridoxine hydrochloride (vitamin B_6), riboflavin (vitamin B_2), thiamin hydrochloide (vitamin B_1), vitamin A palmitate, folic acid, and vitamin D. Quality protected with BHT.

Designing a healthy product

In developing a healthy diet, we want to get enough protein, vitamins, and minerals without getting too much fat, sugar, and sodium or too many calories. We get our energy from the macronutrients consisting of carbohydrates, fats, and proteins. To make effective use of these macronutrients, we need the micronutrients consisting of vitamins and minerals. Dietary fiber binds toxins and helps maintain digestive processes. Nutrition problems stem from malnutrition and overnutrition. Malnutrition results from not getting enough nutrients in the diet to maintain health. Overnutrition results from getting too much of one or more nutrients or an imbalance of nutrients leading to obesity, heart disease, diabetes, cancer, and other diseases.

Reliance on any one food is not a good idea. It is best to eat a wide variety of foods in moderation. The MyPyramid Web site (http://www.

Insert 2.11 Food Guide Pyramid from http://www.mypyramid.gov/.

mypyramid.gov) is a useful tool to help us balance our diets (see Insert 2.11). The pyramid features six colors—orange for grains, green for vegetables, red for fruits, yellow for oils and fats, blue for milk, and purple for meat and beans. Recommendations are for a 2,000-calorie-per-day diet. Six ounces of grains, preferably whole grains, provide a good source of calories from carbohydrates in a normal diet. We should get 2½ cups of vegetables and 2 cups of fruits and their products a day. They are good sources of some vitamins and provide dietary fiber. We should consume the equivalent of 3 cups of milk each day and 5½ ounces of meat or bean products each day to get our high-quality protein, and they are rich in minerals such as iron and calcium. We are encouraged to get most of our fats from fish, nuts, and vegetable oils, limiting solid, saturated, and trans fats. We also should limit our consumption of foods and beverages with added sugars. The previous Food Guide Pyramid was criticized because it emphasized meats, making it difficult for vegetarians to follow. Vegetarians must be careful to get enough of the right kind of protein, minerals, and vitamin B_{12}. Americans did not follow the Food Guide Pyramid very well. MyPyramid is much stricter in its recommendations. It remains to be seen whether Americans will be able to eat healthier with this new pyramid.

Remember this!

- There is no simple way to lose weight.
- Low-carbohydrate diets lead to metabolic imbalances.
- There are ways to lower the fat content in foods, but it is not as simple as just removing the fat.
- Organic food production minimizes the input of fossil fuels and synthetic chemicals.
- Nutraceuticals are foods specifically designed to act as drugs.
- Prolonged fasting can result in severe nutritional deficiencies.
- Eating disorders occur when peer pressure clashes with biological needs.
- Although there are individual dietary needs and restrictions for a select few, in general, our nutritional requirements tend to be quite similar.

- Nutrition labels provide information to assist us in making good decisions to improve the quality of our diets.
- Good nutrition requires an adequate consumption of nutrients without exceeding the recommended allowance of calories.
- Healthy eating demands a balanced diet.

Looking ahead

In the next chapter we'll learn about the many ways in which people make decisions about what foods to eat. In Chapters 4 through 6 we will learn how the foods we find on supermarket shelves are processed or formulated. In Chapters 7 through 9 we will learn how food scientists ensure the quality and safety of foods, develop new products, and how the government regulates them. More detailed information on nutrition and health is presented in Chapter 11.

Answers to chapter questions

Insert 2.4:

Natural: carrots and steak
Organic: soymilk, carrots, and tofu
Whole: corn, carrots, and steak

Inserts 2.8–2.10:

A. Special K
B. 2% milk
C. Lasagna with meat sauce
D. Canned green beans
E. Grilled chicken salad
F. Basil vinaigrette dressing

References

Litt, A. S. 2005. *The college student's guide to eating well on campus*, 2nd ed. Bethesda, MD: Tulip Hill Press.

Further reading

Brunberg, J. J. 2000. *Fasting girls*. New York: Vintage Books.
Mahady, G. B., H. H. S. Fong, and N. R. Farnsworth. 2001. *Botanical dietary supplements: Quality, safety and efficacy*. Lisse, The Netherlands: Swets & Zeitlinger.

Potter, N. N., and J. H. Hotchkiss. 1999. *Food science*, 5th ed. New York: Chapman & Hall.

Roday, S. 2007. *Food science & nutrition*. Oxford: Oxford University Press.

Somer, E. 1999. *Food and mood: The complete guide to eating well and feeling your best*. New York: Henry Holt & Co.

Whitney, E. N., and S. R. Rolfes. 2007. *Understanding nutrition*, 9th ed. Belmont, CA: West/Wadsworth.

Wolinsky, I., and J. A. Driskell. 2008. *Sports nutrition: Energy metabolism and exercise*. Boca Raton, FL: CRC Press, Taylor & Francis Group.

chapter three

Choosing the food we eat

Isaac likes to try new foods. He is always going to new restaurants and trying the latest thing. Nothing is too exotic for him. When he watches television, he is more likely to watch the ads than watch the programs. He drives his friends crazy when he has control of the remote. When he tastes something he really enjoys, he won't shut up until everyone he knows tries it. He's also very vocal about the foods he doesn't enjoy.

Jennifer is taking advantage of an opportunity of a lifetime. She is an exchange student in Ghana. When she first arrived, she was a very picky eater because most of the food was so strange. She longed for the foods she had grown up with and took every opportunity to eat fast food, but the burgers in the fast food chains in Ghana were just not the same as in the chains at home. For the first time in her life she was slimming down. As she adjusted and made new friends, she decided she was missing an important part of the cultural experience. She discovered millet, a crunchy yellow grain, and *nshima*, a mashed corn dish with an interesting flavor. They were delicious. She began to have a great time and to dread going home when the term was over. She's also regained the pounds she lost during the first couple of months.

Kyle avoids all preservatives and processed foods. He reads all the labels of foods he is considering purchasing when shopping. He doesn't want all those nasty chemicals in his body. When he can, he shops at the natural foods market, where he can buy lots of dietary supplements to improve his health. The problem is that he ends up with foods that take a lot of time to prepare; he wants to eat healthy, but he also wants to do other things with his life. He eats at a vegetarian restaurant down the street quite often. The food is good, but it eats into his budget and by the time he walks there and back, it takes almost as much time as it would have if he had just prepared it himself. Maybe he could move in with someone who shares his views on food and loves to cook!

When choosing the foods we eat, we tend to divide them into healthy and unhealthy foods. Like Isaac, Jennifer, and Kyle, we find that our food choices are influenced by many factors––our culture, our budget, and our time. Sometimes our views about healthy eating are distorted, leading to inappropriate food choices.

Looking back

The previous chapters have focused on the safety and healthiness of foods. Some key points covered in those chapters help prepare us for understanding why we choose to eat the foods we eat.

- Healthy eating demands a balanced diet.
- Nutrition labels can help us keep up with the quality of our diets.
- Eating disorders occur when peer pressure clashes with biological needs.
- Low-carbohydrate diets lead to metabolic imbalances.
- There is no simple way to lose weight.
- Fresh foods are more likely to contain harmful microbes than processed products.
- Spoilage is not a good indicator of a safety risk.
- Preservatives are food additives that prevent or retard spoilage.
- Natural chemicals and products are not necessarily superior to artificial chemicals and processed products.

Food choice

Eating can be one of the most enjoyable times of the day or it can get in the way of more important things. Even though the foods we eat can affect our health and happiness, we do not always choose our foods based on logic or even much conscious thought. In this chapter we will explore the reasons people choose the foods they do. It may be a good time to take a serious look at our eating habits and to become aware of the different factors that affect our dietary choices. We have many reasons for choosing the foods we eat, and for many of us, nutrition usually plays a minor role in these decisions. Most of us are busy and are unlikely to spend much of our time analyzing the most nutritious options for our meals. This chapter will look at why we choose the foods we do and how these choices affect the healthiness of our diet.

Safety

We all realize that foods can make us ill. Foods that exhibit an unfamiliar odor, color, or texture are suspect. Some are more adventuresome than others, but almost all of us will reject spoiled foods even though spoiled foods are not necessarily unsafe and unspoiled foods are not necessarily safe. Brand-name foods or those from chain restaurants are assumed to be safe until the brand or chain is linked to a case of food poisoning. Cooking helps ensure that our meats are safe, and rinsing and washing fresh fruits and vegetables helps to make them safer. Other storage and sanitation practices are used during the handling of food to decrease the chances of our becoming ill from contaminated food.

Not all of our efforts to avoid unsafe foods are successful, however, because too many people are sickened and die each year by consuming contaminated foods. Myths like those mentioned in Chapter 1 lead to many of these food-borne illnesses. Some of the myths include the belief that rare burgers and raw eggs are safe to eat, preservatives make foods unsafe, natural foods are safer than processed foods, mishandled foods that look and smell all right are safe, cooked foods are sterile and don't need to be refrigerated, and drinking alcohol protects shellfish eaters from contamination. Although guidelines are available for the proper handling and storage of foods, these practices are not always followed by commercial establishments, and unfortunately, many consumers are not even aware of these guidelines. At other times we are less concerned about safety than we should be. For example, in social settings we might eat food that we wouldn't normally consume when not surrounded by peers. In addition, the consumption of alcohol can cloud our judgment about safe food or handling practices. Even simple practices like washing our hands after using the bathroom or playing with our pets prior to preparing food tend to be ignored when we are in a hurry.

Health

As we learned in the previous chapter, the fundamental function of foods is to provide our bodies with energy in the form of calories. In addition to calories, food provides us with nutrients like protein, vitamins, minerals, and dietary fiber, which are important in maintaining good health. A nutritious diet to most people is one that avoids "bad" foods like breads, burgers, and fries and embraces "good" foods like granola, fruits, and vegetables. Most foods have a nutritional image. Foods like energy bars, yogurt, and fruit drinks have better reputations than others like hamburgers, eggs, and canned vegetables. Nutritional information is provided for a serving of several items in Insert 3.1. By looking at the labels, can you tell the "good" foods from the "bad" ones? Using "good" and "bad" foods to decide on our diet narrows our selection to a point where it can result in nutrient imbalances. Dietitians construct "good" diets by limiting calorie-rich and nutrient-poor foods and increasing nutrient-dense foods. In wealthy countries it is particularly difficult to resist overconsuming because of the many tempting treats available at our fingertips at low prices. Conversely, impoverished populations, without sufficient income to purchase a wide variety of foods, find it difficult to design an economical, healthy diet.

Numerous diet books are available suggesting that we cut back on fats or carbohydrates and consume more fiber. Others suggest creating a

	Product G	

Nutrition Facts

Serving Size (226g)
Servings Per Container – 1

Amount Per Serving

Calories 200	Calories from Fat 70

	% Daily Value*
Total Fat 8g	12%
Saturated Fat 2g	10%
Trans Fat 0g	
Cholesterol 40mg	13%
Sodium 660mg	28%
Total Carbohydrate 14g	5%
Dietary Fiber 3g	12%
Sugars 5g	
Protein 18g	

Vitamin A 90%	•	Vitamin C 25%
Calcium 20%	•	Iron 8%

*Percent Daily Values are based on a 2,000 calorie diet. Your daily values may be higher or lower depending on your calorie needs:

		Calories:	2,000	2,500
Total Fat	Less than		65g	80g
Saturated Fat	Less than		20g	25g
Cholesterol	Less than		300mg	300mg
Sodium	Less than		2,400mg	2,400mg
Total Carbohydrate			300g	375g
Dietary Fiber			25g	30g

Calories per gram:
Fat 9 • Carbohydrate 4 • Protein 4

	Product H	

Nutrition Facts

Serving Size (50g)
Servings Per Container – 1

Amount Per Serving

Calories 210	Calories from Fat 60

	% Daily Value*
Total Fat 7g	11%
Saturated Fat 4g	20%
Trans Fat 0g	
Cholesterol 0mg	0%
Sodium 320mg	13%
Total Carbohydrate 24g	8%
Dietary Fiber 1g	4%
Sugars 14g	
Protein 14g	

Vitamin A 50%	•	Vitamin C 200%
Calcium 6%	•	Iron 0%

*Percent Daily Values are based on a 2,000 calorie diet. Your daily values may be higher or lower depending on your calorie needs:

		Calories:	2,000	2,500
Total Fat	Less than		65g	80g
Saturated Fat	Less than		20g	25g
Cholesterol	Less than		300mg	300mg
Sodium	Less than		2,400mg	2,400mg
Total Carbohydrate			300g	375g
Dietary Fiber			25g	30g

Calories per gram:
Fat 9 • Carbohydrate 4 • Protein 4

	Product I	

Nutrition Facts

Serving Size (40g)
Servings Per Container – 17

Amount Per Serving

Calories 100	Calories from Fat 0

	% Daily Value*
Total Fat 0g	0%
Saturated Fat 0g	0%
Trans Fat 0g	
Cholesterol 0mg	0%
Sodium 5mg	0%
Total Carbohydrate 24g	8%
Dietary Fiber 3g	12%
Sugars 12g	
Protein 1g	

Vitamin A 10%	•	Vitamin C 0%
Calcium 2%	•	Iron 2%

*Percent Daily Values are based on a 2,000 calorie diet. Your daily values may be higher or lower depending on your calorie needs:

		Calories:	2,000	2,500
Total Fat	Less than		65g	80g
Saturated Fat	Less than		20g	25g
Cholesterol	Less than		300mg	300mg
Sodium	Less than		2,400mg	2,400mg
Total Carbohydrate			300g	375g
Dietary Fiber			25g	30g

Calories per gram:
Fat 9 • Carbohydrate 4 • Protein 4

	Product J	

Nutrition Facts

Serving Size (49g)
Servings Per Container – 9

Amount Per Serving

Calories 170	Calories from Fat 10

	% Daily Value*
Total Fat 1g	2%
Saturated Fat 0g	0%
Trans Fat 0g	
Cholesterol 0mg	0%
Sodium 0mg	0%
Total Carbohydrate 40g	13%
Dietary Fiber 6g	24%
Sugars 0g	
Protein 6g	

Vitamin A 0%	•	Vitamin C 0%
Calcium 2%	•	Iron 6%

*Percent Daily Values are based on a 2,000 calorie diet. Your daily values may be higher or lower depending on your calorie needs:

		Calories:	2,000	2,500
Total Fat	Less than		65g	80g
Saturated Fat	Less than		20g	25g
Cholesterol	Less than		300mg	300mg
Sodium	Less than		2,400mg	2,400mg
Total Carbohydrate			300g	375g
Dietary Fiber			25g	30g

Calories per gram:
Fat 9 • Carbohydrate 4 • Protein 4

	Product K	

Nutrition Facts

Serving Size (130g)
Servings Per Container – 3.5

Amount Per Serving

Calories 130	Calories from Fat 50

	% Daily Value*
Total Fat 6g	9%
Saturated Fat 1g	5%
Trans Fat 0g	
Cholesterol 0mg	0%
Sodium 300mg	13%
Total Carbohydrate 21g	7%
Dietary Fiber 8g	32%
Sugars 1g	
Protein 8g	

Vitamin A 15%	•	Vitamin C 0%
Calcium 8%	•	Iron 15%

*Percent Daily Values are based on a 2,000 calorie diet. Your daily values may be higher or lower depending on your calorie needs:

		Calories:	2,000	2,500
Total Fat	Less than		65g	80g
Saturated Fat	Less than		20g	25g
Cholesterol	Less than		300mg	300mg
Sodium	Less than		2,400mg	2,400mg
Total Carbohydrate			300g	375g
Dietary Fiber			25g	30g

Calories per gram:
Fat 9 • Carbohydrate 4 • Protein 4

	Product L	

Nutrition Facts

Serving Size (56g)
Servings Per Container – 8

Amount Per Serving

Calories 180	Calories from Fat 140

	% Daily Value*
Total Fat 16g	25%
Saturated Fat 0g	0%
Trans Fat 0g	
Cholesterol 35mg	12%
Sodium 600mg	25%
Total Carbohydrate 2g	1%
Dietary Fiber 0g	0%
Sugars 0g	
Protein 6g	

Vitamin A 0%	•	Vitamin C 0%
Calcium 2%	•	Iron 2%

*Percent Daily Values are based on a 2,000 calorie diet. Your daily values may be higher or lower depending on your calorie needs:

		Calories:	2,000	2,500
Total Fat	Less than		65g	80g
Saturated Fat	Less than		20g	25g
Cholesterol	Less than		300mg	300mg
Sodium	Less than		2,400mg	2,400mg
Total Carbohydrate			300g	375g
Dietary Fiber			25g	30g

Calories per gram:
Fat 9 • Carbohydrate 4 • Protein 4

Insert 3.1 Nutrition Facts labeling for eleven products. Based on this information, which foods would you classify as healthy and which ones would you classify as unhealthy? Find out what these products are at the end of the chapter. Minor nutrients in products H, J, M, and O are not shown.

Product M

Nutrition Facts

Serving Size (27g)
Servings Per Container – 1

Amount Per Serving

Calories 100	Calories from Fat 10

	% Daily Value*
Total Fat 1g	2%
Saturated Fat 1g	5%
Trans Fat 0g	
Cholesterol 0mg	0%
Sodium 125mg	5%
Total Carbohydrate 24g	8%
Dietary Fiber 1g	4%
Sugars 13g	
Protein 1g	

Vitamin A 8%	•	Vitamin C 10%
Calcium 0%	•	Iron 20%

*Percent Daily Values are based on a 2,000 calorie diet. Your daily values may be higher or lower depending on your calorie needs:

		Calories:	2,000	2,500
Total Fat	Less than		65g	80g
Saturated Fat	Less than		20g	25g
Cholesterol	Less than		300mg	300mg
Sodium	Less than		2,400mg	2,400mg
Total Carbohydrate			300g	375g
Dietary Fiber			25g	30g

Calories per gram:
   Fat 9 • Carbohydrate 4 • Protein 4

Product N

Nutrition Facts

Serving Size (227g)
Servings Per Container – 1

Amount Per Serving

Calories 240	Calories from Fat 20

	% Daily Value*
Total Fat 2g	3%
Saturated Fat 1.5g	8%
Trans Fat 0g	
Cholesterol 20mg	7%
Sodium 110mg	5%
Total Carbohydrate 49g	16%
Dietary Fiber 0g	0%
Sugars 40g	
Protein 7g	

Vitamin A 0%	•	Vitamin C 0%
Calcium 20%	•	Iron 0%

*Percent Daily Values are based on a 2,000 calorie diet. Your daily values may be higher or lower depending on your calorie needs:

		Calories:	2,000	2,500
Total Fat	Less than		65g	80g
Saturated Fat	Less than		20g	25g
Cholesterol	Less than		300mg	300mg
Sodium	Less than		2,400mg	2,400mg
Total Carbohydrate			300g	375g
Dietary Fiber			25g	30g

Calories per gram:
   Fat 9 • Carbohydrate 4 • Protein 4

Product O

Nutrition Facts

Serving Size (62g)
Servings Per Container – 14

Amount Per Serving

Calories 100	Calories from Fat 25

	% Daily Value*
Total Fat 3g	5%
Saturated Fat 3g	15%
Trans Fat 0g	
Cholesterol 10mg	3%
Sodium 50mg	2%
Total Carbohydrate 14g	5%
Dietary Fiber 4g	16%
Sugars 4g	
Protein 3g	

Vitamin A 8%	•	Vitamin C 0%
Calcium 8%	•	Iron 0%

*Percent Daily Values are based on a 2,000 calorie diet. Your daily values may be higher or lower depending on your calorie needs:

		Calories:	2,000	2,500
Total Fat	Less than		65g	80g
Saturated Fat	Less than		20g	25g
Cholesterol	Less than		300mg	300mg
Sodium	Less than		2,400mg	2,400mg
Total Carbohydrate			300g	375g
Dietary Fiber			25g	30g

Calories per gram:
   Fat 9 • Carbohydrate 4 • Protein 4

Product P

Nutrition Facts

Serving Size (52g)
Servings Per Container – 6

Amount Per Serving

Calories 200	Calories from Fat 110

	% Daily Value*
Total Fat 12g	18%
Saturated Fat 0.5g	3%
Trans Fat 0g	
Cholesterol 5mg	2%
Sodium 95mg	4%
Total Carbohydrate 22g	7%
Dietary Fiber 1g	4%
Sugars 10g	
Protein 2g	

Vitamin A 0%	•	Vitamin C 2%
Calcium 6%	•	Iron 4%

*Percent Daily Values are based on a 2,000 calorie diet. Your daily values may be higher or lower depending on your calorie needs:

		Calories:	2,000	2,500
Total Fat	Less than		65g	80g
Saturated Fat	Less than		20g	25g
Cholesterol	Less than		300mg	300mg
Sodium	Less than		2,400mg	2,400mg
Total Carbohydrate			300g	375g
Dietary Fiber			25g	30g

Calories per gram:
   Fat 9 • Carbohydrate 4 • Protein 4

Product Q

Nutrition Facts

Serving Size (35g)
Servings Per Container – 6

Amount Per Serving

Calories 130	Calories from Fat 45

	% Daily Value*
Total Fat 5g	8%
Saturated Fat 0g	0%
Trans Fat 0g	
Cholesterol 0mg	0%
Sodium 65mg	3%
Total Carbohydrate 20g	7%
Dietary Fiber 4g	16%
Sugars 5g	
Protein 5g	

Vitamin A 0%	•	Vitamin C 0%
Calcium 0%	•	Iron 4%

*Percent Daily Values are based on a 2,000 calorie diet. Your daily values may be higher or lower depending on your calorie needs:

		Calories:	2,000	2,500
Total Fat	Less than		65g	80g
Saturated Fat	Less than		20g	25g
Cholesterol	Less than		300mg	300mg
Sodium	Less than		2,400mg	2,400mg
Total Carbohydrate			300g	375g
Dietary Fiber			25g	30g

Calories per gram:
   Fat 9 • Carbohydrate 4 • Protein 4

Insert 3.1 (Continued)

balance among carbohydrates, proteins, and fats, yet others promote the matching up of different foods for different blood types, hair colors, or astrological signs. Meats and dairy products are marked for elimination in some books; others recommend either the abolition or the incorporation of alcoholic beverages. Some diet regimens claim that increasing vitamin and mineral intake is the secret to good health. It is difficult to sell diet books by advocating a balanced diet, chosen from a wide range of sources. The healthiest diet is one that balances macro- and micronutrients. We all know that too little of any particular nutrient is unhealthy, but too much of a nutrient, particularly vitamins A and D, can also be unhealthy.

Weight loss

As discussed in the previous chapter, many of us are unhappy with our weight. Some of us are too heavy and need to shed pounds. Selecting a healthy diet is the first step. Next, we need to find the foods that fit into that diet. Dieters use many methods to lose unwanted pounds: fasting; counting calories; skipping meals; avoiding fat, sugar, and carbohydrates; increasing proteins; taking appetite suppressants or dietary supplements; and exercise.

Fasting involves the complete elimination of food or specific types of food for hours and even days. Extended fasting is voluntary starvation and can have negative health consequences. During fasting, the body's metabolism slows down, making the breakdown of fat less efficient. In addition, fasting leads to ketosis, resulting from the degradation of proteins into forms of energy. Ketosis is characterized by bad breath, although not all forms of bad breath are due to ketosis. The ketone bodies formed in ketosis provide energy but may not be adequate substitutes for glucose for proper brain function. Although many religions practice fasting to cleanse the body and the mind, most nutritionists believe that extended fasting is not a healthy practice for losing weight.

Calorie counters often substitute fruits, vegetables, and cereal grains for fried, highly sugared, and starchy foods. Skipping meals, particularly breakfast, can be self-defeating because breakfast skippers tend to eat "small" snacks throughout the day, which ultimately add up to a greater number of calories. Skipping foods with fat and sugar is a way to reduce calories. There are numerous products on the market that are low in fat and sugar; however, most of these are also low in fiber. Diets low in fiber and fat contribute to hunger pangs because the stomach becomes empty more quickly. In addition, diets with less than 15 to 20% calories from fat are difficult to maintain because they tend to be unappetizing. Increasing fiber content with some fat, preferably from plant sources, will help keep one satisfied without adding too many calories. Avoiding carbohydrates and increasing protein intake is currently the most popular way to lose weight. Foods that are high in protein and low in carbohydrates usually contain fats, so that they are more satisfying than low-fat alternatives, but there are serious problems associated with these diets as discussed in Chapter 2.

It is very common for dietary supplements to promise that the pounds will drop off without cutting back on calories or working out. Unfortunately the phrase "no pain, no gain" is probably closer to the truth. A sensible exercise program, combined with a sound diet is still the most effective and healthiest way to lose weight. Many dietary supplements such as chromium picolinate, the prime ingredient in fat-burner products, have not been shown to be effective in reducing weight. Other compounds such as ephedra, once banned in the United States and now reauthorized, speed metabolism, but the side effects are more dangerous than the benefits. Appetite suppressants do decrease our desire for food, but they are expensive and may also have unwanted side effects. One of the most effective appetite suppressants is sugar. Consuming a little sugar, such as a small serving of fruit about thirty minutes before a meal tends to decrease hunger and leads to a feeling of fullness more quickly during a regular meal.

Our bodies have a built-in mechanism that signals when we have had enough, but there are many tempting treats available that override this feeling of fullness and cause us to eat to excess. Sugared beverages fall into this category, and may be particularly dangerous because they provide excess calories without filling the stomach. Foods high in fats and low in vitamins and minerals can also be consumed quickly to excess before we experience a sense of fullness. In addition, munching all evening at social occasions can be a problem.

The best way of monitoring diet is by counting calories. For approximately every 3,500 calories that our body absorbs more than it burns, we gain a pound of weight. Likewise, for every 3,500 calories our body burns more than it absorbs, we lose a pound. Those who are trying to lose weight should look for foods that are nutrient-dense and restrict high-calorie foods and beverages (particularly alcohol, which provides lots of calories and few other nutrients). For example, people who practice volumetrics or similar techniques find that if they eat more, they gain less weight. This practice stresses the eating of foods high in water and fiber and lower in calories. As a result, the larger volumes fill them up without adding excess calories. To see how food choices affect calorie consumption, see Inserts 3.2 and 3.3.

Foods that help reduce weight are those that are tasty, filling, and low in calories. Fruits, vegetables, and whole grains are high in fiber and water but low in calories. When combined with a protein source and some fat, they can fill up our stomachs and slow the emptying time to delay hunger. The most effective weight-loss diet is one that is low in calories, satisfying,

Insert 3.2 A snack of chips and dip containing 175 calories. (From B. Rolls, *The Volumetrics Eating Plan*, New York: Harper-Collins, 2005. Photo by Michael A. Black. With permission.)

and adaptable. It should also be high in nutrients. Many diets are available that meet these needs, but it is important that we pick the one that is best for us. Guidelines for designing a diet to maintain a healthy lifestyle were provided at the end of the previous chapter.

Weight gain

Others are underweight and would be healthier with some weight gain. Many athletes are interested in bulking-up. Since the best way to lose weight is to decrease calorie consumption, the best way to gain it is to increase calorie consumption. It is ironic that those who wish to bulk-up

Insert 3.3 A veggie platter also containing 175 calories. Which option do you think is healthier—this one or the chips in Insert 3.2? (From B. Rolls, *The Volumetrics Eating Plan*, New York: Harper-Collins, 2005. Photo by Michael A. Black. With permission.)

tend to turn to proteins, the same source of calories for those who wish to lose weight. One reason for the emphasis on proteins is that muscle is primarily protein. It is the combination of impact exercise (e.g., weight lifting) and excess consumption of carbohydrates, however, that is more effective in building muscle. As in dieting to reduce weight, it is recommended that the dieter set realistic weight goals when dieting to bulk-up. A realistic goal for healthy weight gain is 20% of body mass for a young male and 10 to 15% for a young female per year. Such increases in weight should be coupled with resistance (weight) training to ensure the gain is lean body mass and not fat.

Social factors

Although health and safety are important factors that shape what we eat, other factors influence our food choices. Friends and social occasions affect what we eat and thus influence our long-term health. A recommendation of a new food from a trusted friend or family member may be the best introduction to that food, whereas a negative comment may turn us off to a certain product. When in a crowd, it is difficult to maintain our normal diet. If friends are eating fast food or dining at restaurants that feature high-calorie appetizers and main courses, it is hard to eat healthy. Likewise, when everyone around us is eating sensible meals and watching what we choose, it is harder to splurge.

Parties tend to have food and beverages that are high in calories and low in vitamins and minerals. Sporting events tend to increase the need to have tailgate and postgame parties and lots of calorie-laden food. Late-night snacking can also have devastating consequences for the waistline. Foods help people mix and mingle by creating a comfort zone. Finger foods and alcoholic beverages add calories with little compensating nutritional benefits. Pizza is one of the most nutritionally balanced foods, but it is so good it tends to be overconsumed. When we are enjoying the company of others, we are less likely to monitor just how much we are eating and drinking. We also are more likely to ignore our natural mechanisms indicating that we are full. Even in more serious situations like weddings, funerals, and professional receptions, food serves as an icebreaker. Most social occasions offer lower calorie alternatives (e.g., celery and carrot sticks without the dip), which can be consumed slowly, but these items are not usually the most popular ones. Those people who go to one of these activities on an empty stomach are more likely to overeat than those who eat something with a little sugar, fat, or fiber to take the edge off their hunger before going.

Religious influences

Food traditions are associated with many religions: eating fish on Fridays by Catholics, abstinence from pork and other products by Jews and Muslims, avoidance of all meats but seafood by Buddhists, vegetarianism embraced by Hindus, and the potluck suppers of Lutherans and Methodists. Many religious groups abstain from alcoholic beverages, whereas others incorporate them into their ceremonies. Bread and olive oil are considered sacred foods for some groups. Feasts and holidays associated with religions introduce specific foods and traditions. Christians in North America consume turkey and cranberry sauce at Thanksgiving; Jews, unleavened bread during Passover; and Hindus, sweets and puddings in celebration of Divali. In addition, fasting is associated with many religious groups such as the Jews, Muslims, Buddhists, and Hindus. Although there are dietary laws for believers of most religions, not all believers follow them, and sects may adopt their own rules. Some Muslims and Southern Baptists consume alcoholic beverages, while some Hindus eat water buffalo steaks and Zen Buddhists consume very little food mainly consisting of cereal grains. A comparison of some dietary habits of different religions is shown in Insert 3.4.

Kosher laws govern the foods that Jews are permitted to consume. Kosher laws specify types of animals that are permissible and forbidden for food, forbid the consumption of blood, and forbid the consumption of dairy and meat products at the same meal. Among the prohibited animals for kosher foods are wild birds, shark, dogfish, hog, lobster, shrimp, crab, and insects. Strict guidelines must be followed during the slaughter of animals and these processes must be approved by a rabbi to be considered kosher. Salt of a specific grain size (small enough to cover the entire surface but large enough to prevent dissolving within thirty minutes after application), known as kosher salt, is applied to the meat to draw out any remaining blood in the meat. Orthodox Jews serve meat on a separate set of dishes than those on which they serve dairy items. Pareve foods are those that can be consumed with either meat or dairy products. In addition, there are special rules governing the foods that can be consumed during Passover, which include unleavened bread. Kosher practices are far-reaching and prescribed for the way food is prepared and served.

Halal laws govern the foods that Muslims are permitted to consume. Halal refers to permitted foods, and haram to those that are forbidden. Kosher and halal practices have many similarities with some noticeable differences. For example, locusts, shrimp, and lobster are halal. Pork, cats, dogs, birds of prey, carrion, blood, intoxicants, and inappropriate drugs are haram. There is no requirement to separate meat and dairy products in Islam. Slaughter of animals for meat must be done in a humane fashion by a sane Muslim who invokes the name of Allah during slaughter. The blood must be drained from the animal prior to any cutting, but soaking

Insert 3.4 Religious occasions associated with feasting or fasting. (Adapted from P. G. Kittler and K. P. Sucher, *Cultural Foods*, Belmont, CA: Wadsworth/ Thomson Learning, 2000.)

Occasion	Religion	Significance	Associated foods
Ash Wednesday	Christian	Beginning of Lent	Abstinence from certain foods for six weeks until Easter
Christmas	Christian	Celebration of birth of Christ	Varies widely; usually centered around meat like turkey
Divali	Hindu	Darkest night of the year	Many delicacies including *roti*, flatbread with curry
Easter	Christian	Celebration of resurrection	Varies widely; usually centered around meat like ham
Eid al-Fitir	Muslim	Breaking of Ramadan fast	Large feast
Passover	Jewish	Celebration of freedom	*Seder*: chicken soup, matzo balls, unleavened bread
Pravarana	Buddhist	End of the rainy season	Buns and sweets
Ramadan	Muslim	Month of fasting	No food or water between sunrise and sunset
Rosh Hashanah	Jewish	Beginning of the New Year	Apples dipped in honey, sweets, nothing sour or bitter
Sabbath	Jewish	Day of rest and day of prayer	Cooked meals prepared the evening before
Yom Kippur	Jewish	Day of atonement	No food or water from sunset to sunset

and salting required for kosher products is not required for halal. Hunting is permitted if it is for meat but not if it is solely for sport.

Ethnicity

Food is an integral part of a cultural heritage. Many factors contribute to a specific cuisine. Staples are products that form the basis of a cuisine, such as meat and potatoes in the midwestern part of the United States, beans and corn in many parts of Latin America, and rice in many other parts of the world. Specialty foods round out a cuisine, giving it its flair and desirability to people of other cultures. Certain flavors and dishes become associated with a particular cuisine. Frequently these flavors are acquired tastes that are not acceptable to outsiders. When ethnic cuisine becomes popular (Mexican, Chinese, and Italian in the United States, or hamburgers and fried chicken outside the United States), it tends to be a mere semblance of the authentic diet. The most recent wave of ethnic cuisine to hit the United States is from India.

Foods popular in the northeastern United States reflect the immigration of Europeans. Tea, beer, and whiskey come from the United Kingdom; many cheeses, sauces, and spices come from France; pizza and pasta come from Italy. Southern cooking has its roots in "soul food" from Africa. An emphasis on pork products, fried foods, and boiled leafy green vegetables are products of the slave culture. Cajun cuisine, a specialty in Louisiana, features such foods as pralines, beignets, and gumbo. The staples of the central plains are a mix of Native America, central Europe, and Scandinavia. Popcorn, nuts, and many of our vegetables were contributed by plains Indians; sausages and potatoes came from Germany; milk and preserved fish are common foods from Scandinavia. Latinos have had considerable influence on American cuisine, particularly in the southwest. Tacos, tortillas, burritos, and enchiladas are products of Mexico; starchy vegetables such as cassava, and chili-based sauces and jerked meats come from the Caribbean islands; coffee, chocolate, and tropical fruits are part of Central and South American culture. The west coast has seen the influence of Asian cultures. Soy sauce, fried rice, and stir-fried vegetables are just a few of the contributions to the American diet from China; ramen noodles and sushi come from Japan; noodle soups, coconuts, curries, and fermented fish have been introduced from other Asian cultures. Although there are still regional differences, most of these cultures have diffused throughout the American diet. See Insert 3.5 for more information on ethnic foods in America.

Family traditions

Early exposure to foods has a profound effect on our preferences. Foods that we did not like but were forced to eat when we were young are ones we may reject later. Likewise, foods that we enjoyed as a child are those that provide special comfort as we age. Family traditions establish many of our attitudes toward foods and provide a basis upon which we judge new foods. The food preparer in the home may introduce foods that are not common among other families in the region and may fail to prepare some popular items. Family traditions around holidays may differ from the traditions of ethnic groups. Most of us accept unquestioningly that the foods we ate at home when we were young are the ones we are supposed to eat. Most college students say they ate much healthier diets at home than they do at school. Other students seeking a healthier lifestyle reject

Insert 3.5 Ethnic foods in the United States. (Adapted from P. G. Kittler and K. P. Sucher, *Cultural Foods*, Belmont, CA: Wadsworth/Thomson Learning, 2000 and S. T. Herbst, *Food Lover's Companion*, 4th ed., Hauppauge, NY: Barron's Educational Series, Inc., 2007.)

Food	Ethnicity	Description
Arroz con pollo	Cuban	Chicken with rice with seasonings like peppers and saffron
Brunswick stew	Southern	Made from squirrel meat and onions
Burrito	Mexican	Shredded meat, cheese, and/or beans in a flour tortilla
Cracklin' bread	African American	Yellow cornmeal bread with added pork cracklings (crisp fat or skin)
Gumbo	Cajun	Stew with many vegetables particularly okra and tomatoes
Gyro	Greek	Minced lamb in pita bread
Jerky	Native American	Salted, dried meat, originally primarily buffalo
Kasha	Central European	Roasted buckwheat groats (whole or ground)
Kimchi	Korean	Spicy-hot fermented cabbage or turnips
Oatmeal	British/Scottish	Also known as porridge
Pasta	Italian	Wide variety of noodles served with a variety of sauces
Peanut butter	Southern	Blend of ground peanuts, oil, and salt
Pierogi	Polish	Boiled, stuffed dumpling with onions and cabbage
Pita bread	Middle Eastern	Flat, leavened bread with an internal pocket
Sushi	Japanese	With rice, frequently associated with raw fish (*sashimi*)
Taco	Mexican	Shredded meat, cheese, and/or beans in a folded corn tortilla
Tofu	Asian	Curds from soybean milk

the food habits they learned at home and adopt alternative eating patterns such as vegetarianism, low-carb diets, etc.

Advertising

Food companies have large advertising budgets to entice consumers to buy their products. Advertising may play a bigger role in our food choices than many of us are willing to admit. Advertisements like the one shown in Insert 3.6 can be particularly effective at introducing new

Insert 3.6 Food advertisement for banana pudding designed by Laura McKinley, Lauren Hill, Chris Zachary, and Ben Sherrill as part of a class assignment.

food products. If we are not aware that a product exists, we are not going to buy it. Advertisers are particularly interested in communicating with a special group of consumers called *first-adopters* or *influentials*. First-adopters like Isaac are people who are more daring than the general population and are looking for new things to try. They are particularly sensitive to advertising and seek out new products. If a first-adopter tries something that is really good or really bad, friends and family will hear about it. Influentials are individuals whose word is highly respected. If they say something is really good or really bad, people listen and are likely to follow. If a first-adopter is also an influential and endorses a product, it is likely to trigger an increased interest in and sale of that product. Thus, even if we are not directly affected by advertising, it can indirectly influence what we eat and drink. The wide variety of choices available probably would not be as tempting if it were not for advertising. Another role of advertising is to remind us of food we used to consume but have stopped eating for some reason. Such gentle reminders can send us back to the store for old favorites, particularly for those items we enjoyed when we were young.

Time and trends

As society changes, so does the food. Diets today are quite different from those twenty or thirty years ago. If you don't believe this, spend thirty minutes with a parent, grandparent, or some other person over fifty to get their perspective on how eating has changed in the past thirty to fifty years. Many factors have affected what we eat today. Technology has transformed our food supply (see Insert 3.7). The food industry now

Insert 3.7 Photo of food for a family of four in the United States in the 1950s. Notice the absence of processed and formulated foods. Picture is of Steve Czeklinski, a DuPont worker, and his family. (Reprinted with permission of Hagley Museum and Library; photo by Alex Henderson.)

produces more formulated foods with mixtures of ingredients, which reduces the consumption of whole foods. These formulated foods are more convenient with meals prepared in minutes rather than hours. The foods can be eaten on the run, and don't require a sit-down, home-style dinner. We tend to be more health conscious, yet we are heavier than our counterparts of yesteryear. Part of the reason we carry extra weight is because we sit more at our jobs and homes and walk and exercise less. We eat more meals away from home than we once did. Going to a restaurant used to be a once-a-week or even a once-a-month treat. Fast-food restaurants, which featured only hamburgers, fries, or chicken only, now have a wide range of menu options with larger sizes. Ethnic restaurants have taken over the dining scene, offering more options than the meat/ potato/two-vegetable places from years past. Supermarkets are now one-stop shopping centers where we can buy food, drugs, magazines, and even motor oil. The variety for such products as yogurt, fresh salads, sushi, and energy bars are too numerous to list—products that were not even available to our parents.

Economics

Most students are on tight budgets and can't afford to spend much on food. A meal plan can reduce the importance of meal costs. Students looking for bargains often buy canned and dry food because fresh meat, vegetables, and frozen foods tend to be more expensive. Some students are better at budgeting time and money than others. Many students may eat well at the beginning of the month, but are reduced to feasting on ramen noodles toward the end of the month. For students and others who must carefully budget their money, it pays to develop a long-term plan, balancing nutritional concerns with preferences and economics. Items are less expensive when purchased in bulk, but it is no bargain if the leftovers are allowed to spoil because of improper handling or because they are not consumed within a reasonable time. If individuals have compatible tastes and schedules, they may find that sharing meals and expenses will allow them to eat better for less.

Personal philosophy

Our diets are affected by our personal beliefs. Some of us are sensitive to nature and can't bring ourselves to eat meat, but most of us have no such qualms. Others like Kyle eat "organic" foods or those with less packaging because of concerns about pollution and the environment. Some of us boycott companies that advocate things we disagree with or broadcast really dumb television commercials. Many Americans and Europeans avoid all preservatives and processed foods. Our food choices are influenced by our politics, our friends, the media, and many other sources. Many of these influences may be rational, but others are not. Regardless, our beliefs have an effect on what we choose to eat and those choices affect our nutritional status.

Sensory properties

It is the sensory properties of foods that are most likely to affect what we eat and how much. These sensory properties include color, flavor, and texture. They affect what we are willing to put into our mouths, how much we enjoy it, whether we will spit it out or consume more, and even how much benefit we will derive from it. There are studies that show that digestion of food begins in the mouth. Better-tasting food will actually be more likely to be digested than food that does not taste as good.

The first sensory characteristic is color or appearance. The food purchaser is likely to use color to decide whether to purchase items like fresh meat, fresh fruits and vegetables, and foods in see-through containers. Before we put anything in our mouths we tend to look at it to see if it is acceptable. Certain foods are expected to be certain colors. We like our spinach green, but not our meats. Consumers prefer red tomatoes, even though orange tomatoes are higher in vitamin A. Other factors that affect acceptability are blemishes, splotches, bruises, bugs or obvious bug holes, rotten parts, etc. An experiment at the University of Massachusetts determined that the color of fruit-flavored beverages can affect the perception of sweetness. When extra red food coloring was added to a strawberry drink, panelists indicated that the darker red drink was sweeter than the one with normal color. Panelists rated a darker red sample with 4.0% sugar as equivalent in sweetness to a lighter red sample with 4.4% sugar. Color is also related to a product's ability to quench thirst. On the other hand, the appearance of a product may look so good that it increases our expectations beyond what the flavor can deliver. We then become disappointed when we eat it.

Flavor is the combination of the senses of taste and smell. When we say something tastes good, we are usually referring to flavor and not taste. Taste is perceived on the tongue and is primarily confined to sweet, sour, salty, and bitter. Flavor is a much more intense experience than taste alone. We use our sense of smell while we are actually chewing our food, and not just before we put something into our mouth. The unique combination of taste and aroma give a particular food its flavor signature. When we eat a familiar food, we have a set of expectations. If a food does not live up to those expectations, we are disappointed and might not buy it again for a while. If it exceeds our expectations, we may buy it more frequently, but we may also raise our expectations the next time we eat that food. We tend to have higher expectations of fresh foods than frozen or canned foods.

The texture of a food relates to how it feels to our sense of touch— either to the hand or in the mouth. Texture may be the most underrated sensory property, but it is one that leads to the rejection of many foods. Most of us do not like lumpy mashed potatoes or puddings, slimy boiled okra or hot-dog wieners, limp lettuce, watery protein drinks, or mealy apples. A food either feels right or it doesn't. If it is not right we are likely to reject it.

Convenience

Throughout history a large part of people's lives has been consumed with finding food, preparing it, and eating it. Many of us today have ready access to food twenty-four hours a day, seven days a week, and yet we are still looking for ways to make our foods more convenient. In the age of multitasking, we see no problem in eating while watching television, driving down the road, walking to class, or even talking on the phone. Technology has provided us with many options not available to previous generations or other parts of the world.

Convenience comes in packages that have the food portioned into single servings to avoid measuring out what we need. We have packages that have everything we need for a meal for eating directly or for ready preparation. *Speed-scratch* is a term for putting together all of the ingredients of a food such as a cake, so we can experience the satisfaction of preparing the food from scratch without spending extra time and effort. Many packages are developed for use in the microwave, and the package serves as a disposable eating container. Other foods are designed to be eaten with our fingers. These innovations were designed primarily to free us from the drudgery of the kitchen and the confinement of the sit-down meal. For many of us, meals have become just another task to accomplish and scratch off our lists.

Fast-food restaurants provide convenience by delivering a fully prepared meal quickly. Fast-food operations minimize the time between order and delivery of a meal. Thus, the main factor that distinguishes a fast-food operation from other restaurants is the speed of service. Alternate ordering mechanisms at fast-food chains include variation of the number of cash registers, drive-thru windows, and long-line speed-up techniques. Most menus encompass a limited selection of items with emphasis on a single item type. Different strategies are employed to keep prepared meals hot: made-to-order preparation, limited holding times, and hot-food cabinets. Packaging serves many important functions in fast-food restaurants including maintaining temperature, unitization, marketing, item identification, and spill prevention. Nutritional concerns are less important at fast-food restaurants than other food service outlets. The types of foods popular at fast-food outlets tend to be high in fat, high in sodium, and low in fiber, although salads are now provided as an option at many chains.

Others are rejecting the fast-food world and convenience foods. The slow-foods movement started in Italy in 1982 and is finding enthusiasts elsewhere. The slow-foods movement avoids short-cuts in shopping, preparing, and enjoying food. It emphasizes quality––the quality of the ingredients and the food. It stresses the social aspects of enjoying a good meal together with family and friends. For many of us, however, food and meals are not a high priority in our lives, and we prefer to spend that time on more pressing activities.

Pathogenic eating

As mentioned above, mealtime serves as a source of pleasure and offers an opportunity for socializing. Unfortunately, some people develop bad food habits that seriously affect their health. Food scientists and nutritionists stress moderation in food consumption and eating a wide variety of foods. The popular movie *Super Size Me* (2004) illustrated the dangers of a thirty-day McDonald's®-only diet coupled with almost no exercise.

Although it is easy to blame a single factor, like fast-food restaurants, we must take responsibility for our own diets and lifestyles.

Athletes are also at risk, with some estimates indicating that 60% have unhealthy diets. Football linemen may consume massive amounts of food to achieve their weight goals, and wrestlers and gymnasts may sweat off pounds to lose weight. Supplements are sought as the magic answer to gain or lose weight and enhance performance or endurance. Rapid changes in weight place undue strain on our bodies, and the promises of supplements are usually exaggerated with the long-term damage to our bodies ignored.

Pathogenic eating can extend to being overconcerned about healthiness leading to unhealthy habits. *Orthorexia nervosa*, an obsession with healthful eating, was described in *Health Food Junkies* (Bratman, 2000). Such an obsession generally leads to the elimination of "bad" foods, greatly reducing the amount of "good" foods available. Foods may be classified as "bad" due to social concerns, real or perceived safety issues, desire to get closer to nature, and real or perceived food reactions. Likewise, foods may be classified as "good" to eat, right for our type, macrobiotics, or body purification and spiritual gratification. Each of these concerns can be legitimate in the right context. We live in a complex world and must be careful what we put into our bodies. It is when these concerns and strict adherence to a rigid set of rules become more important than our overall health that they become an obsession. The author suggests that orthorexia can lead to social problems when we can't eat what everyone else is eating and start telling others how they are killing themselves. Ironically, orthorexia can also lead to health problems as the elimination of many "bad" foods (sufferers usually have more than a single concern) usually leaves a rather small number of "good" foods available to meet nutritional needs for protein, vitamins, or minerals, etc. This situation generally leads to increased reliance on dietary supplements for nutrients and other "beneficial" agents. The health-food stores in many towns have a fairly small section of whole grains, but a large section of pills and chemicals to address various ills and concerns.

Meal patterns around the world

Meal patterns vary widely across cultures as Jennifer found out during her visit to Ghana. The Spaniards start their day early with a small breakfast, have a big meal at noon followed by a mid-afternoon siesta, and then don't eat again until late in the evening. Indonesians and Southeast Asians eat rice at every meal with many light snacks such as rice cakes between meals. Hungarians have one large meal at noon with small meals and snacks throughout the day. The Chinese have small meals and snacks throughout the day with a big meal served in the evening. Many traditions and rituals are associated with African meal patterns, which are very different from other cultures as Jennifer discovered during her study abroad. Indians have two main meals a day—at mid-morning and in the evening, with much snacking in between.

Likewise, breakfast traditions vary widely between cultures. Australians like a big "breaky" with lots of eggs and meat and Vegemite, a yeast extract that is rarely appreciated by visitors, on their toast. *Akara* (Insert 3.8), a popular breakfast food in many African countries, is made from black-eyed pea meal and resembles a hushpuppy. South American breakfasts tend to be simple, consisting of strong, sweet coffee with milk and breads or rolls with butter and jam. The Japanese wake up with an intensely sour red plum called an *umeboshi* to clear the head and cleanse the mouth, followed by a bowl of *misoshiro* made from fermented bean paste. An Israeli breakfast consists of a wide range of fresh vegetables mixed with meats, cheeses, or eggs and accompanied by hearty hot cereals. Like most Europeans, the Dutch like breads and rolls for breakfast with unsalted butter and jam, but many tend to drink tea instead of coffee.

Cereal grains and their products comprise the most widely consumed source of calories around the world. Wheat breads tend to be favored by most cultures because of their light, spongy texture. Croissants, brioches, and baguettes are French contributions to bread eaters of the world. Australians enjoy damper bread, which tastes like a big biscuit, because it can be easily prepared in the bush and baked over an open fire. *Lavash* is a thin, crisp bread popular in Middle Eastern countries such as Iran,

Insert 3.8 Akara, a deep-fried product made from black-eyed pea flour and enjoyed by Africans. (From K. H. McWatters and B. B. Brantley, "Characteristics of Akara Prepared from Cowpea Paste and Meal," *Food Technology* 36, no. 1 (1982): 66. With permission.)

and *flatbrod* is a crisp rye bread consumed in Norway. Italians consume their grains in the form of pasta. Bulgur, a coarse cracked-wheat grain, is used in the preparation of many Turkish dishes. Rice flour is used to make noodles in Asian countries such as Vietnam.

Fruits and vegetables are great sources of fiber and many vitamins. Fresh berries and rhubarb (a sour red stem) are gathered and enjoyed in Iceland. Tropical fruits such as mangoes, bananas, and papayas are staples in many African countries. Dates and figs are enjoyed in Iran. The *durian* is a fruit enjoyed in Asia that has a sweet and enticing flesh for those who can get past a very offensive odor, so offensive that it is banned in hotel lobbies and on airplanes. Romanians like their fruits preserved in thick sugar syrup in a dish called *dulceata*, and Austrians consume many of their fruits in soups. Beets, cabbage, and potatoes are staples for Eastern European countries such as Ukraine. Cauliflower is a staple in Finland. The *daikon*, a white radish that looks like a white carrot, is prized in Japan. Sauerkraut, made by fermenting cabbage, is a favorite vegetable dish among peoples of northeastern Europe, such as Estonia and Latvia. Tomato sauce is an important ingredient in many Italian foods, and okra is the primary ingredient in African gumbos.

Meats and dairy products provide protein and essential minerals. *Asado* (beef grilled over red-hot coals without flames, is shown in Insert 3.9), is a favorite meal in Argentina, but pork is the most widely consumed meat in the Philippines; lamb is most popular in New Zealand. Hamburgers and fried chicken are America's gifts to the world. *Ceviche* is an uncooked fish popular in Caribbean countries that is "preserved" in lime juice and spices. Czechs prefer their meat cooked until tender and served with a sour cream topping. Bean curd or *tofu* is a meat substitute enjoyed in many Asian countries such as China and Japan. The Danes enjoy specially prepared cheeses that have a nutty, buttery flavor. Cheese dumplings in a flavored milk sauce, known as *ras malai*, are widely consumed on the Indian subcontinent. Ricotta cheese, a mildly flavored white

Insert 3.9 Carne asada (barbequed beef) is an Argentine specialty shown here on an open hearth. (Photo courtesy of Dr. Carlos Margaria, United States Distilled Spirits.)

cheese, is eaten with sourdough bread in Malta. Yogurt and kefir are consumed as liquid beverages in many of the Slavic countries such as Croatia and Macedonia. *Mish*, a skim-milk cheese fermented in earthenware jars, is served with most Egyptian meals. Cottage cheese is a major component in blintzes and knishes.

Not all food is consumed for nutritional purposes. Sweets and fatty foods are enjoyed the world over by those who can afford them. The Greeks consume large quantities of *baklava*, a very sweet pastry usually containing nuts. Morocco boasts of its *makalhara*, a honey-dipped pretzel. Coconut pudding and sweet, sticky rice cakes are favorites of Indonesians. Canadians make delicious pies from pumpkin and rhubarb. Chocolate products are enjoyed throughout the world, but the Belgians and the Swiss may be the biggest consumers. Toasted sunflower and pumpkin seeds are snacks enjoyed by Russians. The Dutch like to dip French fries in mayonnaise, but Americans prefer their fries with ketchup.

Selecting healthy foods

In the previous chapter, the principles for designing a nutritious diet were introduced. The point of this chapter is to show how many things can influence the choices of the foods we eat. Although many of us try to eat healthy foods, there are many temptations and influences that affect our health that we do not consciously consider. Selection of healthy foods should be done in the context of a healthy diet. Obviously we want to pay attention to the safety of our foods to prevent food poisoning. If we are not at the weight we want to be, it is much better to develop a long-term plan to lose or gain weight a little at a time rather than through crash programs that might be more dangerous than our current condition. Planning ahead for the peer pressure associated with social events is the best way to avoid a bad situation. Steering a middle course between letting all inhibitions go, leading to a guilt complex, or becoming the health policeman for everyone is probably the best strategy.

Most religious, ethnic, and family traditions incorporate healthy and unhealthy aspects. An effective strategy is to use basic nutritional principles to incorporate the best parts of these traditions into our weekly routine and minimize the worst parts. Most of us can enjoy celebrations, but enjoy them in moderation. Advertising does not have to coax us into unhealthy habits. If some foods are just too tempting, it is probably best not to have them around. Economic, personal, and political considerations may affect our food choices, but they don't need to rule our diet and affect our health. Anyone who ignores the sensory quality of the diet will soon find it hard to stay on that diet. We can treat ourselves to things we like and try new foods and products, particularly those that are healthier alternatives to foods we should be cutting back on in our diet. Then we can add the new foods that we like to our diet, and discard the ones we don't. It may be harder to design a healthy diet from convenience foods than from whole foods and from fast foods than slow foods, but a careful study of food labels and restaurant Web sites can help us maintain our lifestyle without greatly damaging our health. Finally, we should be adventuresome and try foods from other cultures because, like Jennifer, we may be surprised at what we are missing.

Remember this!

Although many of us try to eat healthy diets, there are many temptations and influences affecting our decisions that we do not consciously consider.

- Meal patterns vary widely across cultures.
- Pathogenic eating can extend to being overconcerned about healthiness, leading to unhealthy habits.
- Fast-food operations minimize the time between order and delivery of a meal.
- Sensory properties include color, flavor, and texture.
- Technology has transformed our food supply, producing more formulated foods with mixtures of ingredients, and moving away from the eating of whole foods.
- Advertising may play a bigger role in our food choices than many of us are willing to admit.
- Staples are those products that form the basis of a cuisine.
- A sensible exercise program combined with a sound diet is the most effective and healthiest way to lose weight.
- If we want to eat a healthy diet, we have an obligation to read labels, analyze our diets, and exercise willpower in our food selections.
- Even though the food we eat can affect our health and happiness, we do not always choose the food based on logic or even much conscious thought.

Looking ahead

This chapter was designed to provide an introduction to food choice. In subsequent chapters we will be introduced to the types of foods we eat and how they are manufactured for safety, a longer life, convenience, and a pleasant eating experience. Chapter 8 describes how new food products and their packages are designed.

Answers to chapter questions
Insert 3.1

1. G. Lean Cuisine® Baked Chicken Florentine
2. H. Zone Perfect® Strawberry Yogurt All Natural Nutrition Bar
3. I. Sunsweet® Gold Label Dried Plums
4. J. Post® Healthy Classics™ The Original Shredded Wheat Spoon Size®
5. K. Kuner's of Colorado Southwestern Black Beans with Cumin & Chili Spices
6. L. Bryan Juicy Jumbos® Franks
7. M. Kellogg's® Froot Loops®
8. N. Breyers® Fruit on the Bottom Mixed Berry Lowfat Yogurt
9. O. Edy's® No Sugar Added Vanilla Flavored Light Ice Cream
10. P. Krispy Kreme Doughnuts® Original Glazed Doughnuts
11. Q. Kashi™ Peanut Butter Chewy Granola Bars

References

Bratman, S. 2000. *Health food junkies.* New York: Broadway Books.
Herbst, S. T. 2007. *Food lover's companion*, 4th ed. Hauppauge, NY: Barron's Educational Series, Inc.
Kittler, P. G., and K. P. Sucher. 2000. *Cultural foods.* Belmont, CA: Wadsworth/ Thomson Learning.
Litt, A. S. 2005. *The college student's guide to eating well on campus,* 2nd ed. Bethesda, MD: Tulip Hill Press.
McWatters, K. H., and B. B. Brantley. 1982. Characteristics of akara prepared from cowpea paste and meal. *Food Technology* 36(1): 66.
Rolls, B. 2005. *The volumetrics eating plan.* New York: Harper-Collins.

Further reading

Barer-Stein, T. 1999. *You eat what you are.* Willowdale, Ontario, Canada: Firefly Books.
Berry, J., and E. Keller. 2003. *The influentials: One American in ten tells the other nine how to vote, where to eat, and what to buy.* New York: Simon and Schuster.
Clydesdale, F. M. 1991. Color perception and food quality. *Journal of Food Quality* 14, 61.
MacDougall, D. B., ed. 2002. *Colour in food: Improving quality.* Cambridge, U.K.: CRC Press Woodhead Publishing Limited.
Meiselman, H. L., ed. 2000. *Dimensions of the meal: The science, culture, business, and art of eating.* Gaithersburg, MD: Aspen Publishers, Inc.
Petrini, C. 2006. *Slow food revolution: A new culture for eating and living.* New York: Rizzoli.
Regenstein, J. M., M. M. Chaudry, and C. E. Regenstein. 2003. The kosher and halal food laws. *Comprehensive Reviews of Food Science & Food Safety* 2, 111.

Riaz, M. N., and M. N. Chaudry. 2004. *Halal food production*. Boca Raton, FL: CRC Press LLC.

Rolls, B., and R. A. Barnett. 2000. *Volumetrics: Feel full on fewer calories*. New York: Harper-Collins Publishers.

Weimann, G. 1994. *The influentials: People who influence other people*. Albany, NY: State University of New York Press.

Whitney, E. N., and S. R. Rolfes. 2007. *Understanding nutrition*, 9th ed. Belmont, CA: Wadsworth.

section two

Commercial food products

chapter four

Processed foods

Martin and Laura were given a class assignment to interview an older person (someone over fifty-five) about what America was like in the 1960s and 1970s, particularly with respect to food. Martin interviewed his mom, who was born in 1961 and grew up in the 1970s; Laura talked to her grandfather, who was born just after World War II. Martin's mom told him all the houses were small but tidy, the yards well kept. All the moms stayed at home and cooked delicious meals, and all the fathers knew best. Politics seemed much simpler then, at least until the Vietnam War came along. Laura's grandfather commented on watching the World Series in the afternoon when everyone was at work and there were far fewer football games on TV. They both said the music was much better back then; you could hum the melodies and understand the words. Martin's mom loved the BeeGees; Laura's grandfather spoke of some guys called the Beatles and the Beachboys.

Life seemed a lot simpler then. Gas was much cheaper, and cars were much bigger. Cruising around town was big. The drive-in restaurants and movie theaters were the places to hang out and make out. When eating fast food, Burger Chef and White Castle were more popular than McDonald's. Laura's grandfather was pleased the burgers were only 15 cents, but Martin's mom pointed out that they were small and not very tasty. The French fries were larger and less crispy, and the serving size was smaller. There was no such thing as super-sizing back then. They had to walk up to a window to order because there was no place to sit down and there were no drive-throughs. Eating out was something you did infrequently. Most meals were prepared and eaten in the home.

The "supermarkets" were very small and most eating out was done at a family restaurant. You couldn't find a deli, bakery, or meat counter. In fact, you usually got your meat at a separate store. There was little fresh produce unless it was in season, and there was a very small frozen food section. With the exception of canned goods, there were not as many packaged foods or as much variety as today. Fresh milk and bread were delivered to the home twice a week by truck. The TV dinners back then were not nearly as good. If they had to go back, they would miss the microwave oven. Even though the kitchens were much simpler back then, mothers spent much of their time preparing most of the family's meals every day. People did not seem to be very concerned about eating healthy. Breakfast was much bigger back then, with bacon and eggs featured almost every morning. Martin's mom was less tempted by chocolate or chips because they were much harder to get, but she did consume lots of whole milk and homemade cookies. Meat with lots of fat was the centerpiece for lunch and supper with lots of starches and vegetables. Almost every meal was topped with a dessert, usually cake, but sometimes only a can of fruit cocktail (not a favorite of Laura's grandfather). Meals were important back then, with the whole family getting together and talking. No one would even think of watching TV or listening to music or the radio during meal-times. Martin's mother and Laura's grandfather both said the family meal is what they missed the most about the good old days. Store-bought ice cream was not big back then, but Dairy Queen and Tastee Freeze were great places to hang out and eat good soft-serve cones.

Looking back

- Meal patterns vary widely across cultures.
- Staples are products that form the basis of a cuisine.
- If we wish to eat healthy, we have an obligation to read labels, analyze our diets, and show willpower in our food selections.
- Even though the food we eat can affect our health and happiness, we do not always choose our food based on logic or even much conscious thought.
- Natural chemicals and products are not necessarily superior to artificial chemicals and processed products.
- Processed foods should be designed to spoil before they become unsafe.
- Fresh foods are more likely to contain harmful microbes than processed products.

What are processed foods and why are they processed?

Processed foods are products that have been preserved so they will not spoil as quickly as the fresh, whole foods (raw materials) from which they were made. Most raw materials are perishable and require careful handling or processing to prevent loss. Foods fresh from the farm, ocean, pond, or other source are only available when they are in season, which can be inconvenient.

The primary reason for food processing is to reduce or eliminate harmful microbes from growing in foods. Some food processes sterilize the product, others kill but do not eliminate microbes, and still others slow or prevent microbial growth. (Can you tell which foods in Insert 4.1

Insert 4.1 Which of the following food products are commercially sterile?	
☐ Chicken of the Sea canned tuna	☐ Kellogg's Pop Tarts
☐ Del Monte canned pineapple	☐ Little Debbie Oatmeal Creme Pie
☐ Frozen Eggos	☐ Pop Secret microwave popcorn
☐ Gatorade	☐ Reese's peanut butter cups
☐ Jeno's frozen pizza	☐ Spam

are sterile?) One type of food preservation actually encourages growth of beneficial microbes to prevent the growth of harmful ones. The major benefit of controlling microbes through processing is to decrease the chances of safety problems and to slow spoilage. Remember that spoiled foods are preferred over unsafe foods. Some processes allow us to store the food at room temperature. These foods are called *shelf stable*.

Another reason for preserving foods is to stop the loss of nutrients. Many of these losses are due to the presence of active enzymes. The same factors that affect microbes affect enzymes. In fact, one reason food processes affect microbes is that the enzymes in the microbe are inactivated. Fresh foods lose their nutritional value as they spoil. Some types of food processing are damaging to vitamins and minerals, but the same conditions that ensure food safety also prevent further loss of nutrients. This chapter introduces us to food processing, describing the types of processing steps used to manufacture foods and the consequences of processing on shelf life, nutrition, quality, and safety. It also introduces the importance of packaging in the manufacturing of foods.

Benefits of processing

Advanced technology has given us many advantages not enjoyed by previous generations. We tend to become spoiled with so many benefits that we forget about the costs associated with technology. Advances in any technology, including food technology, come with associated costs, many of which are not obvious. Food processing and preservation can increase shelf stability of a raw material usually at the cost of affordability, nutrition, and quality. When we buy or eat a processed food, we are trading away a product that may be higher in nutrients, better in quality, and lower in price for one that is less perishable, safer, and more convenient. Some other tradeoffs that are made in food processing include:

- A processed food is more likely to be eaten because a fresh food is more likely to spoil.
- A shelf-stable food is ready to eat when we are, but we need to fit a fresh food into our plans before it spoils.

- Losses of vitamins, minerals, and quality of processed foods tend to be much slower than those in fresh foods.
- Some consumers like the flavor of processed foods better than fresh foods because that is what they are used to.
- Processed foods have less waste while fresh foods usually require trimming and are more likely to have cooking losses.
- Processed foods take more energy to produce but usually take less energy to store than fresh foods.

Processing steps

Many steps occur in a food processing plant to convert raw materials into a processed food. Unit operations are distinct steps common to many food processes. These steps begin as the raw material is unloaded at the plant dock and continue until the packaged product is loaded up on another truck or railcar to the place we will buy it. Most raw materials are perishable and must be processed within a period of hours to days after arrival at the plant. Some common unit operations include materials handling (moving them from place to place in the processing plant), cleaning, separating, grading, pumping, mixing, heat exchanging (adding or removing heat), packaging, and controlling (making sure what is supposed to happen, does happen).

It is important for the food scientist to know the unit operations of a process for a specific product. Each operation provides an opportunity for a problem to develop. An understanding of the potential problems that can occur provides ways to prevent them. The safety and quality of a food product is dependent on the processor's ability to control the individual steps and fit them together properly. The speed and capacity of a process is never greater than the speed and capacity of the slowest operation.

Minor variations in a process can result in major differences in a final product. Think of all of the different types of cheeses or alcoholic beverages. The unit operations for all cheeses are very similar. The major differences among types of cheeses are due to differences in starting materials and addition, subtraction, modification, or rearrangement of a few unit operations. The order of operations is also important. In some processes, the food is packaged before it is processed, while in others, packaging occurs at the end of the process. If careful attention is not paid to the order of operations, late steps can undo positive aspects of early steps.

For example, there are many unit operations in the canning of pimientos. The pimientos are filled (in either the whole, cut, or diced form) into glass jars and sealed (capped) before being processed under pressure with steam. These operations include receiving, holding, washing, peeling, coring, grading, blanching, cutting, dicing, acidification, filling, exhausting, capping, heat processing, cooling, and packing the glass jars into cartons.

Types of food processes

This chapter will focus on processing of fresh raw materials into processed products that are generally recognizable as the original whole food. Later chapters will present information on formulated foods, which involve the mixing of many ingredients; chilled foods, which are perishable and must be refrigerated to maintain quality; and prepared foods, which are ready to eat with or without heating. As food scientists become more inventive with new products, the lines between fresh, processed, formulated, chilled, and prepared foods become blurred. Some foods fit in more than one category, but this separation into groups will help us understand how foods are manufactured.

Raw agricultural materials (whole foods) are processed in many ways. Major types of processing include heating, freezing, drying, concentrating, curing, milling, extracting, and fermenting. Newer technologies are now being tested to preserve foods. Microwaves represent a technology introduced and widely accepted within the last two generations. When you have a chance, ask someone over fifty-five what life was like before the microwave oven. Irradiation is a controversial technology that can be used to either sterilize a food (like canning) or radurize (similar to pasteurize) certain products. Even more exotic processes are being developed that accomplish killing power similar to heat processes but with little or no heat. These processes promise the quality and nutrition of fresh foods with the convenience and shelf stability of processed foods.

Heating

The purpose of heating foods is to kill microbes. The problem with heating is that it can also destroy nutrients and quality. Remember that raw foods are likely to harbor spoilage microbes and pathogens. *Cooking* kills both of these types of microbes but does not sterilize raw foods. Cooking will be discussed in more depth in relation to prepared foods. Blanching is a unit operation in processing of raw fruits and vegetables to inactivate enzymes prior to other processing steps. *Pasteurization* involves a milder form of heating that kills all pathogens (microbes that can make us sick) but does not kill all the spoilage microbes. *Canning* sterilizes the product in the container, usually under pressure. *Aseptic processing* and packaging sterilizes the product before putting it in the package.

Food engineers are responsible for designing safe food processes. If the heat process is not adequate to kill the harmful microbes, the food could spoil too quickly or worse yet become a safety hazard. If the process is too rigorous, the food could lose nutrients and quality. Processing techniques for a specific food are designed with an understanding of the potential microbes present and the properties of the food. For example, many pathogens are not able to grow in high-acid foods, so the heat treatment required to sterilize a high-acid food like tomatoes is not as great as for a low-acid food like green beans. The design of a heat treatment relates to the temperature that will be applied and the time it will be processed. It must also consider the type of heat transfer in the product, the coldest point in the product, the length of time it takes to reach the selected temperature, and the length of time it takes the product to cool down.

Nicholas Appert developed the canning process in the early 1800s in response to a prize offered by Napoleon to help feed his armies more efficiently. Meats, fruits, vegetables, and many formulated foods that are found in cans or jars on supermarket shelves are canned. Raw materials are graded, sorted, and cut or formed before being placed in the container. Fruits and vegetables are usually blanched not only to inactivate enzymes that can hurt product quality, but also to remove oxygen and decrease volume so more product can fit in the container. The can or jar is then sealed and placed in a type of pressure cooker (called a *retort*). Since some foods have acids (naturally occurring or added), they do not take as long to process as those that do not. Canning drives all of the oxygen out of the container, so the microbes of concern are those that can live without oxygen. The most dangerous microbe that can survive in the absence of oxygen is *Clostridium botulinum*, the pathogen responsible for botulism. To test the adequacy of a canning process for a specific product, certain cans are inoculated with a microbe, such as *Bacillus stearothermophilus*, which is more resistant to heat than *Clostridium botulinum*. If *B. stearothermophilus* cannot survive, then botulism is not a concern.

Heat energy is transferred by *conduction* (from one molecule to the next) in solid foods like tuna and puddings. It is transferred by *convection* in liquid foods like chicken broth or evaporated milk, which involves predictable patterns of swirling. Heat is transferred to the food on the outside of the container first and then into the center. The *cold spot* (last point in the container to reach the desired temperature) is the geometric center of the can in conduction heating and below the geometric center in convection heating. The length of time it takes for a process to be complete is based on the cold spot in the can. Mixed foods like pork and beans and clam chowder contain particulates (chunks, which means it is neither completely solid nor completely liquid), which make it more difficult to predict heat-transfer patterns and the cold spot. The larger the container, the longer it takes to reach the proper temperature in the cold spot, and the more nutrients and quality are lost in other parts of the container. One way to speed up the process is to agitate the containers by shaking them or moving them through the heat in a coiled track. Agitation provides a more even heating pattern throughout the container.

In pasteurization and aseptic processing the heating occurs outside the container in a heat exchanger, and the product is packaged into a sterile container under sterile conditions. The higher the temperature, the shorter the time is needed to adequately process the product. This time-shortening effect is dramatic. For example, it takes 30 minutes to pasteurize milk at 63°C (145°F) but only 15 seconds at 72°C (161°F). The high-temperature, short-time (HTST) process results in less destruction to nutrients and quality. Remember that pasteurized products are not sterile and require refrigeration to slow spoilage. Aseptically processed and packaged products are commercially sterile and do not require refrigeration. The most common examples of these are the rectangular milk and juice cartons with the little plastic straw attached that you poke into the foil circle and imbibe. Since the product is added to the container after it has been processed, everything from the packaging material to the filling equipment and the air in the filling room must be sterilized to ensure the final product remains sterile.

Freezing

Freezing is a milder form of preservation than heat treatment resulting in less loss of nutrients and quality. It is also a type of heat exchange, but this heat is being removed from the product, not added to it. Freezing slows growth of microbes but does not sufficiently kill microbes to prevent spoilage or safety problems in thawed product that is not stored properly. The rate of freezing is affected by the freezing equipment and the properties of the food. Although water freezes at 0°C (32°F), other foods have lower freezing temperatures. Sugars, salts, and other components can act as antifreeze. During freezing, water freezes first, concentrating the other components, and freezing occurs from the outside in. The concentration of these components, particularly enzymes, can lead to quality losses if the food does not freeze quickly. Vegetables are frequently blanched before freezing to prevent enzyme damage.

The oldest and slowest way to freeze foods is in still air. It is also the technique most damaging to quality. A better way to freeze foods is to blow air at temperatures below −30°C with high forced air velocities to speed up the freezing process. A food or the package containing the food is placed in contact with a cold surface in indirect-contact freezing. The effectiveness of this type of freezing depends on how much contact is achieved between the food and the freezing unit. Surface dehydration, known as *freezer burn*, can be prevented by quick freezing and proper packaging. Scraped-surface freezers are used for liquids or slushes such as those available at the gas-and-go places on the interstate highways. Individual quick freezing (IQF) involves the immersion of the product into liquid refrigerant. Since there is little contact of the product with air in IQF, there is less chance for oxidation or loss of moisture. Immersion in liquid nitrogen or other ultra-low-temperature liquids (known as *cryogenic freezing*), the quickest and most expensive way to freeze foods, produces the highest quality frozen foods.

Freezing is an excellent method of food processing in countries that have adequate home freezer/refrigerator units, but it places added responsibility on the consumer to handle the frozen food properly. Improper handling of defrosted foods can lead to a safety hazard. Defrosted foods should not be stored above 5°C (41°F) because microbes are free to grow after thawing. Quick defrosting in a microwave or slow thawing overnight in a refrigerator are the safest methods. Defrosted meat and dairy products represent a greater risk of hazard than bakery or other plant products, as they are more likely to be contaminated and more likely to encourage the growth of microbes. Expiration dates for frozen foods assume that the product will be stored properly and are useless if it is not kept frozen. Some products have time-temperature indicators designed into the package to warn the consumer of temperature abuse.

Drying

Removal of water from foods, such as drying of solid foods or concentrating of liquid/semisolid foods and beverages, is another method of preserving foods. Microbes need water to grow. Microbial growth is slowed or halted in dried foods and concentrated beverages. There are many similarities between frozen and dried foods, as the act of freezing removes available water from the food by turning it into ice. Frozen, concentrated juices remove water both ways. Another type of product with less water and less microbial growth is called an *intermediate-moisture food*. Intermediate-moisture foods are solid or semisolid foods that have less moisture (water) available to microbes, but are not as dry as typical dried foods. These foods usually have ingredients added to make the moisture that is present less available, and will be discussed in greater detail in the next chapter.

Dry foods are not as dry as they appear. Water is a very important component of most foods. Even dry flours contain as much as 10 to 15% moisture. There are many ways to remove water from foods (see Insert 4.2 for pictures of three different types of driers). The least sophisticated form of drying is leaving a harvested crop in the field to be dried by the sun such as raisins, dates, and tomatoes. Ovens at low relative humidity (RH) will also remove water but can leave a cooked flavor. By introducing a vacuum, the temperature required to remove water is lower, which achieves the same effect with less cooked flavor. Increased air circulation in cabinets, tunnels, or on fluidized beds can also be used to obtain similar effects. *Spray drying* forces a liquid like milk through a small nozzle to

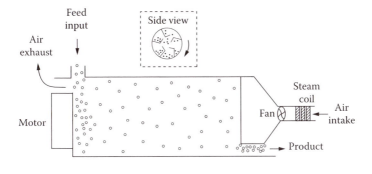

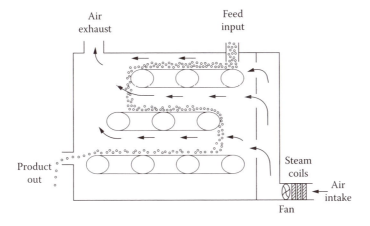

Insert 4.2 Different types of commercial drying equipment. Top is a rotary air dryer. Bottom is a multiple conveyor dryer and next page shows a spray dryer. (From D. R. Heldman and R. W. Hartel, *Principles of Food Processing*, New York: Chapman & Hall/Springer Science, 1997. With permission.)

form a fine mist. The mist circulates in a large conically shaped cabinet with circulating air. The dried powder collects in a container. *Drum drying* involves the rotation of a warm drum or cylinder through a liquid food product. A thin film of the liquid becomes attached to the drum and dries as it is cooked on the drum. A knife blade scrapes the dried product off the surface of the drum into sheets, which are collected in a trough. *Freeze drying* involves freezing a solid or liquid food in a pan in a chamber. The chamber is then placed under vacuum to lower the pressure to a point at which the ice is transformed directly into water vapor without becoming liquid water. This direct conversion from solid to gas is called *sublimation* rather than evaporation.

Different types of drying result in different types of final products. A property that is desirable for dried powders is their quickness at

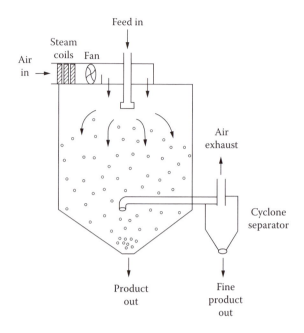

Insert 4.2 (Continued)

dissolving. Spray-dried products tend to be more readily soluble than others. Dried foods that contain sugars are usually *hygroscopic*, which means that they rapidly absorb water. Thus, they appear sticky to the touch. That is why you don't get the sticky mess with sugar-free beverage spills that you do with those that contain sugar. Packaging is critical in protecting a dried food from the moisture in the air.

Concentrating

The removal of water from a liquid food without changing it into a solid is *concentration*. We are most familiar with juice concentrates and syrups as concentrated foods. Most of these foods concentrate sugars, leading to increased sweetness and resistance to microbial growth. Many concentrated juices are susceptible to yeast spoilage or loss of flavor, so freezing is combined with concentrating to preserve quality. Another reason for concentrating liquids is to reduce the size of containers and thus reduce shipping costs. During the evaporation process, delicate flavors also evaporate causing a loss of overall product flavor. These volatile components (known as *essence*) can be captured and added back to the product to give it a fuller flavor. When diluting juice concentrates, the water can add in off-flavors to the reconstituted product.

Curing

Curing of meats is another way of decreasing the availability of moisture to microbes, thus decreasing microbial growth. Curing is achieved by adding preservatives that bind water. The two most common water-binding preservatives are sucrose (sugar) and sodium chloride (salt). Other antimicrobial agents such as sodium nitrate and sodium nitrite are added, which also help maintain the red color. Salt and other preservatives can be added to a meat product by soaking the food in a brine (high-salt) solution. Sugar-curing of hams is achieved by hand rubbing the sugar into the meat. Some products, particularly salt-cured items, are also smoked, which adds flavor, contributes to drying, and can generate antimicrobial agents. Curing also results in typical flavor development in the meats, which many consumers like, while others do not.

Curing of vegetables like onions and potatoes refers to the process of holding the product either in the field or a large ventilated room at high temperatures. Curing helps dry out the surface of the product and permits healing of wounds incurred during harvest. A cured onion or potato is less likely to spoil than one that has not been cured.

Milling

When we eat cereal grains we are eating seeds. Few of us actually eat the whole seed as it is usually hard to digest. Many cereals are milled into flour to make it more digestible. *Milling* separates the seed into many fractions based on the structure of the seed. Flours are produced by dry milling, which is accomplished by adding some water to swell the seed in preparing it for milling. A series of numerous grinding steps further separate fractions without damaging starch granules. Whole flour contains more fractions than white flour, which contains a higher percentage of starch and less fiber. Part of the milling process involves drying with most flours stabilized at 10 to 15% moisture. It is important to store flours in dry conditions as they can absorb water and become susceptible to mold growth.

Wet milling permits the combination of specific components of the grain like starch and protein. Large amounts of warm water are added to the grain, which is then soaked for more than a day. The grain, now swollen with water, is ground and components are separated by solubility properties. Products of wet milling include cornstarch, wheat gluten, and other starch products.

Extracting

Many unit operations in food processes involve the removal of unwanted portions of a raw material. In addition to the removal of water, mentioned previously, other processes remove soil or other objectionable matter from the raw product, while others separate the edible from the inedible portions. *Extraction* involves the removal of part of a substance that is contained within it. Millions of consumers extract soluble components of crushed leaves or ground beans in their homes or offices each day to enjoy fresh cups of tea or coffee.

The liquid and associated solids of a fruit or vegetable are extracted to form a juice. The most commonly processed juice worldwide is orange juice. The flavor of a fresh fruit like an orange and its extracted juice can vary for many reasons such as cultivar (cultivated variety), growing area, season, and growing conditions. Since consumers want the same flavor experience each time they imbibe, juice processors blend juices from different sources to achieve the desired flavor. Juices from different fruits or vegetables can also be blended together to provide unique flavor and color sensations. There are many fruit-flavored beverages on the market that have little juice and few of the associated nutrients. To make sure we are actually getting real fruit juices, we need to read the label carefully. Fresh juices are quite perishable. They need to be refrigerated or undergo further processing to be preserved. The type and length of the process (time–temperature relationships) and storage conditions affect shelf life as well as sensory and nutritional quality. Unit operations of juice manufacture are extraction, clarification, deaeration, pasteurization, concentration, essence addition, and canning, bottling, or freezing. Extraction yield can be improved by the use of pectic enzymes. A Wilmes press, used to extract juice from fruits including the preliminary steps in winemaking, is shown in Insert 4.3.

Oils, or liquid fats, are also extracted from plant parts to provide important ingredients for foods. Removal of the edible oils from the rest of the plant material is usually achieved by physical pressure. After pressing, many unit operations are needed to convert the initial extract into a useable product. Refining and degumming remove unwanted substances from the oil to improve the stability of its flavor and color. Bleaching of the oil removes objectionable odors and colors while increasing its stability. Other deodorization steps improve its acceptability to the consumer. Hydrogenation of an oil can turn it into a solid spread, like margarine, which is easy to spread on toast or biscuits. *Interesterification* rearranges the oil molecules to affect its melting temperature and thus its desirability as a food ingredient. Antioxidants are added to oils to prevent them from deteriorating rapidly.

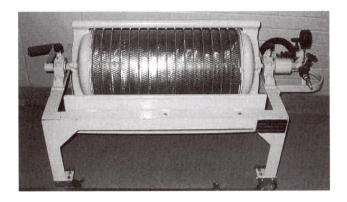

Insert 4.3 Pilot-scale model of a Wilmes press. These presses are used to extract juices for juice and wine manufacture. Pilot-scale equipment is used to produce small batches of a prototype during the development of new products. (Photo by Katherine Erickson.)

Fermenting

Fermentation is the only primary method of food preservation that encourages multiplication of microbes. Among the earliest processed foods are those that were fermented, salted, sun dried, or baked. Fermentation encourages the growth of beneficial microbes to outcompete spoilage and pathogenic microbes. To produce a fermented food, a starter culture of microbes is added to a perishable raw material to change it to a more stable food product. The stability of the product is due to the formation of a natural chemical preservative by the microbe. The most common preservatives generated are lactic acid and ethanol (see Insert 4.4 for chemical structures of these natural preservatives).

The most common types of fermentation include the production of yogurt and alcoholic beverages as well as the rising of dough prior to baking of bread. Yogurt production involves lactic acid fermentation. Alcoholic beverages result from the production of ethanol. Carbon dioxide is the chemical responsible for bread rising. Other lactic acid fermentations include buttermilk, olives, pickles, salami, sauerkraut, sour cream, and vanilla. These lactic acid fermentations are conducted under strictly controlled conditions in commercial processing plants, but they can also be started by leaving the substrate (raw material) in the air or adding some of a previously fermented product to the substrate. Genetic engineering of starter-culture microbes shows potential for more consistent quality of fermented products. Commercial fermentations can be disrupted by bacteriophages (bacterial viruses), which can attack starter cultures.

Insert 4.4 Chemical structures of natural preservatives. Can you identify a source for each?

Irradiating

One of the most controversial forms of food preservation is the use of irradiation. *Irradiation* is a potent killer of microbes and can be used to preserve foods as it induces little or no heat. It is also known as *cold sterilization*. Radioactivity was characterized by the Curies when they were studying uranium and radium, while food irradiation research began in the 1940s in the United States as the Atoms for Peace program. Extensive research has also been conducted in Europe. The types of ionizing radiation used for foods have been selected because of their power to penetrate food tissue without making it radioactive. Chemical changes induced in foods by irradiation appear to be similar to those produced by other food preservation techniques. Radiation can be applied at high doses that will sterilize a product (imagine a raw steak stored in a plastic pouch safely in a cabinet at room temperature) or at a lower dose for *radurization* (the irradiation equivalent of pasteurization). Irradiation kills microbes but does not inactivate toxins formed in the food before irradiation. Low-dose irradiation can be used to replace chemical fumigants to kill insects in imported foods, as an inhibitor of sprouting in onions and potatoes, and to destroy vegetative (active) microbes. Bacteria tend to be more resistant to irradiation than yeasts and molds, but irradiation does not inactivate toxins or undesirable enzymes. The U.S. Food and Drug Administration (FDA) has approved irradiation for bacon, sprouting of potatoes and onions, spices, strawberries, poultry, ground beef, and pork. Although the most promising application of irradiation is to improve the safety of fresh meats, it can also be used to extend the shelf life of fresh fruits and vegetables.

Most food irradiation operations have relied on gamma rays to kill microbes. Gamma rays are also dangerous to human health and thus elaborate safeguards are needed to protect workers in irradiation plants from stray rays. To process large quantities of perishable foods means that either the raw material must be transported to a facility that specializes in irradiation or irradiators must be installed in a processing plant. An alternative technology, electron-beam radiation, can achieve similar results without the health risks associated with gamma irradiation. The electron beam does not penetrate the food as well as gamma rays, so greater doses of energy are required to achieve the same effect.

Opponents of food irradiation consider it to be a nonessential technology that encourages the food industry to cover up practices that encourage contamination, while proponents of the technology indicate that it could prevent numerous cases of food poisoning due to inappropriate handling of foods in the home or food service operations. Part of the difference of opinion with regard to irradiation relates to general beliefs with respect to the safety of foods. Opponents of irradiation tend to believe that foods are inherently safe in their natural state and that they become unsafe when exposed to technology. Supporters of irradiation, including most food scientists, tend to believe that foods are inherently unsafe in their natural state and that a scientific understanding of what causes safety problems leads to development of technology that provides safer foods.

Nonthermal processing

While heat is the most effective means of killing microbes, the benefits are often achieved by sacrificing nutritional and sensory quality. Among the alternatives to heat is high-pressure processing. Microbes are killed in food products put under hydrostatic pressure. While it is not clear how high pressure kills microbes, bacterial spores are more resistant than normal vegetative cells. High-pressure processing is much milder than heat processing because the size and shape of the product is not important, so chunks (particulates) are not a problem. High pressure can be applied in the container, similar to that used in canning or in bulk, and then packaged like aseptic processing. High-pressure processing results in high-quality products, but it is a very expensive process. A pilot-scale high-pressure throttling device is shown in Insert 4.5. Another potential problem is that it does not always inactivate the enzymes that can decrease quality during storage. It will probably not replace conventional processing, but it has potential for high-priced specialty products. A combination of heat treatment and high pressure can help take advantage of the benefits of both techniques.

Other important operations

This section of the chapter has presented many types of food processing with a primary emphasis on preserving fresh, whole foods. Some other

Insert 4.5 Continuous high-pressure throttling may provide an alternative to heating for killing microorganisms. (Photo by Katherine Erickson.)

types of unit operations are generally considered food processes but have not been included in the preceding sections. Three of these operations are extrusion, roasting, and the application of microwaves.

Extrusion is a method for forming a product. Pastas, as well as many ready-to-eat breakfast cereals and salted snacks, are formed by extrusion. Extrusion involves the forcing of a mash of milled product through a small opening. A home pasta machine is a simple extruder. When the mash is heated and forced through the opening under pressure, the release results in evaporation of water and simultaneous heating and drying of the product, and usually results in a puffing of the product. Smacks® and Cheetos® are popular examples of extruded products. Many factors affect the process, including the speed and configuration of the screw that forces the mash forward, the moisture level and speed of the mash fed into the extruder, the temperature and length of the barrel housing the screw, and the size and shape of the opening known as the *die*.

Roasting is a type of dry heating of a raw material that will kill microbes and inactivate enzymes. Both meats and nuts are preserved by roasting. Dry heat is not as effective as wet heat in killing microbes or inactivating enzymes, so the temperature must be higher and the time longer to achieve similar results.

Another form of heat preservation is the use of microwaves. *Microwave heating* is fundamentally different from heating with dry or wet heat. Intermolecular friction is created in the food by a rapid reversal of an electric field, which leads to a change in polarity of water molecules. Microwave heating provides a more uniform way of heating a food than most other thermal processes. Most methods of heat preservation heat the outside of the product first. By conduction the heat then moves from molecule to molecule until the inside is heated. In microwave heating the entire product is heated simultaneously. In the food industry, microwaves are applied to foods moving on a conveyor belt in large tunnels. Microwaves, like gamma rays, are dangerous to humans, but the chambers are sealed to protect workers from their release. Unit operations that use microwaves include solvent removal, concentrating, drying, freeze drying, inactivating enzymes, foaming, puffing, curing, tempering, pasteurizing, thawing, heating, precooking, cooking, baking, and sterilizing.

A novel type of heating, also under active investigation, is *ohmic heating*. Ohmic heating involves the passing of an electric current between electrodes through a food. Most food materials are good conductors of electricity. The method has the advantage, over more conventional heat processing, of heating the food evenly. It is particularly advantageous for foods with particulates (chunks). To do this effectively, the food processor must understand the conductive properties of the substances that make up that food. Once processed, foods preserved with ohmic heating are packaged aseptically. Another way of killing microbes is to zap the food with quick jolts of high-voltage electricity. The jolts don't heat the food, but they do kill microbes. This technique appears to be useful only for clear liquids such as juices, soups, milk, and eggs. Although direct use of electricity to preserve foods has been available for many years, it has not yet found widespread use in the food industry. These techniques may be important processes for the foods of tomorrow.

Consequences of processing

As indicated early in the chapter, foods are processed to keep them from spoiling. To achieve this end, there are some consequences we must consider. Some operations are more severe than other processes. In general, the greater the benefit in extending shelf life and improving convenience derived by processing, the greater the chance that nutrients and quality will be lost. Not all these consequences are bad, as some processes can destroy antinutrients present in foods or improve sensory quality. In this section, we will look at some of the tradeoffs that result from food processing and how they vary by type of process.

Shelf life

Put simply, *shelf life* is the length of time we can keep a product before we need to throw it away. From a more practical point of view, we want enough shelf life in a product so we don't have to rush to eat it before it spoils. While shelf life relates to the safety of a food product, it is most closely associated with spoilage. Since the primary cause of spoilage is microbial, the greater the destruction of microbes or inhibition of their growth, generally, the greater the extension of shelf life. Other major causes of spoilage are enzyme activity and nonenzymic lipid oxidation.

Processing techniques that sterilize a product are the most likely ones to extend its shelf life. Canning, or commercial sterilization, is a very effective way of extending shelf life of foods. Canned foods can last for years at ambient (typical room) temperatures or even in hot warehouses or shelters than at much higher temperatures. It does not matter whether sterilization is done inside (canning) or outside (aseptic processing) the container. Theoretically, irradiation can also be used to sterilize foods by preserving raw foods for storage at room temperature, but it is not currently approved by governmental agencies for many applications. Note also that sterilization usually inactivates enzymes and excludes oxygen, greatly reducing other types of spoilage.

Processing techniques that inhibit microbial growth but do not sterilize the food are also effective at extending shelf life of foods. Frozen foods are stable as long as they remain frozen, and dried foods remain stable as long as they do not become rehydrated. Some yeasts can grow in concentrated or intermediate moisture foods, but growth is usually slow. Combining two types of preservation, such as concentrating and freezing, can be very effective in extending shelf life. Chemical preservation by either curing or fermenting extends shelf life, but many of these products are held under refrigeration to prevent spoilage. While these techniques inhibit microbial growth, generally preventing microbial spoilage, other types of quality losses can occur. Freezing and drying slow enzyme activity but do not necessarily prevent it. Blanching of vegetables can inactivate enzymes, but blanching is not usually performed with other raw materials. Even if enzyme activity is very slow, over a long period of time damage can occur and the product can spoil. Concentration and fermentation can accelerate activity of some enzymes. Nonenzymic lipid oxidation proceeds slowly in foods containing fats. Few of these processing techniques inhibit this oxidation. Oxidation is actually accelerated in intermediate-moisture foods. If a frozen food is stored below its glass-transition temperature (point at which molecules in the food become immobile), enzyme degradation and oxidation are prevented.

Techniques such as pasteurization and radurization (the equivalent operation using irradiation instead of heat), which kill pathogens but do not sterilize the product, have less effect on shelf life. These products are considered perishable and usually require refrigeration. Spoilage is generally by microbial degradation and not by enzymes or lipid oxidation. Other techniques described in this chapter, such as milling, extracting, and extruding, do not generally preserve foods directly, but they are combined with other preservation techniques to extend shelf life.

Shelf life is estimated by food engineers by determining the loss of a positive quality characteristic or appearance of a negative characteristic. The end of the shelf life is when the product becomes unacceptable. In many processed foods, these changes occur in a predictable way that can be calculated using mathematical equations based on chemical kinetics.

Nutrition

Extension of shelf life is usually achieved by sacrificing something else. Frequently nutritional value of foods is lost due to processing, with heat being one of the most destructive forces. Vitamins are lost during heat processing by changing them from active to inactive forms. Generally speaking, the higher the temperature and the shorter the time needed to kill the intended microbes, the less loss of vitamins that occurs. Once a food has been heat processed, however, there is little loss of vitamins or minerals in foods. Minerals and water-soluble vitamins can be lost by *leaching*, which means that they dissolve in water during processing, become separated from the product, and are never eaten. Light and oxygen can also be very damaging to vitamins. Fat-soluble vitamins tend to be more sensitive to the presence of light and oxygen because they tend to be more likely to oxidize. Some of the water-soluble B vitamins are also quite light sensitive. Irradiation tends to affect nutrients in ways similar to light. The stability of vitamins and minerals during storage of foods is affected by pH.

Some processed foods are much less affected by nutrient loss than canned foods (see Insert 4.6). Freezing is a milder form of preservation than canning, but the blanching step (a heat process) will lead to leaching or inactivation. Drying is also milder than canning with the effect on nutrients dependent on the extent of the exposure of the raw materials to heat, light, and oxygen. Fermentation has little or no detrimental effect on the vitamins and minerals and can result in increased availability of these nutrients. Prefermentation steps, such as pasteurization of milk prior to yogurt production, affect nutrient composition. Exposure to light and oxygen are the most damaging aspects of curing, milling, and extraction. Milling and extraction can physically remove nutrients such as minerals, vitamins, and dietary fiber, while the presence of dietary fiber can bind minerals and vitamins, making them less available to us. While hydrogenation of oils improves the spreadability of margarine and other spreads, it results in the production of *trans fatty acids*. Heat and oxygen are destructive in extrusion processes. In addition, packaging can be effectively used to keep light and oxygen away from the product.

Because vitamins and minerals are lost during processing, processed foods have a bad reputation for not being healthy or nutritious. When the nutritional value of processed foods is compared with that of the raw materials from which they are made, it is obvious that there are losses in most but not all cases, and that losses vary by process and product (see Insert 4.7). Most consumers mistakenly believe that nutritional values of fresh foods never change. Food scientists believe that the poor reputation of processed foods is not deserved for the following reasons:

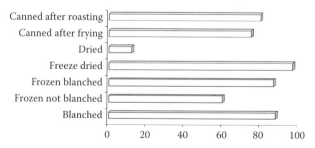

Insert 4.6 Loss of vitamins in processed peppers. (Adapted from S. Martinez, M. Lopez, M. Gonzalez-Raurich, and A. B. Alvarez. "The Effects of Ripening Stage and Processing Systems on Vitamin C Content in Sweet Peppers (*Capsicum annum* L.)." *International Journal of Food Sciences and Nutrition* 56 (2005): 45.)

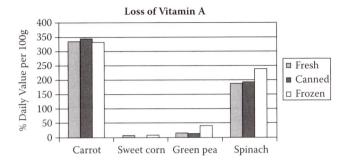

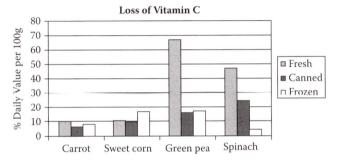

Insert 4.7 Vitamins in fresh foods as affected by processing. (Calculated using data from the U.S. Department of Agriculture National Nutrient Database for Standard Reference, Release 20 (2007), http://www.nal.usda. gov/fnic/foodcomp/cgi-bin/list_nut_edit.pl; and the netrition.com Web site, http://www.netrition.com/rdi_page.html.)

- Some processes are more destructive than others, so it is not fair to lump all processed foods together.
- Raw materials and fresh foods are more likely to lose nutrients during storage than processed foods.
- Cooking (a heat process), particularly boiling, is one of the most destructive processes for vitamins, with fresh foods more likely to be further cooked than processed foods.
- Fresh foods are more likely to spoil and be thrown away than processed ones.

Processed foods (even canned foods) do provide nutritional value. In fact, canning tends to improve the effectiveness of dietary fiber in fruit and vegetable products. As we shall learn in Chapter 11, fiber is one of the most important reasons for consuming fruits and vegetables. In addition, heating of some products, carrots, for example, increases the amount of some vitamins that are available to the body. Other processes such as extrusion can lower the levels of natural toxins and antinutrients present.

Quality

Another thing sacrificed by processing is sensory quality—flavor, color, and texture. Generally speaking, the conditions that affect nutritional value also affect sensory quality. We are all aware of the enticing odors resulting from roasting or cooking of our favorite foods. In many processes, however, heat destroys subtle flavors and can induce off-odors. Heat also tends to fade colors and can lead to unacceptable softening. Light causes oxidation of food components leading to loss of color and development of off-flavors. Oxygen and irradiation also lead to oxidation, with a greater impact on flavor than on color. Texture is not as likely to be affected by light or oxygen as color or flavor.

As with nutritional value, the quality of some processed foods is much less affected than the quality of canned foods. One of the problems with canned foods is that the outside of the product becomes overprocessed before the cold spot in the can is properly processed. Techniques such as aseptic processing, microwaves, and ohmic heating provide more even heat and result in a better quality product. Freezing also has less of an effect on sensory quality than canning, but enzymes can lead to quality losses during storage. Blanching, because it inactivates enzymes, actually helps prevent quality losses during frozen storage. Very small changes due to enzymes that produce off-colors, flavors, and odors can make a big difference over the weeks and months of storage. Even if there is little loss of quality during freezing or storage, major losses can occur during thawing or holding of thawed product at room temperature prior to preparation. Quality losses during drying or storage of dried foods are dependent on how much the raw materials or stored product are exposed to heat, light, and oxygen. Fermentation induces the flavor, color, and texture that we associate with a particular fermented food, for example, with pasteurization of milk prior to yogurt production, which affects some of the flavor of the raw milk. Exposure to light and oxygen are most damaging to quality during milling and extracting, but the oxidation of cured meats provides the characteristic flavor craved by those of us who like them. Heat and oxygen are just as destructive of flavor in extrusion processes as they are of nutritional value. Part of this problem can be solved by adding flavor ingredients late in the barrel of the extruder. Packaging plays an effective role in protecting quality from damage due to light and oxygen.

Processed foods also have the reputation of being poor in quality. Once again food scientists think this reputation is undeserved. Since the effect on quality varies by type of process, certain foods are much less likely to be low in quality than others. It is important to make the appropriate comparisons with stored products and not freshly harvested crops. While it is impossible to see the loss of nutrients during storage, the losses in flavor, color, and texture during spoilage in storage are obvious. As stated previously, microbes are the primary cause for spoilage of fresh foods, and enzymes are the next most likely cause. Prevention of spoilage is a major reason for processing raw materials. Cooking induces changes in flavor, color, and texture, usually in desirable ways, but sometimes in undesirable ways. Processed foods, even canned foods, do provide acceptable quality to some of us, or they would never sell. Some people even prefer the flavor of processed foods to fresh or fresh-cooked products as the quality is what they expect.

Safety

Safety is the most important consideration in the design of any food process. Raw materials should be handled in ways that minimize contamination and growth of microbes. Food is preserved to make foods safer by reducing or eliminating the chances of harmful microbes growing in foods. Proper packaging prevents recontamination. No processing technique should be used as an excuse to avoid proper sanitary practices. A major criticism of irradiated foods is that the processor can pay less attention to proper sanitation, but such reasons are unacceptable for the proper preservation of foods.

Sterile products, like canned foods, are the safest foods we can eat, but they are not usually the most nutritious or highest quality products. Safe foods are only as safe as their handling and preparation. Once a canned food has been opened, it can become contaminated with microbes. If that microbe is a pathogen, and one that can grow in the food, it can grow without competition from other microbes with the food likely to become a safety hazard. Likewise, a frozen food that has become thawed or a dried food that has been rehydrated can support the growth of microbes that are already present, as these techniques slow the growth of microbes with little or no killing. Fermented or cured foods may be safe under specified conditions (some require refrigeration), but they can become safety hazards if stored improperly. Milling and extraction have little effect on microbes, and thus they must be performed under sanitary conditions and the food must be further processed to prevent safety problems.

Packaging considerations

We must remember that the preservation technique is only as good as the package that contains it. The preservation technique slows or stops spoilage, but recontamination of processed foods can restart spoilage or introduce microbes leading to food-borne illness. Food packages are designed to protect the food from microbes, insects, rodents, and physical impact. The main function of a package is to prevent microbial or chemical contamination of a processed food product. Packaging also functions to keep water, odors, and gases in, and water, oxygen, odors, and light out. A primary package is the one directly in touch with the food product. The secondary package contains the primary package(s). The tertiary package contains the secondary package(s), etc. Any part of the primary package in touch with the food must be compatible with the food and must not be toxic.

In the design of a food process, the construction of the package must be carefully evaluated. Considerations include whether (1) the process occurs before or after packaging, (2) the packaging material can withstand the conditions of processing and distribution, and (3) the package is compatible with the product. Canning occurs inside the package, and some frozen product is packaged before freezing. Most other processes, including aseptic processing, pasteurization, concentration, most freezing, dehydration, and fermentation, occur prior to packaging. When packaging comes after processing, great care is needed to prevent recontamination. Any product that is processed by heat in the package needs a package that can withstand the heat (and usually the water associated with the process). Any package that will be exposed to water, in either the processing plant or elsewhere, must not be damaged by water. The package must be able to withstand normal abuse that it could experience during its journey from the processing plant to the consumer. Most packages are designed with tamper-evident seals to prevent unscrupulous humans from deliberately contaminating a product without the consumer's knowledge. Components of the food might interact with the package. Thus, possible interactions must be studied to ensure that no reactions occur that could result in a product hazard or unacceptable product.

Remember this!

- The main function of a package is to prevent microbial or chemical contamination of a processed food product.
- Food is preserved to make it safer by reducing or eliminating harmful microbes.
- The conditions that affect nutritional value also affect sensory quality.
- The higher the temperature of a heat process and the shorter the time needed to kill the intended microbes, the less loss of vitamins that occurs.
- In general, the greater the destruction of microbes or inhibition of their growth, the greater the extension of shelf life.
- Irradiation is a potent killer of microbes that can be used to preserve foods while producing little or no heat in the process.
- Fermentation is the only primary method of food preservation that encourages multiplication of microbes.
- The rate of freezing is affected by the freezing equipment and the properties of the food.

- Design of processing techniques requires an understanding of the microbes and the food.
- Unit operations are distinct steps common to many food processes.
- Food processing and preservation increases the shelf stability of a raw material but usually decreases the affordability, nutrition, and quality of the product.
- Most raw materials are perishable and require careful handling or processing to prevent losses.

Looking ahead

This chapter was designed to provide an introduction to food processes and the types of products that are produced, with a special emphasis on those made directly from raw agricultural materials. Chapter 5 introduces formulated foods, made from mixing ingredients. Chapter 6 describes chilled and prepared foods. We will find how processed food is analyzed for quality and safety in Chapter 7. Chapter 11 explores the nutritional properties of plant products, animal products, and so-called junk foods. The basic principles behind food microbiology are presented in Chapter 12 and food engineering in Chapter 13.

Answers to chapter questions

Insert 4.1:

Sterile: canned tuna, canned pineapple, Spam

Insert 4.4:

Sucrose (table sugar and most sweets)
Ethanol (alcoholic beverages)
Benzoic acid (cranberry products)
Lactic acid (yogurt, sour cream, sauerkraut)
Eugenol (cloves)

References

Heldman, D. R., and R. W. Hartel. 1997. *Principles of food processing*. New York: Chapman & Hall.

Potter, N. N., and J. H. Hotchkiss. 1999. *Food science*, 5th ed. New York: Chapman & Hall.

Further reading

Barbosa-Canovas, G. V., M. S. Tapia, and P. Cano. 2005. *Novel food processing technologies*. New York: Marcel Dekker.

Fellows, P. J. 2000. *Food processing technology: Principles and practices*, 2nd ed. Boca Raton, FL: CRC Press.

Holdsworth, E., and R. Simpson. 2008. *Thermal processing of packaged foods*. New York: Springer Science.

Martinez, S., M. Lopez, M. Gonzalez-Raurich, and A. B. Alvarez. 2005. The effects of ripening stage and processing systems on vitamin C content in sweet peppers (*Capsicum annum* L.). *International Journal of Food Sciences and Nutrition* 56, 45.

Shewfelt, A. L. 1971. *Your future in food science*. New York: Carlton Press.

Shi, J. 2007. *Functional food ingredients and nutraceuticals: Processing technologies*. Boca Raton, FL: CRC Press, Taylor & Francis Group.

Smith, S. J., and Y. H. Hui. 2004. *Food processing: Principles and applications*. Ames, IA: Blackwell Publishing.

Sun, D. W. 2006. *Thermal food processing: New technologies and quality issues*. Boca Raton, FL: CRC Press, Taylor & Francis Group.

chapter five

Formulated foods

Nolan is always on the go. He didn't have much time for meals, but he didn't go hungry. He needed his food fast, and he needed it healthy. Fortunately for him there were a wide range of products that were easy to fix, quick to eat, and nutritious. The most difficult food for him to prepare was a shake; it took some time stirring the high-protein powder into solution. When he didn't have that kind of time, he just ate one of those bars. Besides, it was easier to eat the bar on the run than drink the shake. He couldn't remember the last time he had eaten meat, bread, or vegetables. He just figured he was eating a balanced diet with lots of protein. He could read that right off the labels, when he had the time to read them. The other thing he liked about these products was that they came in lots of different flavors. His favorite bar is Crackling Chocolate Almond, and his favorite shake is Chocolate Marshmallow, but he likes to mix it up, with at least half his meals in another flavor.

Olivia likes to watch her weight, but she isn't much into exercise. Her best friend is her microwave oven. Every evening she selects one of the 300-calorie frozen entrées in the freezer compartment of her refrigerator. She shops for them every two weeks. There are over 20 dinner combinations that she likes, so she buys twelve different ones each time she shops (Saturday nights she goes out). She won't buy anything that takes longer than six minutes to prepare. As soon as she hits the door, she selects the dinner of the evening, rips off the outer package, tosses it in the microwave, and sets the proper time. Six minutes gives her enough time to change her clothes, pour a big glass of chocolate milk (1%), dig a clean fork out of the drawer, and turn on the tube before the beep, beep, beep of the microwave signals that it is supper time.

Paula is poor. She lives in a dorm. She doesn't have much in the way of kitchen appliances. She is not on the meal plan. Early in the month, when she has some money, she lives on macaroni and cheese; when money runs low, she switches to Ramen noodles. Both meals are satisfying without having too many calories. They are easy to prepare in the room: all that is needed is hot water. It is a good thing she likes the flavor of the macaroni and cheese and the noodles. They never seem to get old.

Nolan, Olivia, and Paula have abandoned whole and natural foods for what are called formulated foods. The best way to tell what is in formulated foods is to read the ingredient statement. Many of us depend on formulated foods for our nutrients. Frequently these foods are hard to fit into the guidelines on MyPyramid.gov.

Looking back

Previous chapters focused on food issues we deal with daily. Some key points that were covered in those chapters help prepare us for understanding formulated foods.

- The main function of a package is to prevent microbial or chemical contamination.
- Food is preserved to make it safer by reducing or eliminating harmful microbes.
- Unit operations are distinct steps common to many food processes.
- Although many of us try to eat healthy diets, there are many temptations and influences affecting our decisions that we do not consciously consider.
- Technology has transformed our food supply, producing more formulated foods with mixtures of ingredients, and moving away from the eating of whole foods.
- If we want to eat a healthy diet, we have an obligation to read labels, analyze our diets, and exercise willpower in our food selections.
- Nutrition labels provide information to assist us in making good decisions to improve the quality of our diets.
- Good nutrition requires an adequate consumption of nutrients without exceeding the recommended allowance of calories.
- Preservatives are food additives that prevent or retard spoilage.
- The expiration date represents the food scientist's best guess about how long a food will last before it spoils.

What are formulated foods and why are they formulated?

Formulated foods are products that are mixtures of ingredients. Unlike processed foods, they are not directly recognizable as their original plant or animal sources. Formulated foods are not new. Products like bread, cakes, beer, ice cream, and chocolate bars have been around for a long time, but never before have we had such a wide selection of formulated products.

Formulated foods provide us with flavorful combinations of ingredients in a convenient form, usually preserved or in a stable product. A world with just fresh and processed foods would be a very different one, and for many of us, a much duller one. Most, but not all, formulated foods have been preserved in some way to reduce or minimize microbial growth.

Each of the ingredients is present to perform at least one specific function in that product. One function of an ingredient is as a preservative, to slow microbial growth. Other ingredients may be nutrients, to improve the nutritional value of the product or to replace nutrients lost during processing. Ingredients also function to improve flavor, color, and texture.

Benefits and consequences

Formulated foods that have been preserved in some manner have all of the advantages of processed foods. Like their processed counterparts, they tend to be shelf stable, safe, and convenient. In formulating a food, the food scientist faces fewer restrictions than during food processing. The trade-offs between nutrition and quality are not as significant in formulated foods as in processed foods. Any deficiencies in the nutrients of the raw materials or ingredients can be ignored to create a "fun" or "junk" food, or they can be overcome by adding nutrients to become a "health" or "functional" food. Likewise, ingredients can be added to improve the flavor, color, or texture of a product to enhance its appeal. Formulated foods are frequently safer than processed ones because they contain preservatives. Although *preservative* has become a dirty word in our culture, preservatives function to reduce microbial growth, thus decreasing food spoilage and increasing food safety. Therefore, leftovers from formulated foods with preservatives are less likely to spoil than those from fresh or processed foods. Finally, many formulated foods fit in nicely with our fast-paced culture. They tend to be ready to eat or require minimal preparation. They can be eaten with our fingers on a slow walk or a fast ride, leaving little or no mess other than the package, which can be easily discarded.

There are problems with formulated foods, however, particularly from a nutritional standpoint. Three components in formulated foods that improve their flavor are salt, sugar, and fat. When consumed in moderation, each of these components makes a valuable contribution to our diet. However, when consumed in excess, they can be detrimental to our health, and can decrease our desire to eat more nutrient-dense foods. Formulated foods that contain fruit or vegetable ingredients may convey the false idea that they provide the benefits of whole fruits and vegetables. Even the "health" or "functional" foods that Nolan lives on tend to provide high levels of only a few nutrients and do not provide a balanced diet. Many of them also have high levels of salt, sugar, and fat. The health-conscious consumer of formulated foods has an obligation to carefully read food labels to make sure they are indeed getting the vitamins and minerals they need. Although the food industry is frequently blamed for making processed foods "tasteless," it would appear the major problem with many formulated foods is that they are too tempting.

Formulation steps

Many steps occur in a food manufacturing plant to mix ingredients and transform them into tempting foods and beverages. Just as in processed foods, there are distinct unit operations common to many types of food formulations. These steps begin as the ingredients are unloaded at the plant dock and continue until the packaged product is loaded onto another truck or railcar for transport out. Most ingredients are shelf stable, but many companies do not want the added expense of maintaining large warehouses for storing supplies. Some common unit operations include materials handling, pumping, mixing, heat exchanging, packaging, and controlling.

In the production of yogurt, for example, there are many unit operations (see Insert 5.1). All ingredients must be received and properly stored. The milk must be standardized to the proper fat content. The fat content of the yogurt will be the same as the fat content of the milk because it will not change during fermentation. Pasteurization of the milk occurs after the addition of the dry ingredients (such as sugar). In the production of yogurt, pasteurization must be more vigorous than for regular milk because it must eliminate undesirable microbes that can compete with the starter culture microbes that are added to produce the yogurt. The fruit is put into the bottom of the container with the pasteurized milk on top. As the product is incubated, the starter culture organisms multiply and the liquid milk slowly develops a tart flavor and a thick consistency. When the incubation period is complete, the containers are chilled, ready for shipment to supermarkets or other points of sale.

Formulated products

This chapter will focus on the formulation of ingredients into food products. The previous chapter emphasized the operations that are used to process foods. The next chapter will present information on chilled foods (perishable, requiring refrigeration to maintain quality) and prepared foods (ready to eat with or without heating). These distinctions are made to help us understand how foods are made. Most of us don't think much about whether a food has been processed or formulated, chilled or prepared. In many cases these categories overlap. However, an understanding of how foods are formulated and their importance in modern diets is critical to understanding food science.

Baked goods

Think of the vast array of breads, cookies, crackers, cakes, and specialty items such as chocolate éclairs that are available in the marketplace today. All are baked goods. The differences in these products are the result of the

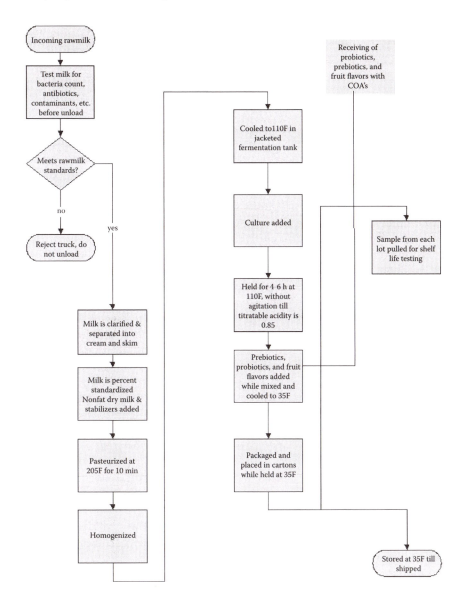

Insert 5.1 Schematic for producing a drinkable yogurt. (Prepared by Jesse Crosswhite and Brooke Boretski as part of a class assignment.)

ingredients that provide unique colors, flavors, and textural sensations during eating. Some of these sensations are a direct result of the ingredient, but others develop during the heating process known as baking.

Most baked goods rise using a leavening agent. Breads and baked goods rise from carbon dioxide (CO_2) bubbles producing a light, fluffy texture. There are two ways of generating CO_2 gas in baked goods––microbially or chemically. Yeast produces CO_2 microbially; baking soda or baking powder produces it chemically. Most wheat is milled into flour, and wheat is the best grain for making bread. The most dramatic action during baking is the rising of the flour, which begins before heating and is due to the expansion of CO_2 gas. In addition to gas expansion, baking results in coagulation of proteins, gelatinization of starch, and evaporation of the water. All these changes contribute to how a baked product feels in the mouth. Other actions that are occurring during baking are the development of the flavors, hardening of the crust, and darkening of the color. Many factors affect the quality of baked goods. The most important factors are temperature and time. As little as 15 seconds can make a noticeable difference in flavor and texture. Other factors affecting the quality of the finished loaf include the size, the shape, and the heat transmission properties of the baking pan and the evenness of the temperature in the oven.

Crackers contain little sugar and about 10 to 20% fat. The dough used is generally low in water. The preparation of crackers begins by fermenting a small amount of dough with yeast that is contaminated by bacteria to form a "sponge." During the fermentation process, the sponge becomes more acidic. The acidified sponge is neutralized by the addition of sodium bicarbonate, regular dough, and other ingredients. The dough is then rolled into large sheets, which are cut and *docked* (little holes punched in them). Docking is necessary to prevent the cracker from separating into layers. From here the crackers are baked (less than 3 minutes), cooled, broken into appropriate sizes, and packaged. See Insert 5.2 for a picture of how cracker dough is processed, and what it looks like after cutting and docking.

Pasta and noodles

Pasta and noodle products have become staples of the American college student diet. Popular for their ease of preparation, low cost, and satisfying quality, these products are the food of choice for many. Noodles and pasta provide good nutrition at a low price. They are high in carbohydrates and reasonably low in fats; however, they do not supply a complete, balanced diet unless supplemented with protein, minerals, and fiber.

By American regulations, noodles must have at least 5.5% egg solids, but pasta is generally made from flour and water only. Traditionally, Italian pasta was made from semolina, coarse granules that are low in starch and high in protein, ground from hard durum wheat. However, the merging of European Community laws and regulations has overturned these restrictions. To make pasta, the granules are then made into a paste by adding water, and the paste is formed into dough. The dough is extruded into the

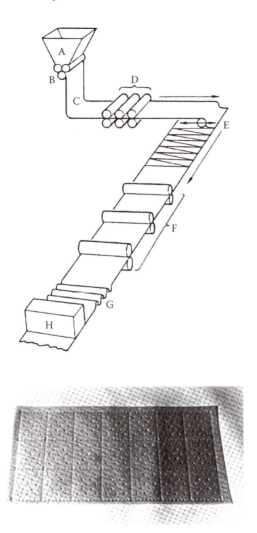

Insert 5.2 Top: cracker dough processing consisting of (A) dough hopper, (B) forming roll, (C) dough web, (D) reduction rolls, (E) lapper, (F) final reduction, (G) relaxing curl, and (H) cutter-docker. Bottom: cracker dough after cutting and docking. (From R. C. Hoseney, *Principles of Cereal Science and Technology*, 2nd ed. St. Paul, MN: American Association of Cereal Chemists, Inc., 1994. With permission.)

desired shape and dried to about 12% moisture. Instant noodles (the kind that Paula and many of her friends turn to at the end of the month) are prepared by frying the noodle dough after it has been steamed.

Noodles and pasta vary by size, shape, color, and light transmission. The sizes and shapes are formed by forcing the dough through the holes of the die at the end of the extruder. During extrusion the hot dough rapidly expands, causing rapid evaporation of moisture and a puffing of the product. The key to maintaining quality and retarding spoilage is in controlled drying. Most noodle and pasta products Americans are familiar with are hard and dry; noodles from certain Asian countries are more commonly fresh, refrigerated pasta products. Although pasta and noodle products require little preparation, timing is of utmost importance. When not cooked long enough, pasta and noodle products are tough, but overcooking leads to stickiness. Once again, as little as fifteen seconds can lead to a noticeable loss in quality. The popular instant (Ramen) noodles have been cut, waved, steam-cooked, portioned, and fried prior to packaging. During the frying process, they lose moisture and pick up fat.

Jams and jellies

Since jams, jellies, preserves, and similar products are made from fruits, they are considered to be healthy. Most consumers are surprised when they learn that these products are about 70% sugar. Although many people think sugar from fruits is better than refined sugar (usually from sugar cane or sugar beets), the body can't tell the difference between the two. Jams and preserves use whole fruit during their manufacture, which provides fiber; jellies are made from fruit juice, which usually contains no fiber. Jams and jellies usually have more sugar added during formulation, while preserves do not. Acid and pectin are also added to fruit gel formulations. The resulting gelling action is due to the unique combination of sugars, acids, and pectins. These three ingredients form an extensive network that is at once solid and flexible.

Beverages

One of the most important nutrient requirements is water. We are urged to drink six glasses of water daily. Although most of us consume large quantities of liquid refreshment, many of us look to sources other than plain water to supply this nutrient. The problem with getting water from beverages is that many contain caffeine or alcohol, which are diuretic (induce urination), causing elimination of more water than one would anticipate.

Juices are a healthy beverage because they contain some vitamins and minerals from the original fruit. The unit operations of juice manufacture are extraction, clarification, deaeration, pasteurization, concentration, essence addition and canning, bottling, or freezing. The liquid portion of the fruit is removed from the solid part during extraction. Extraction yield increases by the addition of pectic enzymes, which release more of the liquid portion of the fruit. Clarification removes the pulp to make a clear juice. Pectic enzymes are also used to clarify the juice by removing the pulp, which makes them clear. Deaeration removes the air to reduce the chances of off-flavor development. Pasteurization kills harmful microbes but does not sterilize the juice.

Concentration involves the removal of water so the nutrients, colors, and flavors can be put into a smaller container, making it easier to ship and store. During the concentration process some of the delicate aromas evaporate. These delicate aromas, known as essence, can be captured and returned to the concentrated juice to give it a fuller flavor. The type and duration of the juicing process and the conditions of storage affect flavor, color, texture, nutritional quality, and shelf life. The most commonly manufactured juice worldwide is orange juice, but there are many other fruits and some vegetables that are also sources of juice. Combinations of juices from different sources are known as juice blends. Just because a beverage has a fruity name and fruity properties doesn't mean that it contains juice. Many fruit-flavored beverages have little juice and few nutrients; others have more vitamins and minerals than fresh-squeezed juice. Once again, it is important to read the label to know what you are getting.

A popular drink on college campuses is the carbonated beverage. In fact, carbonated beverages are the most widely consumed liquid products in America. The unit operations of carbonated beverages include mixing, adding carbon dioxide, and packaging. Sweeteners and acidulants provide the tantalizing tastes of carbonated beverages. Natural and artificial flavorants provide a wide range of appealing aromas in carbonated beverages: cola, root beer, cream soda, lemon-lime, and orange. Although many people associate the colors with the flavors, flavorants for most of these products have no color and the colorants have no flavor.

Carbon dioxide provides the tingle that tickles the throat as these beverages pass through our mouths. Although carbonated beverages are widely criticized as being junk foods, there appears to be a place in the diet for these beverages when consumed in moderation (i.e., 24 to 30 fluid ounces a day). Caffeine can be mildly addictive, but many carbonated beverages are caffeine-free. The color of a carbonated beverage is not a good indicator of caffeine content because some dark root beers are caffeine free, whereas some clear lemon-lime sodas provide a strong caffeine jolt. Caffeine and some acidulants can interfere with calcium absorption. The acidulant of most concern is phosphoric acid. A frequent cola drinker, who drinks no milk, consumes few dairy products, and does not take a calcium supplement, should be concerned. We can tell if our favorite beverage contains phosphoric acid by checking the ingredient statement. Many noncarbonated beverages and juice-type drinks provide little or no nutritional advantages over carbonated beverages.

Alcoholic beverages differ in ethanol content, fermentation substrate, and method. Ethanol is the alcohol that causes intoxication when consumed in excess. Fermentation is the process that turns sugar into ethanol. Wines are formed by fermentation of grapes and are named after the grape variety (chardonnay, zinfandel, cabernet sauvignon), region of origination (burgundy, Chianti, port), or added characteristics (sherry, sparkling, vermouth). Although similar processes can produce similar types of products from other fruits, wine purists do not consider those made from peaches, plums, or other fruits as true wines.

One of the more popular yet destructive beverages on college campuses is beer. The unit operations in beer making include filtering, brewing, fermenting, aging, filling, sealing, pasteurizing, labeling, and packing. Grain is mixed with water to form a mash. The water-soluble components are extracted from the solids in the mash during brewing. Aging helps develop the flavor of the beer. The addition of hops imparts distinctive flavors and aromas that are associated with many brews. Brewers use different techniques to produce flavor and body variations in beer. For example, draft beer is unpasteurized, resulting in a better flavor with a shorter shelf life, whereas light beer is fermented from a mash with less solids.

Hard liquor is produced by the distillation of low-alcohol fermented products. Brandy is produced from wines or other fermented juices; gin from rye; rum from molasses or cane solids; tequila from the roasted stems of the agave plant; vodka from grains and potatoes; and whiskey from grains. Neutral spirits are distilled to 190 proof or higher and are then added to wines to fortify (increase the percentage of alcohol present) them. Liqueurs or cordials are made from brandy that has been sweetened with sugar syrup and flavored.

Confections

Confections are sweet, tempting treats that offer more satisfaction than nutrition. Confections are based either in chocolate or sugar. In both types of confections, sugars are important ingredients. Minor differences in the unit operations result in major differences in the final product. One of the most important variables is whether the sugar is in the crystalline form or not. Those confections containing crystalline sugar include rock candy and fudge; those with noncrystalline sugar include peanut brittle, caramel, gumdrops, hard candy, jelly beans, marshmallows, and taffy.

Manufacturing chocolate is a complex process that begins with the raw white to pale purple cacao bean. Raw beans are fermented, which removes the pulp, kills the germ, and modifies both color and flavor. Fermented beans are then cleaned, roasted, winnowed, milled, and ground to form the chocolate liquor. Winnowing removes the outer coating of the bean and separates the germ from the rest of the seed. The chocolate liquor is more than half fat. Although most people associate chocolate with caffeine, it actually contains very little. Removal of the fat (cocoa butter) from the liquor leaves cocoa powder. White chocolate is made using the cocoa butter with no cocoa powder.

The unit operations in the manufacturing of chocolate include the combining of ingredients, mixing, refining, conching, tempering, and molding, or enrobing. *Conching* is a very slow mixing process that changes the very grainy cocoa mixture into a fine smooth-textured mixture. Tempering involves stirring and heating to permit fat crystallization. Although many consumers think chocolate is sweet, it is actually very bitter due to the presence of tannins and theobromine (a molecule that is similar in structure to caffeine but does not have the stimulating properties of caffeine). If you don't think chocolate is bitter, buy some baking chocolate (100% natural chocolate with no sugar added) and take a nice big bite! Another secret of chocolate's success as an ingredient is that it melts at body temperature. This melt-in-the-mouth property adds to the overall enjoyment of chocolate confections.

Many baked goods, because of their high sugar content, are also considered confections: cakes, cookies, éclairs, and blintzes. The mixing technique is important and strongly influences the texture of the final product. During mixing, air is incorporated into the cake batter to help provide a light, fluffy texture. In some cakes, air is pumped directly into the batter using a high-speed mixer. Commercial boxed mixes contain surfactants to allow the added water to combine with the shortening present in the mix. Ready-to-eat cakes can require several mixing steps including creaming in which the sugar and fat are mixed prior to adding water. The primary reason for creaming is to incorporate air into the batter. Baking powder is added as a chemical leavening agent. Full flavor development occurs during baking. Cookie batter is made from soft wheat and is either molded or cut after extrusion. The proper mixing of air into the batter, as with cakes, is essential for a proper cookie texture. Creaming is usually part of the cookie process. Cookie baking occurs in long ovens on a slow-moving conveyor belt. Baking melts the shortening, fully dissolves the sugar, and can gelatinize the starch.

Sausages

Many of us associate sausages with breakfast; however, sausages are much more than breakfast products. They include the pepperoni on our pizza, the frankfurter in our hot dog bun, and the salami in our sandwich. Most of these products originated in Germany, where they remain a staple of the German diet. They come in fermented or unfermented form (see Insert 5.3).

Probably the most popular form of sausage in America is the frankfurter or wiener. The unit operations of wiener production include forming an emulsion, filling, twisting, heating, smoking, cooking, peeling, and

Insert 5.3 Processed meats. (From the U.S. Department of Agriculture Photo Gallery (http://www.ars.usda.gov/is/graphics/photos/k7223-7.htm). Photo by Scott Bauer.)

packaging. An *emulsion* is the dispersion of one liquid in another; with the wiener, water is dispersed in fat. To form the emulsion, cuts of meat, water, dry mustard, and spices are blended together in a huge grinder until smooth. The dry mustard helps bind the water to the fat to give the final product its juiciness. The comminuted (ground) meat, which is mushier than raw hamburger, is then stuffed into a casing and allowed to gel and become solid. Twisting is done to form the individual frankfurters. Heating, smoking, and cooking are the primary preservation methods for this product, and also give it its characteristic flavor. Once cooked, the artificial casing is peeled off.

Unit operations for link sausage are similar to those of wieners, except the casing is left on the sausage. Wieners and link sausage originated as a salvage operation to use up as much of the slaughtered hog as possible (an old saying goes, "they ate everything but the squeal!"). The sausage stuffing was meat that could not be recovered as distinct cuts, and the cleaned intestine was used for the casing. Today artificial casings have replaced natural intestine casings. Wieners were originally exclusively hog products, but beef, chicken, and turkey have been added to replace all or part of the pork meat, primarily to reduce the fat content.

Variations of sausage processing include fermentation and curing. These products were the result of attempts to preserve the sausage when refrigeration was not as widely available as it is today. Thus, bologna, salami, bratwurst, knockwurst, and frankfurters are all sausage products. Curing with salt or nitrates also helps prevent microbial growth. The nitrates give frankfurters and bologna its typical red color. They help prevent the growth of the bacterium *Clostridium botulinum*, the cause of deadly botulism. However, they are not as effective in preventing the growth of *Listeria monocytogenes*, the cause of listeriosis, which can grow at refrigerated temperatures.

Frozen desserts and entrées

The frozen food aisle is the favorite location in the supermarket for consumers who are more interested in convenience and quality than price. The frozen food aisle has something for everyone. This aisle is the first place Olivia visits. The entrées provide her with balanced meals that are convenient, tasty, and low in calories. Even with all the choices available, these meals can still become boring after a while. Besides, three hundred calories at a single meal is probably not enough to sustain a healthy diet over a long period of time.

Ice cream is an American favorite. The unit operations of ice cream start with pasteurizing and homogenizing the milk, followed by mixing the ingredients, aging, whipping air into the mix, freezing, and hardening. Ice cream is a frozen foam; that is, a gas dispersed in a liquid. The gas is air; the liquid is cream. The incorporation of air into the mix is the step that gives ice cream its smooth and creamy texture. This step increases the volume of the original mix; the extra volume is called *overrun*. Without the air, the ice cream would be hard and grainy. When churning homemade ice cream, air is being incorporated into it. Next time you make it, try freezing it without the churning and see if you like it better with or without the air.

Frozen entrées, whether they are of the low-calorie or full-fat variety, are complex foods with many ingredients and components. Some ingredient statements are shown in Insert 5.4. Can you name the products listed? Many are full formulations of ingredients that are completely mixed together, but others have several components that are prepared separately and then placed into the package. Most of these items have been at least partially prepared, and most are designed to be heated prior to consumption. The packaging usually is designed with the idea that the product will be heated in a microwave oven. There are even foil shields, particularly for frozen pizza, that direct the microwaves to a particular location to provide better crisping. Some packages are dual-ovenable, meaning they can be heated in a microwave or a conventional oven. Before we try placing a container in either a microwave or conventional oven, we should read the directions carefully.

Functional foods

Functional foods, also known as *nutraceuticals*, are described as "foods that have been modified or formulated to have specific physiological or

Insert 5.4 Ingredient statements for frozen entrées. Can you guess these products from their ingredient statements?
Answers are at the end of the chapter.

R.

INGREDIENTS: COOKED BROWN RICE, VEGETABLES (MUSHROOMS, ONIONS), WATER, ROLLED OATS, MOZZARELLA CHEESE, PASTEURIZED MILK, CULTURES, SALT ENZYMES), NONFAT MILK, CORNSTARCH, NATURAL FLAVORS, ANATTO (VEGETABLE COLOR), BULGHUR WHEAT-HYDRATED, CHEDDAR CHEESE (PASTEURIZED MILK, CHEESE CULTURE, SALT ENZYMES) SOY PROTEIN CONCENTRATE CONTAINS LESS THAN 2% OF SALT, PARSLEY, VEGETABLE GUM, AUTOLYZED YEAST EXTRACT, WHEAT GLUTEN, DRIED GARLIC, MILKFAT, APOCAROTENAL (VEGETABLE COLOR) SPICES, CITRIC ACID, NATURAL FLAVORS, BARLEY MALT, NATURAL BUTTER FLAVOR, ANATTO (VEGETABLE COLOR), DRIED ONION, DRIED MUSHROOMS.

S.

INGREDIENTS: LINGUINE (WATER, SEMOLINA, EGG WHITES) HALF-AND-HALF (MILK, CREAM) COOKED CHICKEN BREAST MEAT (CHICKEN BREAST MEAT, WATER, SEASONING (SALT, DEHYDRATED GARLIC, DEHYDRATED ONION, DEHYDRATED PARSLEY, SPICES, CITRIC ACID, PARTIALLY DEHYDROGENATED SOYBEAN OIL) ISOLATED SOY PROTEIN, FLAVOR (SALT, YEAST EXTRACT, MALTODEXTRIN, CHICKEN SKIN, DEHYDRATED CHICKEN MEAT, DEXTROSE, FLAVOR, SOY SAUCE (SOYBEAN, WHEAT, SALT), CHICKEN BROTH, SMOKE FLAVOR, GRILL FLAVOR (FROM VEGETABLE OIL), MODIFIED CORNSTARCH, CORN SYRUP SOLIDS, CITRIC ACID), MODIFIED RICE STARCH, CANOLA OIL, SODIUM PHOSPHATES) PEAS, PARMESAN CHEESE (CULTURED MILK, SALT, ENZYMES), BACON (CURED WITH WATER, SALT, SUGAR, SMOKE FLAVORING, SODIUM PHOSPHATES, SODIUM ERYTHORBATE AND SODIUM NITRITE), BUTTERFAT, CANOLA OIL, HAM BASE (HAM CURED WITH WATER, SALT, DEXTROSE, SODIUM PHOSPHATES, SODIUM ERYTHORBATE, SODIUM NITRITE), SUGAR, SALT, DRY SOY SAUCE (WHEAT, SOYBEANS, SALT, MALTODEXTRIN), NATURAL FLAVORING, POTATO STARCH) SALT, SPICE OATEM, MODIFIED CORNSTARCH, BLEACHED WHEAT FLOUR, XANTHAN GUM. CONTAINS: MILK, EGG, SOY, WHEAT INGREDIENTS.

T.

INGREDIENTS: ENRICHED FLOUR (WHEAT FLOUR, MALTED BARLEY FLOUR, NIACIN, REDUCED IRON, THIAMINE MONONITRATE, RIBOFLAVIN, FOLIC ACID), LOW MOISTURE MOZZARELLA CHEESE (CULTURED PASTEURIZED MILK, SALT ENZYMES), SKIM MILK, TOMATOES (TOMATOES, WATER, TOMATO PASTE), MARGARINE (PARTIALLY HYDROGENATED SOYBEAN AND COTTONSEED OILS, WATER, SALT, VEGETABLE MONO AND DIGLYCERIDES, NON-FAT DRY MILK SOLIDS, SOY LECITHIN, ARTIFICIAL FLAVOR BETA CAROTENE, VITAMIN A PALMITATE), WATER, MODIFIED FOOD STARCH, SMOKED FLAVORED PROVOLONE CHEESE (CULTURED PASTEURIZED MILK, SALT, ENZYMES AND NATURAL SMOKE FLAVOR) ASIAGO CHEESE (PASTEURIZED CULTURED MILK, SALT ENZYMES), SUGAR, YEAST, CONTAIN 1 PERCENT OR LESS OF PARMESAN CHEESE (PASTEURIZED CULTURED PART SKIM MILK, SALT, ENZYMES), ROMANO CHEESE (PASTEURIZED CULTURED COW'S MILK, SALT ENZYMES) DOUGH CONDITIONER (DIACETYL, TARTARIC

T. (Continued)

ACID, GUAR GUM, ACTIVE MALT FLOUR, CALCIUM PYROPHOSPHATE, SOY
LECITHIN, ASCORBIC ACID, ENZYME), DRIED WHOLE EGG, SHORTENING
(PARTIALLY HYDROGENATED SOYBEAN AND COTTONSEED OILS) SALT,
DEXTROSE, BUTTER POWDER, (BUTTER (CREAM, SALT, ANATTO EXTRACT)
NONFAT DRY MILK, MALTODEXTRIN, BUTTERMILK, PARTIALLY
HYDROGENATED SOYBEAN OIL, SALT, SOUR CREAM (CULTURED CREAM,
NONFAT DRY MILK), DISODIUM PHOSPHATE, NATURAL AND ARTIFICIAL
FLAVORS, LACTIC ACID, CITRIC ACID, COLOR), DEHYDRATED SWEET CREAM
(SWEET CREAM, NONFAT MILK, AND LECITHIN), CORN STARCH, BUTTER
POWDER (BUTTER, NON-FAT MILK SOLIDS, SODIUM CASEINATE, BHT ADDED
TO IMPROVE STABILITY) SODIUM ACID PYROPHOSPHATE, SODIUM
BICARBONATE, DEHYDRATED PARSLEY, DOUGH CONDITIONER (WHEAT
STARCH, L-CYSTEINE HYDROCHLORIDE, AMMONIUM SULFATE) AS CURBIC
ACID (CONTAINS 10.4% SAUSAGE AND PEPPERONI).

U.

INGREDIENTS: CORN SYRUP, ENRICHED BLEACHED FLOUR (WHEAT FLOUR,
NIACIN, REDUCED IRON,THIAMINE MONOHYDRATE, RIBOFLAVIN, FOLIC
ACID), VEGETABLE SHORTENING (PARTIALLY HYDROGENATED SOYBEAN AND
COTTONSEED OILS), SUGAR, WATER, COCOA, WALNUTS, HIGH FRUCTOSE
CORN SYRUP, WHEY, EGGS, SOY LECITHIN, EGG WHITES, SALT, LEAVENING
(BAKING SODA, SODIUM ALUMINUM PHOSPHATE) COLORS (CARAMEL
COLOR, RED 40) CORN STARCH, NATURAL AND ARTIFICIAL FLAVORS, SORBIC
ACID (TO RETAIN FRESHNESS).

---

nutritional effects" (Schneeman, 2000). Nutraceuticals are foods specifically designed to act as drugs. The nutrients often found in nutraceuticals
and fortified foods are calcium, B vitamins, ascorbic acid, and antioxidant
vitamins. Ingredients designed to give an energy burst include caffeine,
L-carnitine, ginseng, guarana, and taurine. Chromium picolinate is used
for weight loss. Although some of these ingredients may indeed provide
health benefits at certain levels, the levels in foods may be too low to offer
any benefit. When the levels are high enough to provide the benefit, the
food or beverage may not be palatable. Some of the concerns facing the
developers of these products are bioactivity, customization, expense,
health claims, packaging requirements, physiological targets, quality control measures, sensory characteristics, stability, and time release.
Bioactivity means that the ingredient actually works; *customization* means
that it is compatible with the other ingredients in the product. See Insert
5.5 for some nutritional labels from these products.

High-protein products, like the ones Nolan consumes, are very popular. Protein supplements are available as whole proteins or individual
amino acids; they come in pills, powders, and potions. Nonfat dried milk,
soy products, and other nonanimal sources are the primary components

Product R

Nutrition Facts

Serving Size (71g)
Servings Per Container – 4

Amount Per Serving

Calories 110	Calories from Fat 25

	% Daily Value*
Total Fat 3g	5%
Saturated Fat 1.5g	8%
Trans Fat 0g	
Cholesterol 20mg	7%
Sodium 560mg	23%
Total Carbohydrate 16g	5%
Dietary Fiber 3g	12%
Sugars 1g	
Protein 6g	

Vitamin A 0%	•	Vitamin C 0%
Calcium 6%	•	Iron 0%

*Percent Daily Values are based on a 2,000 calorie diet. Your daily values may be higher or lower depending on your calorie needs:

		Calories:	2,000	2,500
Total Fat	Less than		65g	80g
Saturated Fat	Less than		20g	25g
Cholesterol	Less than		300mg	300mg
Sodium	Less than		2,400mg	2,400mg
Total Carbohydrate			300g	375g
Dietary Fiber			25g	30g

Calories per gram:
    Fat 9 • Carbohydrate 4 • Protein 4

Product S

Nutrition Facts

Serving Size (340g)
Servings Per Container – 1

Amount Per Serving

Calories 530	Calories from Fat 200

	% Daily Value*
Total Fat 22g	34%
Saturated Fat 12g	60%
Trans Fat 0g	
Cholesterol 110mg	37%
Sodium 1380mg	57%
Total Carbohydrate 50g	17%
Dietary Fiber 4g	16%
Sugars 9g	
Protein 33g	

Vitamin A 8%	•	Vitamin C 6%
Calcium 25%	•	Iron 10%

*Percent Daily Values are based on a 2,000 calorie diet. Your daily values may be higher or lower depending on your calorie needs:

		Calories:	2,000	2,500
Total Fat	Less Than		65g	80g
Saturated Fat	Less Than		20g	25g
Cholesterol	Less Than		300mg	300mg
Sodium	Less Than		2,400mg	2,400mg
Total Carbohydrate			300g	375g
Dietary Fiber			25g	30g

Calories per gram:
    Fat 9 • Carbohydrate 4 • Protein 4

Product T

Nutrition Facts

Serving Size (113g)
Servings Per Container – 2

Amount Per Serving

Calories 320	Calories from Fat 140

	% Daily Value*
Total Fat 15g	23%
Saturated Fat 6g	30%
Trans Fat 0g	
Cholesterol 40mg	13%
Sodium 580mg	24%
Total Carbohydrate 34g	11%
Dietary Fiber 1g	4%
Sugars 5g	
Protein 12g	

Vitamin A 8%	•	Vitamin C 0%
Calcium 25%	•	Iron 4%

*Percent Daily Values are based on a 2,000 calorie diet. Your daily values may be higher or lower depending on your calorie needs:

		Calories:	2,000	2,500
Total Fat	Less than		65g	80g
Saturated Fat	Less than		20g	25g
Cholesterol	Less than		300mg	300mg
Sodium	Less than		2,400mg	2,400mg
Total Carbohydrate			300g	375g
Dietary Fiber			25g	30g

Calories per gram:
    Fat 9 • Carbohydrate 4 • Protein 4

Product U

Nutrition Facts

Serving Size (12g)
Servings Per Container – 12

Amount Per Serving

Calories 10	Calories from Fat 10

	% Daily Value*
Total Fat 2.5g	4%
Saturated Fat 0.5g	3%
Trans Fat 0g	
Cholesterol 5mg	2%
Sodium 30mg	1%
Total Carbohydrate 8g	3%
Dietary Fiber 0g	0%
Sugars 5g	
Protein 1g	

Vitamin A 0%	•	Vitamin C 0%
Calcium 0%	•	Iron 2%

*Percent Daily Values are based on a 2,000 calorie diet. Your daily values may be higher or lower depending on your calorie needs:

		Calories:	2,000	2,500
Total Fat	Less than		65g	80g
Saturated Fat	Less than		20g	25g
Cholesterol	Less than		300mg	300mg
Sodium	Less than		2,400mg	2,400mg
Total Carbohydrate			300g	375g
Dietary Fiber			25g	30g

Calories per gram:
    Fat 9 • Carbohydrate 4 • Protein 4

Insert 5.5 Nutrition labels corresponding to ingredient statements in frozen entrées shown in Insert 5.4.

of these powders. These items can usually be purchased in other parts of the supermarket at much lower prices. One big problem with protein powders is getting them to dissolve in cold milk. Consumption of large amounts of animal proteins adds excess fat and cholesterol to the diet. In addition, too much protein and not enough carbohydrates can contribute to the development of ketoacidosis. The side effects of large doses of individual amino acids have not been tested as extensively as high levels of complete proteins. However, the consumption of individual amino acids may lead to metabolic imbalances because the body needs to match up amino acids to synthesize proteins. As we learned in Chapter 2, excessive intake of proteins for energy can lead to excretion of urea in the urine. One of the long-term consequences of urea excretion is kidney failure. The popularity of high-protein, low-carbohydrate shakes and bars comes out of a mistaken belief that increased protein consumption helps build muscle. Actually, increased carbohydrate consumption, when combined with a rational exercise program, is best for bulking up and building muscle mass. Nolan would do well to hook up with a qualified sports dietitian to design an appropriate diet and exercise program to meet his nutritional needs and physical goals.

Types of ingredients and their functions

What really makes food formulation different from food processing is the wide range of ingredients that are used. In selecting the ingredients for a formulation, the food scientist must consider the quality of the food, its safety, its stability, and its cost. Ignoring any one of these aspects could doom it in the marketplace. When we are concerned about what is in our food, the best place to start our research is by looking at the label. All the ingredients are listed there. It is important not to jump to conclusions about ingredients. Just because they have weird-sounding or unpronounceable names doesn't mean they are dangerous. Each ingredient has at least one, and frequently more than one, function in a product. Food scientists responsible for designing food products must understand how each ingredient functions and which ingredient is best for the specific product.

Flours and grains

Worldwide, individuals take in more of their calories from flours and grains than any other source. The primary function of flours and grains, in fact, is to provide calories. They are also important in providing the basic structure and texture of a product. Flours and grains are the main ingredients in baked goods; however, they can also be found in soups, beverages, confections, and many other products.

Although the largest component of flour is carbohydrates, flour proteins are also important. Wheat flour is best for making bread. The reason wheat flour is superior to other flours is because its primary protein is gluten. *Gluten* is elastic, which allows it to bend, expand, and stretch rather than break. Thus the CO_2 gas bubbles become trapped in pockets formed by the gluten network, creating the light, fluffy texture that bread lovers enjoy. Bread flour, particularly white flour, is rather bland; the delicate bread flavors form during baking. Unique bread flavors are a result of the addition of other flours like rye, potato, soy, oats, and others. Most of these breads contain only a small amount of the other flour for flavor, while maintaining a high level of wheat flour for a light, fluffy texture. In general, the less wheat flour the coarser and heavier the texture. For instance, compare cornbread with wheat bread. Whole-wheat flour contributes flavor but at the cost of lightness. The use of whole-wheat flour or addition of other whole grains such as rye increases the dietary fiber in a product but changes the product quality. Be careful not to judge a loaf of bread by its color. Some breads that appear to be whole wheat merely have caramel coloring added to turn them brown. To make sure, check the fiber content on the Nutrition Facts part of the label.

Whole grains include bulgur, rye, oats, buckwheat, amaranth, millet, barley, and many others. They make excellent cereals, or they can be ingredients in casseroles, soups, pasta dishes, and other products where grains are important in providing texture. Flours or meals can be produced from these grains, as well as corn, soy, peanut, potato, black-eyed pea, and others. Some of these grains are used sparingly in baked goods, and many are added as thickeners in products like soups, salad dressings, and beverages. Grains also serve as the fermentation substrate for many alcoholic beverages.

Starches extracted from flours are very important ingredients in many products. Different starches have unique properties. Most are added to foods to provide the proper texture. Starches swell in the presence of water. Selection of the right starch is critical in product performance. They are primary components of flours in breads, cakes, cookies, breakfast cereals, and other baked goods. They also show up on ingredient statements in salad dressings, gravies, soup mixes, puddings, and many other products. Some starches are stable when frozen and thawed; others are not. Most starches will gel after boiling, but some won't. Starches can be modified to cook and gel quickly (quick grits or oatmeal), which improves speed and convenience in food preparation. The structural changes in starch during processing and cooking are shown in Insert 5.6. A main problem with cooked and cooled starches is retrogradation, which manifests itself as staling. Staling is the result of the starch separating from the water in the product. Staling produces off-flavors and poor texture.

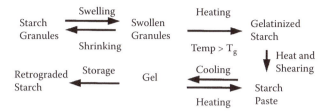

Insert 5.6 Changes in the form of starch resulting in changes of textural properties in foods. (From J. Jane, "Starch: Structure and Properties," in *Chemical and Functional Properties of Food Saccharides*, ed. P. Tomasik. Boca Raton, FL: CRC Press, Taylor & Francis Group, 2004.)

Fruits and vegetables

Fruits and vegetables are often eaten whole—fresh, frozen, canned, or dried. Fruits and vegetables are also important ingredients in many other products. Fruits form the fillings in many pies and baked goods; contribute to the flavor of many beverages, ice creams, and other desserts; serve as a topping for cereals; and appear in jams, jellies, preserves, and roll-ups. Vegetables are major components of many soups and frozen entrées, but they also appear in pies (pumpkin and sweet potato), cakes (carrot), and juices.

In addition to providing nutrition and flavor, fruits and vegetables add color and texture to products. Fruits, when used as ingredients, usually are accompanied by extra sugar. Although dietary fiber is an important nutritional quality of fruits and vegetables, many times the fiber is removed to improve the functionality of the ingredient or its appeal to the consumer. For example, in discussing juices above we talked about clarification to remove the pulp from juices like orange and apple to make them more appealing to consumers. Unfortunately, the clear juices that are more popular with students than those with pulp also have less dietary fiber. Fresh fruits and vegetables that will be used as ingredients are usually processed to keep them from spoiling. Fruits and vegetables can be processed into juice concentrates, preserved with sugar, canned, frozen, or dried until ready for use.

Dairy and eggs

Dairy products are excellent sources of high-quality protein and calcium. They are also of great interest to food scientists because of their functional properties (contributing directly to a specific function of an ingredient in a food product). Many formulated foods were developed by taking advantage of the functional properties of milk and egg proteins. For example, yogurt, pudding, cheese, and fondue all owe their texture to milk proteins, and omelet and meringue texture is due to the functional properties of egg proteins.

The two main protein fractions in milk are casein and whey. It is casein that coagulates and comes out of solution to form the smooth and creamy texture of yogurt and sour cream or the many different textures of cheeses. Gels are formed from fluid milk in products like puddings and other milk-based desserts. Most gels have extensive cross-linking of the proteins and will not reverse when reheated. Another gel, gelatin, will change back into liquid form upon reheating. Egg yolk protein acts as an emulsifier, keeping water and fat from separating in products such as mayonnaise. Egg whites are known for their ability to foam, providing the light, fluffy texture of cakes, meringues, and shakes.

Plant proteins

Although complete proteins come from animal products, many consumers get their proteins from plant sources. Plant proteins are not complete proteins. Plant proteins from one source, like corn, can be nutritious if they are balanced with complementary proteins from another source, like beans. For use as ingredients, plant proteins must first be isolated and purified. The isolation process separates the small quantities of protein in plant cells from the nonprotein material—the cell walls, carbohydrates, fats, and acids. Purification must be rigorous enough to remove all unwanted material for the protein to function properly, yet mild enough to maintain the desired functional properties of the protein. Plant proteins can function as emulsifiers, water binders, and gelling agents. Textured vegetable proteins replace meat protein in veggie burgers and other meat substitutes for the consumer who doesn't want to eat meat, and doesn't want to give up some of their favorite products. Plant proteins can be modified by chemical, enzymatic, or physical processes to meet the ingredient's functional requirements. Sources of plant proteins include grains such as corn and oats and oil-rich seeds such as soybeans and peanuts.

Fats and oils

Although many of us are trying to cut back on our fat and oil intake, they are still important ingredients in many of our favorite products. Fats and oils are extracted from plant and animal products. Fats and oils belong to a chemical group called *lipids*. Fats are solid at room temperature, and the oils are liquid at room temperature. Rendering, pressing, extracting, refining, modifying, and forming are common unit operations for fat and oil products. Rendering melts an animal fat and separates it from other components. Pressing physically removes the oil from a plant seed. Organic solvents such as hexane are also used to extract oils from water in the seed.

Refining removes undesirable compounds, bleaching removes pigments, and deodorizing reduces off-odors and flavors. Hydrogenation improves the functional properties of the fat or oil, but the process increases the level of trans fatty acids. More information on trans fatty acids is available in Chapters 6 and 11. Butter, chocolates, cooking oils, margarine, mayonnaise, peanut butter, and salad dressings are examples of products with fats and oils as primary ingredients.

So why do we want to put fats and oils into food products? First, there is the issue of satiety (feeling full). Fats and oils slow down the stomach-emptying time. The longer the stomach-emptying time, the longer it is before we become hungry again. Another reason for adding fats and oils is that they enrobe flavors and contribute to mouthfeel. That means that fats and oils act as carriers of flavors. Our food may not taste as good without the fat and oil. In addition, when fats and oils are heated, they turn to liquid in the mouth and contribute to the juiciness of meats and other entrées, serving as lubricants in the mouth. Think about a dry biscuit in your mouth. Then think about the difference some butter or margarine makes. Finally, fats and oils can either serve as flavor precursors or can become oxidized. A *precursor* is a flavorless compound that changes form to either a desirable or an undesirable flavor during mixing, heating, or storage. *Oxidation* is a series of reactions that leads to off-flavors, generally described as tasting like cardboard or like fresh-mowed grass.

Sweeteners

Sugar can be a major source of calories for anyone with a sweet tooth. However, sugar is much more than a sweet source of calories. Sugars contribute to the color and texture of food products. Table sugar, or sucrose, can caramelize. Other sugars, such as fructose (primary sweetener in honey), glucose (blood sugar), and lactose (milk sugar), are *reducing sugars* and can turn brown at lower temperatures. Sugars also contribute to the way a food feels in the mouth (mouthfeel). Sugars also provide bulk. If you take away the sugar in a food that is 10% sugar, you have 10% less food. It requires much less artificial sweetener to provide the same level of sweetness that sugar provides. In addition, the form of sugar (crystalline or noncrystalline) can affect the structure of a confection. Sugar has other functional properties relating to the sponginess of cakes and the creaminess of ice creams, in addition to adding a sweet flavor to these products.

Artificial sweeteners can be used to replace sugars. Artificial sweeteners are high-intensity compounds; therefore, much smaller amounts are needed to provide the same level of sweetness as sugar. As a result, there is a major reduction of calories when artificial sweeteners are used. For example, a 12-ounce can of regular cola contains 39 grams of sugar (196 calories). A sugar-free product with aspartame (NutraSweet®) contains one

Insert 5.7 Sweetness of sugar substitutes relative to sucrose. (Adapted from L. O. Nabors, ed., "Alternative Sweeteners: An Overview," in *Alternative Sweeteners*, New York: Marcel Dekker, Inc., 2001, chap. 1.)

Sucralose	600
Saccharin	300
Stevioside	300
Acesulfame potassium	200
Aspartame	180
Sucrose	1
Maltitol	0.9
Sorbitol	0.6

calorie. Other artificial sweeteners include saccharin, sucralose, sorbitol, and acesulfame K. Each sweetener has its advantages and disadvantages. For example, aspartame breaks down when heated, making it unsuitable for baked goods; saccharin and acesulfame K have a bitter aftertaste; sorbitol works well in baked goods, but it can cause flatulence. Some of these problems can be overcome by mixing the alternative sweeteners. For more information on artificial sweeteners see Insert 5.7.

Fat replacers

Replacing the fat in products is of even more interest than replacing the sugar. Unfortunately, individual fat replacers cannot perform all the functions of lipids. Flavor and mouthfeel are the two functions that are the most difficult to replace. Protein particles can produce a similar mouthfeel and reduce the calories by half; carbohydrates can increase the viscosity. Neither, however, is as stable in high-temperature frying and cooking. Sucrose esters (like Olestra®) are chemically similar to fats and have similar stability but are not digested and absorbed by the body. Sucrose esters have the full lubricating properties in the mouth as real fats. Unfortunately, because they are not digestible, the lubricating properties can cause side effects such as diarrhea and abdominal cramping if consumed in excess. Another option to reduce caloric content is to simply remove the fat without adding a fat replacer. But such removal can have unexpected consequences. For example, most low-fat peanut butter products have more calories per serving than the full-fat versions.

Other fat replacers include gums and hydrocolloids, which are carbohydrates similar to dietary fiber. They provide similar textures to foods, while replacing the bulk of the fat so the low-fat product contains fewer calories. The gums and hydrocolloids are either not absorbed by the body or are absorbed at a much lower rate than the fats. Unfortunately, these ingredients do not provide the lubrication or the flavor properties of fat. Starch and protein-based products can provide bulk and some lubrication, but not flavor. Many companies are working on synthetic fat replacers that will have all the advantages of fat without the problems, but none are currently approved by the U.S. Food and Drug Administration (FDA).

Although fat reduction is a good thing for most of us, there are difficulties. In general, many of the functional properties of a fat can be maintained even with the elimination or replacement of over half of the fat in a product. Complete elimination of the fat, however, makes it difficult to match the flavor and texture of the full-fat product. Another potential problem with low-fat products deals with satiety. Since fats slow stomach-emptying time, consumers may eat more low-fat product to fill them up and thus still not reduce their calorie intake.

One day Olivia went into a yogurt shop and said to the guy operating the soft-serve machine, "I'm really glad you have sugar-free/fat-free yogurt, but couldn't you do something to make it taste more like the real thing? It just doesn't feel the same way in my mouth." If she had taken a food science course, she would have known that the sugar and the fat have functional properties that contribute to the mouthfeel of her favorite yogurts. See Insert 5.8 for more information on fat replacers.

Flavors and colors

Flavor and color frequently make the difference between a food we enjoy and one we just won't eat. Some ingredients are added to foods to enhance the flavor and color. Most flavor ingredients that are added to food are colorless, yet color can influence perception of flavor as mentioned in Chapter 3.

As we will learn in more detail in Chapters 7 and 14, flavor is the combination of the senses of taste and aroma. We have talked about bitter, sweet, sour, and salty tastes and the ingredients responsible for them. Most flavor ingredients (flavorants) are volatile compounds (evaporate rapidly at room or body temperatures) that contribute to the aroma of food products. Flavorants can be natural or synthetic. Early attempts at artificial flavoring were only suggestive of the real thing, but recent advances have been much more sophisticated. In many products, a flavor-impact compound is responsible for much of the characteristic aroma of that product; however, the aroma of most products is the combination of a wide range of aromatic compounds. Flavorants may provide or add to the flavor impact of a product, enhance or mask other flavors, or provide a background for other flavors present. Monosodium glutamate is a flavor enhancer that tends to bring out flavors not normally perceived by a consumer.

Makers of sports beverages, breakfast cereals, and other products know that many consumers are more likely to buy a product if it is brightly

Insert 5.8 Types of fat replacers found in foods. (Adapted from C. C. Akoh,
"Fat Replacers: A Scientific Status Summary,"
Food Technology 52, no. 3 (1998): 47.)

Chemical base	Subtype	Composition	Functionality
Lipid	Sucrose fatty acid polyester (Olean & Olestra)	Transesterified and interesterified sucrose by fatty acids	Provides mouthfeel and lubrication of fats without being absorbed
	Sucrose fatty acid esters	Monoglycerides, diglycerides, and triglycerides of sucrose	Excellent surfactants and emulsifiers
	Structured lipids	Mixed chain-length fatty acids	Specific applications
Protein	Simplesse	Microparticulated whey protein	Baked goods, salad dressings, and mayonnaise
Carbohydrate	Gums	Negatively charged high-molecular-weight polymers	Stabilizers, gelling agents increases viscosity
	Starches	Varying sources	Slippery mouthfeel
	Maltodextrins	Polymer of mixed-length saccharides	Water binding, viscosity, smooth mouthfeel
	Polydextrose	Polymer of glucose, sorbitol, and citric or phosphoric acid	Bulking agent and texturizer

colored, so they add artificial colors to these products to enhance their appeal. These colorants are the same ones used in the home to color Easter eggs and cake icings. Natural colorants are isolated from fruits, vegetables, flowers, or insects and added to food products. Most food manufacturers prefer artificial colorants because they are brighter, more stable, and easier to handle than natural colorants. Natural colorants do not appear to be any safer than artificial ones.

Stabilizers

Stabilizers are added to formulated foods to keep them from breaking down. Peanut butter with the oil floating to the top of the jar and a milk-chocolate bar with an uneven white coating are examples of food breakdown that is not related to microbes. Important functions of stabilizers include preventing the separation of food components and maintaining product structure.

Emulsifiers are an important group of stabilizers. Many formulated foods are emulsions, wherein oil is dispersed in water (sauces and soups) or water is dispersed in oil (margarine and mayonnaise). Ordinarily water and oil don't mix, but emulsifiers can dissolve in both water and oil. With part of the emulsifier dissolved in water and the other part dissolved in oil, it can keep the water and oil bound together. Some natural peanut butters do not have added emulsifiers and are more likely to separate. Egg yolk contains lecithin, a natural emulsifier that helps keep mayonnaise together.

Gums represent another important type of stabilizer. *Gums* are water-soluble polysaccharides that help thicken liquid and semisolid foods. They thicken by binding the water present and swelling up. Each gum possesses unique properties that are useful in formulating food products that we enjoy. For example, gum tragacanth is stable over a wide pH range, agar forms nice gels, guar gum is an excellent thickener at low concentrations, and alginate works well when combined with calcium as found in milk products.

Preservatives

In addition to being resistant to breakdown, formulated foods are also resistant to microbial spoilage. Preservatives are food additives that prevent or slow spoilage. Although we don't think of them as preservatives, table sugar (sucrose) and table salt (sodium chloride) are the two most widely used preservatives. Despite their bad reputation, preservatives extend shelf life and lower costs, which can increase profits for food companies and lower prices for consumers. Much food is wasted because it looks, smells, and tastes bad: molded bread, spoiled chicken, sour milk, rancid oils, and slimy wieners. Preservatives help reduce this waste. See a picture of a food not protected by preservatives in Insert 5.9.

Many spices are added to foods to prevent microbial growth and rancidity as well as to add flavor. Garlic, nutmeg, oregano, and thyme are known to protect foods from harmful microbes. Cloves, cumin, and sage slow the growth of molds, and basil and rosemary are antioxidants. Chemical compounds added to foods to prevent the growth of molds and yeasts include sorbic, benzoic, and acidic acids or their salt forms. Sulfur dioxide, nitrates, and nitrites are used to prevent the growth of bacteria that can cause illness or spoilage. Natural antioxidants such as vitamins A, C, and E or synthetic antioxidants such as BHA (butylated hydroxyanisole) and BHT (butylated hydroxytoluene) prevent fatty foods and oils from developing rancidity.

Insert 5.9 Moldy pear. (Photo provided by Dr. Larry Beuchat, Center for Food Safety, University of Georgia.)

Remember this!

- Formulated foods are products that involve the mixing of ingredients.
- Minor variations in a process can result in major differences in a final product.
- Noodles and pasta provide good nutrition at a low price.
- Most wheat is milled into flour, and wheat is the best grain for baking bread.
- Carbonated beverages are the most widely consumed liquid products in America.
- In selecting the ingredients for a formulation, the food scientist must consider the quality of the food, its safety, its stability, and its cost.
- Fats and oils are extracted from plant and animal products.
- Individual fat replacers cannot perform all functions of lipids.
- Stabilizers are added to formulated foods to keep them from breaking down.

Looking ahead

This chapter was designed to provide an introduction to formulated foods, made from mixing ingredients. Chapter 6 describes chilled and prepared foods. Chapter 7 explores the role of food scientists in measuring foods for quality and safety. Chapter 8 describes the design of food products.

Answers to chapter questions
Inserts 5.4 and 5.5

 R Gardenburger Original
 S Stouffer's® Corner Bistro Chicken Carbonara
 T RED BARON® Stuffed Pizza Slices SINGLES 2 SUPREME PIZZAS
 U Little Debbie Fudge Brownies

References

Akoh, C. C. 1998. Fat replacers: A scientific status summary. *Food Technology* 52(3): 47.
Hoseney, R. C. 1994. *Principles of cereal science and technology*, 2nd ed. St. Paul, MN: American Association of Cereal Chemists, Inc.
Jane, J. 2004. Starch: Structure and properties. In *Chemical and functional properties of food saccharides*, ed. P. Tomasik. Boca Raton, FL: CRC Press, Taylor & Francis Group, chap. 7.
Nabors, L. O., ed. 2001. Alternative sweeteners: An overview. In *Alternative sweeteners*. New York: Marcel Dekker, Inc., chap. 1.

Further reading

Akoh, C. C. 2008. Fat-based fat substitutes. In *Fatty acids in foods and their health implications*, 3rd ed., ed. C. K. Chow. Boca Raton, FL: CRC Press, Taylor & Francis Group, chap. 17.
Damodaran, S., K. L. Parkin, and O. R. Fennema. 2007. In *Fennema's food chemistry*, 4th ed. Boca Raton, FL: CRC Press, Taylor & Francis Group.
Fellows, P. J. 2000. *Food processing technology: Principles and practices*, 2nd ed. Boca Raton, FL: CRC Press.
Lindhorst, T. K. 2007. *Essentials of carbohydrate chemistry and biochemistry*, 3rd ed. Weinheim, Germany: Wiley VCH.
Man, D. and A. Jones. 2000. *Shelf-life evaluation of foods*. Gaithersburg, MD: Aspen Publications, Inc.
Potter, N. N. and J. H. Hotchkiss. 1999. *Food science*, 5th ed. New York: Chapman & Hall.
Schmidl, M. K. and T. P. Labuza. 2000. *Essentials of functional foods*. Gaithersburg, MD: Aspen Publishers, Inc.
Schneeman, B. O. 2000. Relationship of food, nutrition, and health. In *Essentials of Functional Foods*. Gaithersburg, MD: Aspen Publishers, Inc.
Shahidi, F., C-T. Ho, S. Watanabe, and T. Osawa. 2003. *Food factors in health promotion and disease prevention*. Washington, DC: American Chemical Society.
Smith, S. J. and Y. H. Hui. 2004. *Food processing: Principles and applications*. Ames, IA: Blackwell Publishing.
Webb, G. P. 2006. *Dietary supplements & functional foods.* Ames, IA: Blackwell Publishing, Ltd.

chapter six

Chilled and prepared foods

Rob loves to shop, particularly for food. Shopping is a social occasion for him. He knows most of the employees at the supermarket by their first names and has been known to kill a couple of hours shopping for food for the week. He particularly likes the perimeter of the store because that is where the good stuff is. He likes fresh produce, high-quality meats, and chilled and frozen foods. The middle aisles are particularly boring with all those cans, boxes, and cartons. He knows the food on the perimeter is more expensive, but he is very selective and usually gets good value for his money.

Shaundra's grandparents had four peach trees in their back yard. When she visited them in the summer she had delicious fresh peaches and loved them. She can't understand why anyone would buy a fresh peach at a supermarket because the flavor is so lame. She also can't understand why anyone would buy cut-up fruits and vegetables. They aren't anywhere near the quality of the ones her grandparents grew and prepared. If she can't have the best, she doesn't want the rest.

Tanya is concerned about her weight. She avoids the dining hall on campus and fast foods but still finds herself gaining weight. She is very busy and does not have a lot of time to cook. She needs something that is quick, easy, and filling. She's big on frozen dinners and prepared sandwiches from the supermarket deli. She has tried many diets, but they either have unappealing foods or don't fill her up. She wishes there were some magic pills that would do the trick.

Rob, Shaundra, and Tanya are familiar with chilled and prepared foods. Rob takes time shopping and gives his selections thought and attention. Some up-front time helps him save time later while still eating a healthy, satisfying diet. Shaundra is more demanding in her standards for fresh fruits and vegetables, and so she probably does without them many times. She is not willing to sacrifice quality for convenience. Tanya is interested in convenience and does not think about her food selections enough to eat healthy and maintain her weight. Chilled and prepared foods can help us design healthy diets, but we must evaluate them just like any other food.

Looking back

Previous chapters focused on food issues we deal with daily. Some key points that were covered in those chapters help prepare us for understanding chilled and prepared foods.

- Noodles and pasta provide good nutrition at a low price.
- The main function of a package is to prevent microbial or chemical contamination of a processed food product.
- Food is preserved to make it safer by reducing or eliminating harmful microbes.
- The conditions that affect nutritional value also affect sensory quality.
- Unit operations are distinct steps common to many food processes.
- Most raw materials are perishable and require careful handling or processing to prevent losses.
- Fast-food operations minimize the time between order and delivery of a meal.
- Technology has transformed our food supply to produce more formulated foods, leading to the eating of fewer whole foods.
- If we want to eat a healthy diet, we have an obligation to read labels, analyze our diets, and exercise willpower in our food selections.
- Nutrition labels provide information to assist us in making good decisions to improve the quality of our diets.
- Good nutrition requires an adequate consumption of nutrients without exceeding the recommended allowance of calories.
- Healthy eating demands a balanced diet.
- Preservatives are food additives that prevent or retard spoilage.
- The expiration date represents the food scientist's best guess about how long a food will last before it spoils.

What are chilled and prepared foods and why are they necessary?

As the pace of society increases, most consumers are looking for ways to reduce food preparation time and increase convenience. Chilled and prepared foods are part of the food industry's answer to meet these consumer needs. Chilled foods tend to be fresh and perishable, requiring refrigeration. Prepared foods are those that are ready-to-eat or ready-to-heat then eat. Prepared foods can come in commercial packages, or they can be purchased and consumed at a retail outlet such as a cafeteria or restaurant. Because chilled and prepared foods have generally been exposed to less harsh preservation measures than their processed counterparts, they have a reputation for being of a higher quality. The cost of higher quality and increased convenience may be an increased safety risk.

Distribution

To understand perishable products like chilled and prepared foods, we must understand distribution. Distribution is everything that happens to a product from the time it is produced to the time it is consumed. Although perishable and shelf-stable products have similar distribution steps, many of the unit operations are of little consequence to stable products, but they are critical for maintaining the quality of chilled and prepared foods. To monitor potential problems, many producers use a systems approach to distribution—looking at how individual operations interact within the overall context of the system rather than focusing on specific steps. A systems approach is based on the idea that the whole is greater than the sum of its parts. Since chilled foods tend to be minimally processed raw materials, distribution of these raw materials prior to packaging must also be carefully studied. Although many chilled foods may be consumed without heating, many prepared foods are heated either shortly before or after purchase. Such heating may serve to decrease microbial load, increasing the safety of the product, or provide a false sense of security to the consumer.

Chilled foods

As you walk around a modern supermarket, you will notice that the higher-value items tend to be located around the perimeter, with the lower value staples located in the internal aisles. Chilled foods tend to be higher-value items. If we take a walk around Rob's favorite supermarket, we will find the produce section across from the deli as he enters the store. At the back of the store are the luncheon meat coolers, the fresh fish and meat counters, and around the corner are the dairy product cases.

Whole fresh fruits and vegetables

For pure eating pleasure it is hard to beat a fresh peach, mango, or watermelon at the peak of the growing season. Unfortunately, it is difficult to find a ripe one in the supermarket that will deliver on flavor. Fresh fruits and vegetables are picked live and continue to live and respire through the distribution system. *Postharvest physiology* is the scientific discipline that follows biological changes in a fresh fruit or vegetable as it proceeds from harvest to consumption. Postharvest technology is the term that incorporates the unit operations employed in slowing down these changes to extend the shelf life of an item.

The plant and its detached fruit undergo trauma during harvest. Ethylene gas is emitted from the wound and leads to the formation of the wound scar. Some fruits, like apples and bananas, will continue to ripen off the plant. These fruits are harvested before they reach full ripeness because they deteriorate rapidly after being detached from the plant and are unlikely to withstand shipment if picked when fully ripe. Other fruits like watermelon, oranges, and strawberries must be allowed to ripen on the plant. These fruits tend to be harvested closer to their peak level of ripeness. Fresh fruits are sorted and graded in a packinghouse prior to shipment to market. Many fresh fruits, such as peaches and cherries, are shipped in refrigerated conditions to slow decay and respiration, which lead to spoilage. Other fruits, particularly tropical fruits like bananas and mangoes, are not refrigerated because storage at low temperatures reduces the flavor and can accelerate decay. Some fresh fruits such as apples and pears are stored in controlled atmospheres (high carbon dioxide, low oxygen) to slow ripening so they can be distributed year round. Fresh vegetables undergo many of the same biological processes as fresh fruits and are preserved by similar technological processes. Some consumers believe tomatoes and squash are vegetables, but they are considered fruits. Tomatoes ripen off the vine and should not be refrigerated; squash do not ripen after harvest and need refrigeration.

In most distribution schemes, fresh fruits and vegetables are loaded and unloaded many times in their journey from harvest to the retail store. Any rough handling will cause bruising and lower the quality at purchase and consumption. The distribution system tends to value appearance and shipping stability over flavor, which is why Shaundra is frequently disappointed in the flavor of the fresh peaches that look so good in the supermarket. For the distribution scheme of a fresh peach, see Insert 6.1.

Packaged salad vegetables and cut fruits

One of the objections to fresh fruits and vegetables is that they take so long to prepare. Fresh-cut items make them more convenient to consumers who want to eat healthy but are always on the run. Cutting involves wounding of the plant tissue. As at harvest, ethylene is generated and wound healing begins. In cut lettuce the wound scar may not be noticeable, but the white blush of cut carrots and brown ends of cut green beans are objectionable to consumers. Fresh-cut salad vegetables have become more popular than fresh-cut fruits. It is usually more difficult to preserve the flavor of cut-up produce than whole produce. Since cut fruits and vegetables are usually more perishable than their whole counterparts, they are usually shipped whole from the growing area and cut and packaged close to the consumers. Sophisticated modified atmosphere packaging (MAP), which is low in oxygen and high in nitrogen or carbon dioxide, slows the deterioration process. Once the consumer opens the package, however, any benefits of the MAP are lost. Refrigeration of cut products is critical to slow spoilage. Many of the whole fruits sensitive to low temperatures are less susceptible to chilling when cut.

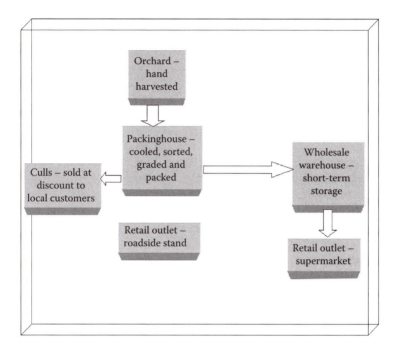

Insert 6.1 Possible handling and distribution scheme for fresh peaches.

Fresh meats

Fresh meats are also called muscle foods because the edible portion of most meat, including fish, is from the muscle tissue of the animal. Unlike fresh fruits and vegetables, which are living tissue, meat is dying tissue. The study of the reactions involved in the conversion of muscle to meat is called *postmortem physiology*. A series of these complex reactions proceeds to rigor, which produces tough meat. Thus, for most meats, rigor needs to be resolved before the meat can be marketed and sold. Textural problems, known as cold shortening, can develop in beef, lamb, and pork if the carcass is cooled too rapidly after harvest. Beef and pork carcasses are hung at low temperatures to allow for the resolution of rigor and the development of flavor. Red meats are actually purple right after harvest. It is only in the presence of oxygen that the meat becomes the bright red color associated with fresh hamburger and steak. Inappropriate lighting in a retail case causes quality loss due to increased temperatures. Although many supermarkets are equipped to produce the appropriate cuts from partial carcasses, some chains prefer *case-ready meats*. Case-ready cuts are prepared and packaged at a central processing facility and then shipped to the store for display.

Food scientists are wary of consuming raw meat because any harvest operation is likely to contaminate the surface of the meat. In nondiseased animals, the inside of the muscle is assumed to be sterile; however, if the surface is contaminated, any cutting, grinding, or puncturing of the surface will inoculate the inside of the muscle and contaminate it as well. Meats spoil rapidly even when refrigerated. Spoilage is characterized by the development of off-odors, dull colors, and slime. The shelf life of meats is affected by the initial numbers of microbes in the meat, storage temperature, and relative humidity. Use of MAP in meats can slow the spoilage process considerably and thus extend the shelf life. *Drip*, the release of liquid from the tissue, is also a major problem with chilled meats. Rapidly cooling the meat and holding it at low temperatures reduces drip. Absorbent materials in the bottom of retail packages, known in the trade as *diapers*, help reduce the level of visible drip in the package.

Fresh fish and seafood

Fresh fish and seafood are very perishable. A fresh-caught fish has little or no aroma, but foul aromas can develop in a relatively short period of time due to the breakdown of amines and related compounds in the muscle tissue. Immediate chilling, preferably close to the freezing point of water, helps preserve fish flavor. MAP in nitrogen atmospheres can slow unpleasant aroma development but can enhance the chances of botulism. Because most refrigeration systems are not kept at the freezing point, storage on ice helps keep the temperature down. At the same time, ice can serve as a medium for the spread of spoilage and harmful microbes. Expedited handling minimizes loss of quality and safety problems.

Contrary to popular belief, *sushi* means "with rice," not "raw fish." *Sashimi* refers to raw fish. Sashimi can be part of the filling in a sushi roll in one of four forms: *futomaki* (large rolls), *hosomaki* (thin rolls), *temaki* (hand rolls) and *uramaki* (inside-out rolls). Many other ingredients can serve as fillings for sushi rolls. The basic ingredients in sushi rice include nori sheets, short-grain rice, soy sauce, and wasabi paste. Raw fish used as ingredients in sushi rolls include eel, mackerel, salmon, snapper, tuna, and yellowtail. Sushi rolls are very popular and have a record of safe consumption. Even so, raw fish spoils rapidly, and any raw meat product can contain pathogenic bacteria. See Insert 6.2 for pictures of sushi products.

Deli meats

Fresh meat cuts can serve as either raw materials for processed deli meats, such as smoked turkey, cured ham, and roast beef, or as formulated products such as bologna and salami. In general, these products have been cooked and are sold as cold cuts in ready-to-eat form. In the processed products, the original texture of the muscle structure is obvious;

Insert 6.2 Sushi is prepared daily and found in a refrigerated display case of a special section of most large supermarkets. (From http://office.microsoft.com/en-us/cl:part/results.aspx?qu=sushi&sc=20.)

in formulated products the meat is usually comminuted (finely ground) and then restructured. Some of these products, like salami and bratwurst, are fermented to provide unique flavors, colors, and textures. The distinct color of bologna, frankfurters, and similar products is due to the addition of nitrites, which stabilize the myoglobin in the meat. Nitrites also protect the product from the growth of *Clostridium botulinum,* the microbe responsible for botulism. Deli meats are usually packaged in atmospheres of little or no oxygen to slow the growth of spoilage microbes; however, *Clostridium botulinum* is only able to grow in the absence of oxygen.

In recent years there has been some concern expressed about the consumption of deli meats because they are known to contain nitrosomines and *Listeria monocytogenes*. Nitrosomines, a carcinogen, are formed by the interaction of nitrites with free amines in the meat. Nitrites can be found naturally in many items, including lettuce and other vegetables. Listeriosis is primarily a concern for the immunocompromised (the young, elderly, pregnant, or people with AIDS). Moderate consumption of deli meats is unlikely to increase the risk of developing cancer, and healthy consumers between the ages of 5 and 60 who are not pregnant have little to fear from *Listeria*. The U.S. Department of Agriculture (USDA) recommends that all luncheon meats be cooked before they are served, generally defeating the purpose of cold cuts. Maintaining adequate refrigeration of perishable products like deli meats is necessary, and expiration dates should be strictly observed. Daily consumption of these products is probably not a good idea. See Insert 6.3 for a fact sheet on *Listeria* from the FDA.

During wholesale distribution to retail outlets, deli meats are reasonably stable if they are kept refrigerated. Once the package has been opened, however, the protection offered from the low-to-no-oxygen environment

Insert 6.3 Listeria Fact Sheet found at the U.S. Food and Drug Administration (FDA) Web site (http://www.fda.gov/womens/getthefacts/listeria.html).

You can get very sick from some ready-to-eat foods if you wait too long to eat them. The sickness is called Listeriosis. A germ called *Listeria* causes it. It's unusual because it can grow at refrigerator temperatures where most other foodborne bacteria do not. You can't see it, smell it, or taste it. Only heat can kill it, but if heated food cools, *Listeria* may grow again.

Who's at risk?
- Pregnant women and their unborn babies
- Older adults
- People with cancer, AIDS, and other diseases that weaken the immune systems.

How can I reduce my risk?
- Keep your fridge set at 40 degrees Fahrenheit or colder.
- Use precooked and ready-to-eat foods as soon as you can.

Danger: Avoid these foods
All pregnant women and others at risk should not eat certain foods.

Don't eat soft cheeses:
- Mexican-style soft cheeses including: queso blanco, queso fresco, queso de hoja, queso de crema and asadero
- feta, brie, Camembert, blue cheese, and Roquefort
- cheeses made from raw milk.

You don't have to cut all cheeses from your diet. Cheese can be a good source of protein, vitamins, and calcium when you are pregnant.

These cheeses are safe to eat: hard cheeses (such as cheddar and Swiss); semi-soft cheeses such as mozzarella, pasteurized processed cheeses such as slices and spreads, cream cheese, and cottage cheese.

Don't eat refrigerated smoked seafood right from the package:
- This includes salmon, trout, whitefish, cod, tuna, or mackerel. They might be called "nova-style," "lox," "kippered," "smoked," or "jerky." It is safe to eat smoked seafood if it is cooked in its dish, like a casserole.

Canned seafood is safe. Examples are canned salmon or tuna in a pouch.

Don't eat refrigerated pâtés or meat spreads. Canned meat spreads and pâtés are safe.

Don't drink raw (unpasteurized) milk or eat foods that contain raw milk.

Be careful with these foods
- Ready-to-eat foods: Hot dogs, cold cuts, lunchmeats, deli counter meats, and other ready-to-eat foods. Eat these foods only if they're reheated until steaming hot. Even cured meats such as salami must be heated.
- Meats and seafood: Cook these all the way through. Stay away from rare meat and seafood.
- Leftovers: Reheat all until steaming hot.
- Fruits and vegetables: *Listeria* can grow on some fruits and vegetables. Do not buy sliced melon. Wash all fruits and vegetables with water. Scrub hard produce such as cucumbers and melons with a clean produce brush.

How can I keep my kitchen safe?
- When you buy pre-cooked or ready-to-eat foods that go in the fridge, use them as soon as you can.
- Clean your fridge often.
- Make sure that the fridge always stays at 40 degrees F or colder. Use a refrigerator thermometer.
- Read labels. Follow instructions on foods that must be kept in the fridge or have a "use by" date.
- Wash your hands with warm soapy water after you touch raw foods. Wash any knives or other tools you used with hot, soapy water before you use them again.

You can get sick from two to 30 days after you eat food. Pregnant women can start early labor if the infection spreads to the unborn baby. Tell your doctor right away if you get any of these symptoms:
- Fever and chills
- Headache
- Upset stomach
- Throwing up

To learn more:

U.S. Food and Drug Administration (FDA)

Center for Food Safety and Applied Nutrition

1-888-723-3366

www.cfsan.fda.gov

is lost and the meats will spoil fairly rapidly. It is generally best not to keep freshly cut products more than 5 to 7 days at refrigerated temperatures. Leftovers can be kept for longer periods if properly packaged and frozen. Freezing can change the texture of the meat product and decrease its desirability, however.

Fluid milk and soy alternatives

Milk is collected on dairy farms under sanitary conditions and pumped into a tanker truck. Milk in a truck could come from several farms. Thus, a sample of the milk is collected from each farm before it is pumped into the truck. Once the truck arrives at the processing plant, a sample of the milk is collected and sent to the lab to test for added hormones, antibiotics, pesticides, and other undesirable compounds. If the milk passes the test, it is pumped from the truck into large refrigerated holding tanks until ready for further processing. If the milk sample fails, the load is rejected and the samples from each farm are tested to see which farm is responsible for the rejection.

Fluid milk goes through many steps before it is packaged. *Clarification* is the process of removing undesirable components such as pus and cells from the milk. *Pasteurization* is a heating process that kills all harmful microbes but does not kill spoilage microbes. It is necessary to remove some of the fat, usually by centrifugation, to produce skim milk, 1%, and 2% milk. *Homogenization* then breaks the fat globules into very small droplets to prevent the fat from separating from the milk. Thickeners and colorants can be added to skim milk to improve its acceptability. The fat can be used for cream or half-and-half products. After these steps, milk is pumped into sterile glass or plastic containers. In many manufacturing plants, plastic containers are formed on-site just prior to being filled. Dairy companies have designed more user-friendly packages, making them easier to open, pour, and put in drink holders.

For consumers who don't wish to refrigerate their milk, there is what is called "brick-pack" milk. These shelf-stable packages come in different sizes, the most popular holding eight ounces of milk and the other holding a liter. The eight-ounce size comes with a small straw on the side that can be poked through a silver patch on the top of the carton. These products are processed aseptically. The milk is commercially sterilized before being put into the containers, killing all spoilage and harmful microbes. Aseptic processing frequently sacrifices nutrition and flavor for shelf stability.

Many people are lactose intolerant. Lactose is the primary sugar in milk and is found in many other dairy products. A lactose-intolerant person experiences bloating, gas production, diarrhea, and other intestinal discomfort when consuming too much lactose. One way to avoid this discomfort is to avoid any products containing milk. Since milk and dairy products are among the best sources of calcium, this may not be the best strategy. Other alternatives include consumption of low-lactose milk, consumption of an enzyme called β-galactosidase (found in products such as Beano™) just prior to a meal, consumption of low-lactose dairy products like yogurt and many cheeses, or consumption of a cow's milk substitute such as soy milk. Soy milk is made by water extracts of soy proteins and fats through a complex process. Soy milks are high in protein and lack lactose, but the characteristic soy flavor is present. Although some consumers like this flavor, most consumers do not. For this reason, these milks are usually flavored with something like vanilla or chocolate to help mask the strong soy flavor.

Spreads

Our favorite bread products are usually enhanced by a spread of some sort like butter or margarine. The fats in these spreads provide lubrication in the mouth. Think of a fresh biscuit. Although the flavor of the biscuit is tempting, it can become dry in the mouth and hard to swallow without a little (or a lot) of our favorite spread. Just like oil in an engine, the spread helps lubricate the biscuit to make chewing and swallowing a better experience. Butter is a by-product of milk production and is composed primarily of milk fat. Margarine is made from vegetable oil and contains an emulsifier to mix the oil and water. By definition a fat is solid at room temperature and an oil is liquid at room temperature. Spreadability is a critical quality factor for a spread. Since butter is perishable at room temperature, it must be refrigerated. When it is removed from the refrigerator it does not spread well because the fats are too solid due to the predominance of saturated fats. Vegetable oils are liquid because they are more unsaturated. To achieve a solid and spreadable margarine, the oils can be hydrogenated. *Hydrogenation* adds hydrogen to the unsaturated double bonds in the oil to produce a margarine that is solid but not as solid as butter, and can be spread easily even when just removed from the refrigerator. Although margarine provides excellent spreadability, there are problems associated with hydrogenation. Hydrogenation produces trans fats, which appear to contribute to heart disease as much as saturated fats, if not more so. Regulations now require the product label to display the trans fat content in addition to the total and saturated fat content. In addition to margarines and similar spreads, hydrogenated oils and trans fats are in many popular products.

    In addition to spreadability, the color and flavor of spreads are important functional properties. Without added color, margarines are generally not acceptable to consumers. When margarines were first introduced, regulations did not allow color to be added to them by the manufacturer. The consumer was supplied with a food color packet to add to the product to improve its color. Now β-carotene (provitamin A) is usually the color of choice in margarines. Butter has a distinctive character that is hard to duplicate in margarine. Consumers introduced to butter before consuming margarine are rarely pleased with the margarine flavor; those introduced to margarine before butter usually prefer the flavor of margarine. Brands of margarine tend to have a greater variation in flavor than brands of butter. Butter flavor can be greatly influenced by the feed of the cows producing the milk, so butter flavor may vary by season. In a desire for low- or no-fat foods, the fat content has been lowered or eliminated in some spreads. These spreads may perform well when applied to a bread product, but they may not have certain functional properties we have come to expect. For example, Rob bought a no-fat spread one day and was very surprised when the spread would not melt and soak into his bread toasted in a conventional oven. The spread just formed a little puddle on top of the unheated and toasted bread without penetrating its surface!

During distribution and storage, the biggest problem with spreads is development of rancidity. The fatty acids in fats and oils oxidize and the flavor becomes unacceptable. Unsaturated fatty acids are more susceptible to oxidation than saturated fatty acids. Since margarines are higher in unsaturated fatty acids than butter, margarines are more susceptible to the development of rancidity. Adding antioxidants such as BHA and BHT helps slow the development of off flavors due to rancidity. Butter tends to be more susceptible to spoilage by microbes. Maintaining low temperatures during handling and distribution can slow the development of rancidity in margarines and spoilage in butter.

Prepared foods

Prepared foods are foods that are ready to eat or ready to heat then eat. In our fast-paced society many of us want convenience and we want it now. Some consumers see prepared foods as an answer to the drudgery of meal preparation; others see them as a technological blight on consumption of healthy, flavorful foods.

Salads and sandwiches

Prepared salads can be found in the produce or deli section of many supermarkets. Unlike the fresh-cut products described earlier in the chapter, these salads are ready to eat right out of the package. Fresh fruits and vegetables deteriorate rapidly as they senesce and die. They also exude fluids. In salads with mixed items, the items exuding the fluids become limp or wilted, and the items absorbing the fluids become soggy. Cutting accelerates deterioration and leakage. Due to the lack of proteins, fresh-cut fruits are not as susceptible to the growth of harmful microbes as other products, but they can become contaminated during preparation, particularly if the preparation area of the fruits and vegetables is not separated from the preparation area of the raw meats. Contamination of fresh fruit and vegetable salads with harmful microbes is very serious because these products are rarely heated to kill the harmful organisms. Some salad products contain other ingredients such as liquid dressings, croutons, cooked meat or seafood, or eggs and mayonnaise. Salads tend to be mixtures of distinct ingredients in a nonhomogeneous product rather than a blending of ingredients in a continuous, homogeneous product. The more distinct the ingredients are, the greater the possibility for ingredient interactions and deterioration. In addition, different ingredients have different storage requirements, leading to quality problems. One way to handle such problems is to have separate packages for particular ingredients inside the larger package. This approach sacrifices some convenience for better stability and quality. Although most consumers are afraid of products containing mayonnaise, and believe it can increase the chances of food poisoning, mayonnaise actually inhibits the growth of microbes. It is important that any salad with high-protein ingredients such as eggs, milk, or meats be kept refrigerated.

Sandwiches come in many shapes and sizes, but they are all characterized by the presence of a filling in a bread product. Chilled sandwiches have a short shelf life and are usually prepared daily. Refrigeration helps prevent growth of harmful microbes, but promotes bread staling. Thus, there is usually a trade-off between safety and quality. Long-term storage of sandwiches in refrigerated vending machines sacrifices quality to provide safety. Kiosks that sell sandwiches without refrigeration sacrifice safety to provide quality.

Pasta products

Refrigerated food sections of most supermarkets feature fresh pasta. These products are vacuum packaged to preserve freshness. Of particular concern is the migration of moisture from the high-moisture product such as the sauce to the low-moisture product, the pasta, because it leads to soggy pasta. One way to get around this problem is to keep the pasta separated from the sauce. Because pastas are intermediate-moisture products, they are not susceptible to most microbes except molds. But the low- or no-oxygen package environment (vacuum package) prevents the mold growth. Fresh pasta, however, is susceptible to staling. The shelf life of pastas is about two to three weeks. Marinara, cheese, and meat sauces have specific requirements. Many convenient fresh pasta products are available (see Insert 6.4).

Insert 6.4 Example of fresh, refrigerated pasta available at supermarkets. (This photo is made available as a courtesy by Nestle USA and www.buitoni.com.) Photo by Colin Rose.

Prepared entrées

Prepared entrées are available at most supermarkets and many specialty shops. They tend to be high in protein and fats from meats, eggs, and cheeses. As such, they are usually very perishable due to microbial growth. They need to be kept hot or cold, outside the optimal temperature growth range for microbes. Prepared entrées that are held at warm temperatures on steam tables tend to dry out and lose valuable nutrients, particularly vitamins. Consumers who purchase prepared entrées and hold them at room temperatures for extended periods of time (more than ninety minutes) are increasing their chances of food poisoning. When entrées are packaged with other components, the best conditions for one component may not be the best for the others. For example, breads or rolls generally do not store well at the refrigerated temperatures required for casseroles or other perishable items.

Food service

Food service provides meals or meal components for consumption on-site. Among the various types of food service outlets are sit-down restaurants, cafeterias, fast food, catering, and vending machines.

Casual dining restaurants

Typical restaurants provide convenience by saving food preparation but not by saving time. Although most consumers assume that the meals are prepared on-site from scratch, many items may have been prepared elsewhere and heated or poured out of containers prior to serving. Many items, from appetizers to main entrées to desserts, are prepared in processing plants under carefully controlled conditions and delivered to the restaurant. Many of these items are delivered frozen. Restaurants find the advantages of frozen products are their stability and safety. Although there may be some sacrifice in flavor or other quality characteristics, most consumers are sufficiently pleased with the product's quality. Safety problems can come from items that are prepared on-site but insufficiently cooked or from cross-contamination of uncooked foods with raw meats. Sit-down restaurants usually place a premium on presentation. A congenial atmosphere and flair in placement of the item on the plate enhances the eating experience. Many foreigners complain that most American restaurants have large portion sizes, which encourages overeating.

Cafeteria

In a sense, cafeterias were the first fast-food restaurants because ordering occurs at the same time as service. The prepared foods are displayed with hot foods kept on steam tables. Cafeterias can be self-serve or can be served by cafeteria personnel. Safety becomes a concern if hot foods are not kept hot (above 140°F) and cold foods are not kept cold (below 40°F). Loss of heat on a steam table can compromise quality and safety. The display of so much tempting food tends to lead to overfilling of plates and overconsumption of food.

The unlimited supply of food in college dining halls is frequently cited as a major cause of the freshman fifteen, particularly by first-year women students. Some possibilities for cafeteria eating are shown in Insert 6.5. Litt's *The College Student's Guide to Eating Well on Campus* (2005) offers some suggestions to help overcome the temptations of the dining hall:

- scoping out all the choices before making selections will give us a better chance for healthier choices,
- selecting and consuming at least one fruit each at lunch and supper,
- choosing a wider range of foods that represent all the food groups,
- avoiding casseroles and stews for more plain foods, and
- saving dessert for the second time through the line since we might be full and more likely to pass it by.

Insert 6.5 Brian, Kyle, Jennifer, and Ursula went to the cafeteria to plan a class project. Brian had a cola and pepper steak over steamed rice. Kyle tried the vegan option with chili garlic tofu and carrot juice. Jennifer had the chicken alfredo with broccoli and reduced-fat milk, splurging with a dish of chocolate chip cookie dough ice cream for dessert. Tanya went to the salad bar to get two helpings each of fresh carrots and red cabbage with a helping each of mushrooms, cucumber, black olives, and green peas. No dressing was going to touch her lips. How healthy is each of these meals based on what you have learned in the book this far? These values were calculated using data from The Ohio State University Office of Campus Dining Services Web site, http://diningservices.osu.edu/menus/.

Student	Brian		Kyle		Jennifer		Tanya	
Entrée	Pepper steak		Chili garlic tofu		Chicken alfredo		Salad bar	
Beverage	Cola		Carrot juice		Milk		Water	
	Amount	%DV	Amount	%DV	Amount	%DV	Amount	%DV
Calories	1017		950		967		119	
Carbs	173 g		150 g		111 g		16 g	
Vitamin C	61 mg	102	56 mg	93	44 mg	73	43 mg	72
Iron	9 mg	50	6 mg	33	6 mg	33	0 mg	0
Fiber	3 g	12	6 g	24	6 g	24	2 g	8
Protein	41 g	82	21 g	42	45 g	90	3 g	6
Fat	20 g	31	29 g	45	38 g	58	5 g	8
Cholesterol	82 mg	27	0 mg	0	141 mg	47	0 mg	0
Calcium	82 mg	8	163 mg	16	621 mg	62	57 mg	6
Sodium	2692 mg	112	2860 mg	119	1926 mg	80	298 mg	12

Fast foods

Fast foods have become America's scapegoat for its obesity problem. Fast food offers the opportunity to get a high-calorie meal in a short time, but a judicious consumer does not need to avoid fast foods to eat a healthy diet (for some comparisons see Insert 6.6). Unlike most casual dining restaurants, their fast-food counterparts usually offer a form of portion control. Not everyone needs to supersize his or her order.

The main factor that distinguishes a fast-food operation from other restaurants is the speed of service. Different ordering mechanisms at

Insert 6.6 Comparison of four typical meals consumed at fast-food restaurants. Note that regular cola and lemonade add calories and increase carbohydrate consumption, but meals with diet soda are lower in calories and carbohydrates. Where possible, medium sizes were ordered. Portion size matters. Compare these values to those in Insert 6.5.

Restaurant	McDonald's		KFC		Taco Bell		Schlotskzy's Deli	
Entrée	Big Mac		Breaded drumstick and thigh		Beef soft taco		Small original	
Sides	Medium fries		Biscuit and cole slaw		Nachos and hot sauce		Sour cream & onion chips	
Beverage	Diet cola		2% milk		Medium cola		Medium lemonade	
	Amount	%DV	Amount	%DV	Amount	%DV	Amount	%DV
Calories	920		980		860		1026	
Carbs	31 g		69 g		119 g		122 g	
Vitamin C	12 mg	20	15 mg	25	5 mg	8	NA	
Iron	31 mg	172	5 mg	25	5 mg	25	NA	
Fiber	8 g	32	4 g	16	10 g	40	4 g	16
Protein	29 g	58	46 g	92	22 g	44	31 g	62
Fat	49 g	75	57 g	88	35 g	54	39 g	60
Cholesterol	75 mg	25	200 mg	67	60 mg	20	85 mg	28
Calcium	27 mg	3	400 mg	40	250 mg	25	NA	
Sodium	1220 mg	51	2250 mg	94	1505 mg	63	2163 mg	90

Note: Calculated from the Web sites:

McDonald's: http://www.mcdonalds.com/app_controller.nutrition.index1.html

Kentucky Fried Chicken: http://www.yum.com/nutrition/menu.asp?brandID_Abbr=2_KFC

Taco Bell: http://www.yum.com/nutrition/menu.asp?brandID_Abbr=5_TB

Schlotzsky's Deli: http://www.schlotzskys.com/nutrition.html

fast-food chains include the following: variation of the number of cash registers, drive-thru windows, and long-line speed-up techniques. Most fast-food menus have a limited selection of items with emphasis on a single item type. Most chains prefer to have all the food prepared outside of the restaurant so that heating, assembling, and servicing are the major activities.

Different strategies are employed to keep prepared meals hot and include made-to-order preparation, limited holding times, and hot-food cabinets. Packaging serves many important functions in fast-food restaurants, including maintaining temperature, unitization, marketing, item identification, and spill prevention.

Convenience, image, and price drive fast-food sales. The primary reason a consumer goes to a fast-food restaurant is to save time. Meals that can be eaten without cutlery are preferred, particularly for drive-thru orders. Specialization is a major attractor to a specific chain. Fast-food chains employ a wide range of marketing techniques. Low-price specials stimulate sales but cause stress on inventory and slow service. Fast-food restaurants get blamed for food poisoning outbreaks, but many of these reports are unfounded. A primary reason for blaming restaurants is that most people blame the last meal they consumed for a stomach illness, as we learned in Chapter 1. Most food-associated illnesses, particularly those that would be associated with fast foods, take 12 to 24 hours to develop. Although fast-food restaurants employ many minimum-wage workers, they have excellent systems to maintain sanitation and safety. Since they serve so many meals, however, when a serious mistake is made, many consumers are likely to become ill.

Catering

Food catering businesses have the same challenges as other food service outlets plus the added challenge of needing to transport the food from the place of preparation to the place of consumption. The major safety concern is keeping hot foods hot and cold foods cold. Some of the conditions that maintain safety can degrade quality, particularly in keeping hot foods hot. Shelf-stable foods such as breads, cakes, and cookies present little problem, but main entrées and frozen desserts can present a challenge. Caterers want to present the food at top quality at the proper time. Traffic delays, set-up time by personnel at the consumption location, and long-winded speakers at pre-banquet meetings are all factors that caterers must take into consideration with respect to serving times. Facilities for heating and reheating, short-term storage, and meal assembly at the consumption location must also be considered. Food must be secured properly in the transportation vehicle to provide proper presentation to the consumers. In a sense, home delivery systems such those as for pizza or

Meals on Wheels are also catering businesses. These delivery systems use specialized, hot-food containers to keep the meals from getting cold.

Vending machines

It seems we are never very far away from a vending machine in America. Although some vending machines have become very sophisticated, dispensing hot coffee, meals, fresh fruit, cold sandwiches, and exotic ice cream, it is more common to see them stocked with either cold drinks or sweet or salted snacks. Carbonated beverages, sugared snacks, and their salted counterparts are ideal for vending machines because they are well packaged, not very perishable, and require no preparation prior to eating (see Insert 6.7). Furthermore, they are easily consumed right out of the

Insert 6.7 Typical snack and beverage vending machine setup. (Public photo extracted from http://www.flickr.com/photos/73416633@N00/432654745/.)

package. Minimally processed juices, fruit cups, sandwiches, and entrées with meat are perishable and require refrigeration. Failure of the refrigerated vending equipment can lead to rapid growth of microbes, compromising the quality and safety of the product. Some of these items may be heated in the vending machine or in a microwave oven. Improper heating of foods that have not been properly refrigerated can give the purchaser a false sense of security about its safety. Frozen items require even more monitoring for proper temperature conditions. Thawing and refreezing will damage the quality of the product before it threatens its safety.

Remember this!

- Convenience, image, and price drive fast-food sales.
- Safety becomes a concern if hot foods are not kept hot and cold foods are not kept cold.
- Food service provides meals or meal components for consumption on-site.
- When entrées are packaged with other components, the best conditions for one component may not be best for the others.
- Prepared foods are those foods that are ready to eat or ready to heat then eat.
- Fats provide lubrication in the mouth.
- Pasteurization is a heating process that kills all harmful microbes but does not kill spoilage microbes.
- Maintaining adequate refrigeration of perishable products like deli meats is necessary, and expiration dates should be strictly observed.
- Expedited handling minimizes loss of quality and safety problems.
- Food scientists are wary of consuming raw meat because any harvest operation is likely to contaminate the surface of the meat.
- Sophisticated modified atmosphere packaging (MAP—low in oxygen and high in nitrogen or carbon dioxide) slows the deterioration process.
- Distribution is everything that happens to a product from the time it is produced to the time it is consumed.

Looking ahead

This chapter was designed to provide an introduction to food processes and the types of products that are produced, with special emphasis on those made directly from raw agricultural materials. Chapters 7 through 9 cover how food scientists ensure the quality and safety of food products, design new products and processes, and work in government to regulate foods and beverages.

References

Litt, A. S. 2005. *The college student's guide to eating well on campus,* 2nd ed. Bethesda, MD: Tulip Hill Press.

Further reading

Brody, A. L. 2001. *Case-ready meat packaging*. West Chester, PA: Packaging Strategies.

Florkowski, W. J., R. L. Shewfelt, B. Breuckner, and S. E. Prussia. 2009. *Postharvest handling: A systems approach,* 2nd ed. San Diego, CA: Academic Press/Elsevier.

James, S. J., and C. James. 2002. *Meat refrigeration*. Boca Raton, FL: CRC Press, Taylor & Francis Group.

Kays, S. J., and R. E. Paull. 2004. *Postharvest biology*. Athens, GA: Exon Press.

Kerry, J., J. Kerry, and D. Ledward. 2002. *Meat processing: Improving quality*. Cambridge, U.K.: CRC Press, Woodhead Publishing Limited.

Lamikanra, O. 2002. *Fresh cut fruits and vegetables*. Boca Raton, FL: CRC Press.

Man, D., and A. Jones. 2000. *Shelf-life evaluation of foods*. Gaithersburg, MD: Aspen Publications, Inc.

Stringer, M., and C. Dennis. 2000. *Chilled foods—A comprehensive guide*. Cambridge, U.K.: CRC Press, Woodhead Publishing Limited.

Swatland, H. J. 2004. *Meat cuts and muscle foods*, 2nd ed. Nottingham, U.K.: Nottingham University Press.

Functions of food scientists

chapter seven

Quality assurance

Nolan grew up in the city and thought it would be cool to spend spring break with his friend Greg on his dairy farm. At first he enjoyed the drive through the countryside, the fresh air and the fresh, home-cooked meals. He thought it was pretty dull in the evenings though, and he wasn't really eager to get up so early to milk cows. Greg's dad urged him to try some raw milk, assuring him that there was "nothing like it for flavor and nutrition!" Greg seemed to enjoy it, so although he resisted at first, Nolan finally got up his courage late in the week and drank a whole glass. It was rich and full-flavored. It sure didn't taste like any kind of milk he had ever tried before. It wasn't as great as Greg's dad claimed, but he could see how you could come to really like it. The next morning he was sick with stomach cramps, nausea, vomiting, diarrhea, and a fever. He was sure it was the milk, but Greg and his dad insisted that all that talk about the only safe milk being pasteurized milk was hogwash. They had been drinking raw milk all their lives and they never got sick! It took Nolan about three days to recover. He now knew what it was like to live on a farm. It might be all right for others, but it was definitely not for him. He'd also steer clear of raw milk, regardless of what Greg said.

Laura bought her favorite chocolate bar at the snack shop on her way to chemistry class. It was the last bar in the box. She was running late, so she threw it into her book bag for later. She forgot all about the candy bar, but later, while looking for a highlighter, she rediscovered the treat. She didn't have lunch and was incredibly hungry. She tore off part of the wrapper and bit in. Instead of the usual great flavor she was expecting, this bar took her taste buds to a new level. It was awesome! Somehow it was sweeter but not too sweet, and the chocolate was more chocolaty. The flavors exploded in her mouth, sending her into spasms of delight. She had never tasted anything quite like it. Was she hallucinating? She took a second bite. It was as good as or better than the first! She closed her eyes and savored every little sensation. She thought about sharing it with a friend, but it was so tempting, so satisfying, and so delicious, she could not resist eating the whole thing. After a quick look to see if anyone was looking, she licked the wrapper clean. The next day, Laura bought the same kind of bar at the same place. It was the first bar in a brand new box. She threw it into the same bookbag and waited until the same time she had eaten the candy bar the day before. She could hardly wait for the first bite! When she ripped open the wrapper and took the first bite, she was bitterly disappointed. It was nothing like what she had the day before. Thinking back, it was probably as good as it usually was, but it was still disappointing. She bought several bars of the same brand at several different places, but it was never as good as the one on that magical day. She has just about stopped eating that bar because it just can't compete with perfection. Every now and then she buys one but is still disappointed.

Emily likes fruit juice. It is part of almost every meal she eats. She avoids soft drinks because of the sugar and the caffeine. She wants something healthier. Since she is on a tight budget, she usually buys whatever fruit beverage is cheapest. Then she learned in her food science class that not every fruit-flavored beverage is that nutritious. As part of an assignment, she chose to find the juice that gave her the most nutrition for her money. She went to the juice aisle in the supermarket and was surprised to find that many of the products had little juice. She narrowed it down to five products (see Insert 7.1 for the comparison). She finally chose the orange juice enriched with vitamin C because she wanted real juice, and it was the least expensive of the 100% juice products. What would you choose? Why?

Nolan, Laura, and Emily are interested in the safety and quality of their food. Food scientists employed by food companies are responsible for ensuring the products are safe and the quality is consistent. These food scientists work in the Quality Department to design and oversee laboratory tests to make sure that Nolan's milk has been properly pasteurized, that Laura's chocolate bars are consistent in quality, and that Emily's juice product nutrition label is accurate. In this chapter we will see what is involved in measuring the quality and ensuring the safety of foods.

Looking back

Previous chapters focused on food issues we encounter in our daily lives and with the types of food products we encounter in the marketplace. The following key points that were covered in those chapters help prepare us to understand quality assurance.

- Distribution is everything that happens to a product from the time it is produced to the time it is consumed.
- Expedited handling minimizes loss of quality and safety problems.
- The conditions that affect nutritional value also affect sensory quality.
- Food processing and preservation increases the shelf stability of a raw material, but usually decreases affordability, nutrition, and quality of the product.
- Most raw materials are perishable and require careful handling or processing to prevent losses.

Insert 7.1 Nutritious fruit beverage comparison.

Name	% juice	Serving size	Calories	Sugars	Vitamin A	Vitamin C	Calcium	Iron	Sodium
						% daily values/serving			
White House Apple Juice from concentrate	100	7 oz (207 ml)	110	23	0	0	0	0	1
Juicy Juice Score Spikin Strawberry	10	8 oz (240 ml)	80	18	0	10	0	0	1
Ocean Spray Cranberry and Georgia Peach	100	8 oz (240 ml)	140	34	0	130	2	2	1
Diet V-8 Splash Tropical Blend	8	8 oz (240 ml)	10	2	100	100	0	0	1
Tropicana Plus enriched orange juice	100	8 oz (240 ml)	110	23	10	150	10	0	1
Whipple Snapple Strawberry Banana Smoothie	15	10 oz (297 ml)	160	38	0	0	0	0	3

- Sensory properties include color, flavor, and texture.
- Spoilage is not a good indicator of a safety risk.

What is quality and why does anybody care?

Perhaps the most important function of a food scientist is achieving and maintaining the quality of a product. Like Phaedrus in *Zen and the Art of Motorcycle Maintenance* (1974), we find that we know what quality is, even if we can't define it. Many authors have tried to define quality for products and services, but it is difficult to find one definition to meet all needs. The simplest definition, "an absence of defects," may be good for a laptop, lamp, or textbook, but we want more when shelling out big bucks for a sound system, a set of wheels, or a gourmet meal on a hot date.

When it comes to food, defects include blemished apples, bruised peaches, mushy bananas, green eggs, iridescent ham, slimy wieners, stale cereal, lumpy mashed potatoes, and crunchy macaroni and cheese. Truly fine dining is the intertwining of pleasant flavors and textures with an excellent presentation. Quality is thus more than the absence of the bad; it is also the presence of good to great to exquisite. Even that description oversimplifies the situation because different foods and flavors appeal to different consumers. What one consumer might consider good quality, another might reject; some like their salsa mild, while others like it hot.

To help make sense of this complex situation, we can place the factors involving individual preference into a category called *consumer acceptability* and the factors involving price into a category called *value*. That leaves *quality* to describe properties of food that can be measured by food scientists. A favorable experience in the mouth leads to consumer acceptability, while quality at a given price relates to value. This chapter will define quality, describe how it is measured by food scientists, explain what is done in the food industry to ensure safety and consistent quality, and relate quality to acceptability and value.

Quality characteristics

Quality has been defined as meeting or exceeding the expectations of the consumer. A more detailed definition involves the characteristics of a product that differentiate it from similar products and are important in determining acceptability by the consumer. Thus, the quality of a given product is defined by characteristics that vary from one brand to another within a category (differentiation) and that make a difference to consumers (degree of acceptability). The two types of characteristics we use to judge the quality of a product are sensory and hidden. *Sensory characteristics* are those easily detected by the five senses, and *hidden characteristics* are important to consumers but are not readily detectable by them.

Hidden characteristics include safety and nutrition. The packaging also contributes to maintaining the quality of the food and its acceptability to the consumer. We take on faith that all commercial products we buy or meals we order out are safe. Much of the rest of this chapter will focus on the work food companies and food-service institutions do to make sure their products are safe for the consumer. The consumer can use the safety and nutritional quality information on a product label to decide whether or not to buy it.

Although we use all five senses when evaluating the sensory quality of our foods, food scientists tend to reduce sensory quality to three types of characteristics: flavor, color, and texture. Flavor combines the senses of taste and smell. Color, or appearance, involves the sense of sight. Texture involves the sense of touch, both with our hands and in our mouth. Finally, through our sense of hearing, we can experience the sizzle of a steak, the crunch of a raw vegetable, and the snap, crackle, and pop of a breakfast cereal.

When we say we *taste* a food, we are actually evaluating its flavor. The five main tastes are bitter, sweet, salty, sour, and umami. There may be other taste sensations, but these five are the most easily identifiable. Flavor includes both aroma and taste. We perceive aroma in two ways: directly through the nose (orthonasally) before we put food into our mouth and through the back of our mouth (retronasally) while we are chewing. To take full advantage of the retronasal contribution of flavor, we must have proper airflow through our mouth and nose. That is why when we have a cold, pinch our nose, or eat with our mouth tightly closed, we say the food does not taste as good, but it is aroma and not taste that is affected. Much of what most of us think of as food quality is how we perceive its flavor. Would we be tempted to eat so much chocolate, fatty meat products, potato chips, fresh-squeezed juice, or other tempting treats if they did not emanate an enticing aroma and follow it up with a pleasant taste?

Most consumers think that color is overrated as a quality characteristic, indicating that they are willing to give up appearance for flavor. In reality, color tends to be the most important factor in deciding what is purchased and consumed. Although we now know that appearance is not a good indicator of a safety hazard, most of us are unwilling to eat food that just doesn't look right. For example, many consumers refuse to eat bananas that have brown spots; tomatoes, more orange than red, indicating a high vitamin A content, are rejected by consumers who want them to be red; meat cooked in a microwave oven is not as popular as when it is cooked in a conventional oven because the meat is not as likely to turn brown. Food manufacturers add artificial colors to breakfast cereals, beverages, candies, and maraschino cherries because bright colors sell and dull colors do not. Color does not usually get the respect it deserves as an important quality characteristic.

Texture is another characteristic that is generally underrated, but many consumers will refuse to eat a food strictly on the basis of its texture. Raw oysters and boiled okra, a southern delicacy, are rejected by many consumers because they have a tendency to become slimy. One reason consumers reject bananas with brown spots is that they find them too mushy, even though they are sweeter than those without the spots. Most of us prefer our carrots and apples crunchy and our peaches and cherries soft. We don't like lumps in our grits or mashed potatoes or soft spots on our oranges or fried chicken. We don't want our maple syrup or orange juice to be too thin and watery or our milk to be thick. We also have textural sensations known as *chemical feeling factors*, which are not part of flavor; for example, pungency (onions), heat (peppers), cool (menthol), and astringency (puckering). All of these textures that we feel, whether with our hands or in our mouths, are very important to us.

As mentioned before, safety and nutrition factors cannot be readily detected by the consumer but are important in helping the consumer make a decision as to whether or not they will buy or eat a product. Without a sophisticated analytical lab, consumers must take the word of the manufacturer or the preparer that the product is safe and nutritious. All quality characteristics, whether they are sensory or hidden, must be measurable and must be measured within the context of a quality system within the processing plant, which has a responsibility to do everything possible to provide safe, nutritious, and wholesome foods.

Measuring quality

Measurement of sensory and hidden quality characteristics is not as easy as we might think. In this section only a brief overview will be provided. Later chapters will provide greater detail. For a measurement to be useful it must be accurate, precise, sensitive, and relevant. Many measurements are made using instruments. Thus, the reliability of the instrument must also be considered. Measurements are only estimates of reality. *Accuracy* of a measurement is how close the estimate is to the real value. *Precision* is how close that estimate is each time it is measured. *Sensitivity* of the estimate is how effective it is at detecting very small amounts. Sensitivity is easy to determine by diluting the sample until the characteristic can no longer be detected. *Relevance* is how important the measurement is in relation to consumer acceptability. This property is the most difficult to determine because measuring consumer acceptability is difficult. *Reliability* of the instrument refers to the consistency of measurements over time. Food scientists can easily determine the precision of a measurement by measuring the same characteristic of the same product numerous times. Reliability of the instrument needs to be checked periodically by trained personnel. The installation, competence of the operator, proper calibration, and proper equipment maintenance can affect day-to-day readings.

Measuring the safety of a product is done primarily by microbiological testing and analysis to determine the number and types of microbes present in a select sample. To ensure the safety of a product, the presence of pathogens (microbes that cause food poisoning) must be determined. Canned products should be sterile and pasteurized products should be free of pathogens. Most formulated products contain very low levels of pathogens (not a health risk), but raw meats are likely to contain harmful levels of pathogens. Food scientists are much more concerned about pathogens present in a product that is not likely to be cooked than one that must be cooked. Even though no pathogens are detected in a product, they may be present. Thorough knowledge of the product, such as how it will be stored and prepared prior to consumption and the types of pathogens likely to be present, aid food scientists in deciding what further testing must be done. In some cases, prior to performing the analysis, the food sample is placed in an enrichment broth to allow damaged pathogens to recover. These pathogens may be present at such a low level that regular testing might miss them. In other cases, even more sensitive testing may be needed. In the absence of a serious health hazard, the absence of detectable pathogens may be sufficient. Although most food safety problems are directly related to microbial contamination, other hazards could be chemical or physical. Chemical hazards are unauthorized compounds or compounds present at unsafe levels. Physical hazards could be stones from a field or pieces of metal from processing machinery. Tests are also conducted to ensure that chemical or physical hazards are not present.

Calories per serving is another of the most fundamental measurements of nutritional quality. However, some of us consume too much of a good thing and must count the calories to ensure that we do not gain weight. Calories come from carbohydrates (4 calories per gram), proteins (4 calories per gram), and lipids (9 calories per gram). Fiber is also a carbohydrate, but it does not contribute calories. Estimates of caloric energy components as well as fiber are also required for the nutritional label.

Nutritional labels must be accurate at the time the food is consumed. Since all nutrients are chemical, analyses are done to determine the chemical composition. Vitamins and minerals must be estimated and translated into the percentage of daily values. A food scientist must be able to measure the nutritional composition of a food, and trace and allow for possible changes or deterioration during its shelf life (handling and storage). Since nutrient loss tends to be small in most fully processed foods (canned, frozen, and dried), measuring the nutritional composition is not difficult. However, it is much more difficult to estimate the projected vitamin and mineral content, at the time of consumption, of highly perishable items such as fresh meat, fruit, and vegetables. It is not enough to determine how much of a nutrient is present; food scientists must also consider the nutrients' bioavailability, which is how much of it is absorbed through the intestinal tract and actually reaches the blood stream, as well as whether it is in the proper form that the body can use.

Sensory quality of foods can be measured by physical, chemical, or sensory testing. Color is best measured by an instrument called a *colorimeter*, which measures the physical properties of reflected light and translates those readings into values that can be related to consumer perception of color. Visual defects in foods can be detected by using machine vision systems with video cameras. Textural properties, such as viscosity of beverages and liquid foods, gumminess, hardness, adhesiveness, chewiness, and fracturability of foods, can be determined using other physical measurements. Taste is defined using chemical processes to measure sugar concentration for sweetness, acid concentration for sourness, sodium chloride for saltiness, and for specific bitter compounds such as caffeine. The aromatic part of flavor can be separated out of the volatile compounds using chromatographs. A liquid extract is injected into a gas or liquid chromatograph, which separates individual molecules by retention time based on properties like molecular weight. In some cases a simple physical or chemical method can give a very good estimate of the sensory quality of food that is accurate, precise, and sensitive.

With respect to flavor, often the estimate isn't accurate according to the consumer's perception. There are many types of sensory tests to satisfy all of the taste quality specifications of a product. They range from determining if there is a difference between two brands to finding the best formulation (a scientific version of a recipe). Tests may also make comparisons between competitor's products or between experimental formulations. It is important to use human beings to actually *taste* the product. These types of sensory tests have trained panelists who are much more aware of and sensitive to specific tastes and aromas than a typical consumer. A unique method, known as a preference test, involves finding the "best" product. The panelists for this kind of test are typical consumers. We'll learn more about these sensory tests in the final chapter. Some instruments that measure sensory characteristics are shown in Insert 7.2. Colorimeters can plot product color in a three-dimensional space; the Instron Universal Testing Machine can be adapted to test many textural properties, and the Electronic Nose can differentiate products by their aromas.

It is critical that a sealed package remain sealed because without an intact package the quality of the product is compromised. Can seams are evaluated every hour on every line in every canned food processing plant. Glass container seals are evaluated. All tamper protection devices are checked. Gas permeability and water vapor transmission rates of plastic packages are important to maintain the desired conditions inside

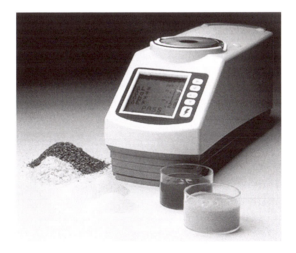

Insert 7.2 Examples of instruments for measuring sensory quality character-istics of texture, color, and flavor. (Images used with permission from Instron, HunterLab, and Alpha MOS.) Top: Instron Universal Testing Machine Model 5582. Bottom: Hunter-Lab CLFX Food CMYK_300 Color Flex Spectrophotometer.

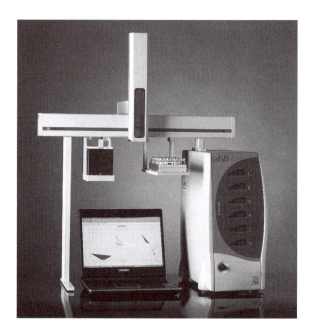

Insert 7.2 (Continued) Alpha MOS Fox Electronic Nose Analyzer.

the package. Packages that will be subjected to rough handling between leaving the processing plant and being purchased in the supermarket are tested for burst strength. In addition, quality department personnel check drained weights to ensure they are within a certain range. A drained weight that is below what is stated on the package violates federal regulations. A drained weight that is above the stated amount is costing the company money.

Evolution of quality management

Quality has always been important for food products. Before mass production and distribution, chances are you knew where your food came from and could complain directly to the provider. Whether the person was a member of the family, a next-door neighbor, or someone from the nearest town, their reputation and ability to earn a living were at risk; therefore, it was important to be careful. As food production and manufacture became more mechanized, it was necessary to develop ways to ensure good quality.

Quality control (QC) was developed to make sure that a product is safe, wholesome, and of good quality. Consistent quality is achieved by inspecting the final product. Because inspecting every item (opening the can or package) is not only impossible but impractical, only some of the finished items are statistically sampled. A quality control manager usually runs the QC programs. Statistical quality control (SQC) was incorporated into quality assurance (QA), which considers all parts of the process. The idea is that if the process is done correctly, then the final product will be safe and of good quality. A director of QA is generally responsible for a Quality Assurance Program. Quality management (QM) takes this idea one step further in that it looks at what the consumer wants and needs (including changes in consumer preferences) and modifies the process accordingly. Total Quality Management (TQM) encompasses the entire processing company and is the responsibility of the chief executive officer (CEO) of that company. Each change in the way food companies viewed and managed quality has affected the type of products available to us in the supermarket.

Quality control

QC involves inspecting the product to make sure it meets standards and that procedures are in compliance with all laws and regulations. When someone bakes bread, churns ice cream, or cooks meatloaf for a family, a party, or a small group, the preparer tests the quality of the product (usually orally) before serving. In parts of the food industry, inspection of every item is still practiced. Nearly every piece of fresh fruit on a grading table is inspected by humans, and every carcass of fresh meat in a slaughterhouse is inspected by a U.S. Department of Agriculture (USDA) inspector. To ensure fresh fruit in the supermarket is edible and reasonably consistent in quality, the fruit graders remove fruit that is underripe, overripe, bruised, damaged, or unsightly. A USDA inspector visibly checks every carcass and marks for removal any that have visible signs of disease. Inspectors can stop a production line for a closer look, if necessary, by using a bell or buzzer. Diseased carcasses cannot be sold to an American consumer in any way (whole, cut up, or further processed). Laws and regulations cover all aspects of food processing, from farm production to waste disposal at the processing plant.

Quality is more than what we can see. As mentioned earlier, many food quality measurements require destruction of an item. Since no food company can make money if its entire production line is destroyed, statistical sampling is conducted to help determine the quality of the raw material and the finished product. Raw materials that do not meet the specifications of the processing plant are either rejected or sold at a reduced price. Tables and sampling schemes show the QC manager how many samples are required to assess quality. Then the quality of the samples is compared to a specification or standard to ensure acceptability. If there are no problems with a product that is finished during a certain period of time (one hour, two hours, or a shift), it can be certified by the QC manager for release and sale. However, if problems are detected, further sampling is conducted to determine the extent of the problem. If it is a safety issue, the product must either be destroyed or reprocessed. If the problem is with flavor, color, or texture, then the product may be sold at a reduced price under a different label.

Quality assurance

QA considers the whole process when evaluating quality. The emphasis on QA is to get it right the first time. By keeping the process operating correctly, the product should consistently be of good quality. Testing is still important in QA, but the testing occurs at various points in the process to make sure the operations are working. As we read earlier, the QA manager is responsible for the administration of quality testing and reporting; however, everyone working in the food processing plant also has a responsibility for product quality. The QA manager is also responsible for understanding how the system works and identifying areas for improvement. QA begins when receiving raw materials and ingredients, continues through every unit operation during processing, and extends beyond the plant through distribution and other activities of the company that can make a difference in final product quality. A constant study of the whole process by QA personnel and the people who operate the machinery leads to continuous improvement of product quality.

Quality management

While quality control and quality assurance focus on the characteristics of the product that define quality, QM defines quality in terms of consumer acceptability. QM incorporates everything in QA, but develops standards based on what consumers want. Understanding the consumer is difficult; therefore, many techniques are necessary to gather information. Quality function deployment provides a systematic way of matching product characteristics to consumer desires. Other approaches involve use of the marketing and sales departments to identify consumer wants and needs. Consumer profiles are then incorporated into the quality measurements adopted by the company and monitored by the quality department. In TQM, quality principles are extended to every aspect and activity of the company; therefore, ideally, every employee becomes a quality inspector and strives for continuous improvement. Workers get together daily or weekly for meetings called Quality Circles to discuss operations and ways to improve them.

All but the smallest food companies have shifted from QC to QA. Many companies have incorporated aspects of QM although they may still refer to their quality program as QA.

A few food companies have adopted TQM, but it is still somewhat controversial and has been difficult to apply consistently. There are many versions of TQM. Some versions focus on eliminating defects with a strong emphasis on measurement. These versions are useful in assuring safety and preventing products from having poor flavor, color, or texture. They are, however, not very useful in separating products with a range of characteristics, such as great flavor versus just acceptable flavor. Other versions stress the less measurable aspects of quality and attempt to evaluate degrees of excellence in a product. For the 14 basic principles of one popular version, see Insert 7.3.

Quality management is being recognized by organizations around the world. Standards have been developed by the International Organization for Standardization (ISO). ISO 9000 is a family of standards that incorporates many guidelines for specific processes to ensure uniform quality and safety of products that come from different countries. A company can become ISO certified by properly modifying its operation and going through a thorough inspection. Some companies will only buy ingredients for their products from ISO-certified processors.

Insert 7.3 Deming's basic principles. (Adapted from W. E. Deming, *Out of the Crisis*, MIT Press, Cambridge MA, 2000.)

---

1. Develop a realistic statement of objectives and purpose for the organization.
2. Learn the new philosophy.
3. Understand the purpose of inspection.
4. End the practice of basing decisions solely on the lowest price.
5. Constantly improve the system of service and production.
6. Institute training.
7. Teach and institute leadership.
8. Eliminate fear. Foster trust. Provide an environment that encourages innovation.
9. Optimize all activities toward the objectives and purpose of the organization.
10. Eliminate slogans and gimmicks. Treat employees as integral parts of the organization.
11. Eliminate quotas for production and management by objectives. Learn how to improve the systems.
12. Remove barriers that compete with pride of workmanship.
13. Encourage education and self-improvement for all members of the organization.
14. Implement all changes to transform the organization.

Statistical process control

An essential element of QA and QM is statistical process control (SPC). SPC involves the measurement of the effectiveness of a process or unit operation. It is based on the idea that by doing everything right in each phase of the process, the final product will be good. Measurements are made on the product at the end of a process to ensure that it is operating within guidelines. Statistical sampling techniques provide the number and frequency of measurements needed for a particular operation. A Shewhart control chart provides an operator with a visual indication of whether the process is in control or not (see Insert 7.4). A trend of points, either up or down, indicates if a process is about to go out of control. Adjustments to the operation can then be made to prevent problems before they occur. When all operations are in control, less inspection of the final product is needed than in QC. Thus quality is assured for QA and QM through SPC.

Advances in technology have led to the development of in-line sensors. Such sensors measure physical characteristics of the product like color, viscosity, or temperature. They may also measure chemical composition such as percent sugars, percent acids, or percent salt. Technology is moving rapidly in improving accuracy, precision, and sensitivity of these measurements. Long-term reliability of a sensor is sometimes a problem, and the types of measurements easily made by sensors are not always relevant to the consumer acceptability of the product.

Hazard analysis and critical control point

Hazard Analysis and Critical Control Point (HACCP) is a means of ensuring microbial safety in a product. This method started out as a way

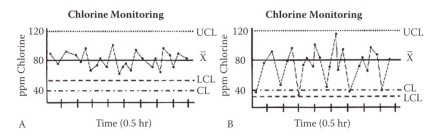

Insert 7.4 Shewhart control charts for monitoring chlorine concentration in a fresh-cut produce operation. Part A illustrates an operation that is in control with all values falling between the upper control limit (UCL) and lower control limit (LCL). Part B shows an operation that is out of control, as four values fall below the LCL, but no value falls below the critical limit (CL). (From W. C. Hurst, "Safety Aspects of Fresh-Cut Fruits and Vegetables," in *Fresh-Cut Fruits and Vegetables*, ed. O. Lamikanra, Boca Raton, FL: CRC Press LLC, 2002, chap. 4. With permission.)

to meet the challenges of providing foods for astronauts in space and has been adapted to programs here on earth. HACCP seeks to (1) destroy, eliminate, or reduce hazards, (2) prevent recontamination, and (3) inhibit growth of harmful microbes and limit toxin production. Hazard analysis identifies anything that could be harmful in a product. Types of hazards include biological (microbes or insects), chemical (pesticides or toxins), nutrition related (antinutrients), or physical (stones, metal, or glass). Hazards are classified as either severe or moderate. Unit operations have critical control points that are monitored and controlled to prevent hazards.

In developing a HACCP plan we must do the following:

- Assess all hazards
- Identify critical control points
- Establish critical limits
- Monitor requirements
- Take corrective action when in noncompliance
- Develop an effective record-keeping system
- Develop a verification system

HACCP has been applied to all meat slaughtering and packaging operations, but it has not replaced the inspection of every carcass by USDA inspectors. For a chilled food HACCP system, see Insert 7.5.

HACCP has been so successful in identifying and minimizing hazards in processed foods that advocates have proposed that it be extended to sensory quality as well. Some scientists insist that it should be reserved for safety problems only. HACCP involves a systems approach to safety. Systems approaches characterize related operations within a process, view the whole process as a series of integrated steps, and study how the steps interrelate with respect to their fundamental properties. Quality measurements within a QA or QM system benefit greatly from a systems approach similar to HACCP with regards to quality characteristics.

Sanitation

One of the major obstacles to food quality and safety is product contamination from either microbes or chemicals. The key to preventing contamination is sanitation. Sanitation, in short, is keeping things clean.

Sanitation begins with the production of the raw materials, before anything ever gets to the processing plant. Although it may be impossible to keep plants and animals completely free of microbes, any procedure that reduces the contact of raw materials with soil or animal feces will reduce the presence of microbes. Growing of plants on plastic minimizes contact with soil and pathogens associated with organic fertilizers. Efforts that

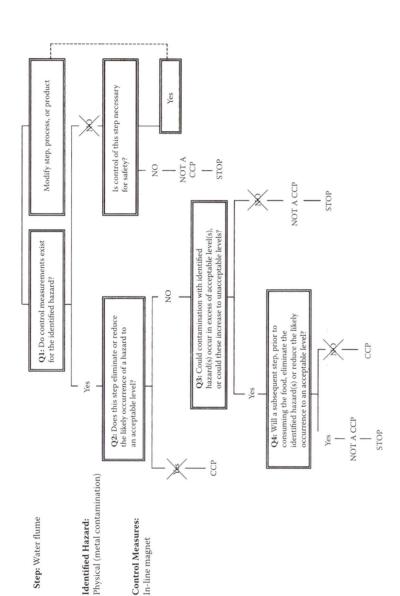

Step: Water flume

Identified Hazard:
Physical (metal contamination)

Control Measures:
In-line magnet

Insert 7.5 Decision tree for applying HACCP to determine critical control points for a fresh-cut produce operation. (From W. C. Hurst, "Safety Aspects of Fresh-Cut Fruits and Vegetables," in *Fresh-cut Fruits and Vegetables*, ed. O. Lamikanra, Boca Raton, FL: CRC Press LLC, 2002, chap. 4. With permission.)

minimize the opportunity for animals to roll in dirt or excrement reduce chances of contamination. Keeping harvested plant parts and live animals in clean environments during transport to and storage at the plant before processing are also important steps in limiting contamination.

Many things need to be done to keep the processing plant clean. Control of pests, such as insects and rats in warehouses and processing areas, is very important. There is a constant need to keep floors, horizontal surfaces, and food surfaces clean and free of dust. Employees who handle food must be sure their hands and clothes are clean. Most processing plants operate for 14 to 18 hours and then are completely cleaned and sanitized before the next day, when operations begin again. Some equipment can be cleaned by circulating chlorinated water or some other sanitizing solution through it. Other equipment must be completely broken down, scrubbed with warm soapy water, rinsed, sanitized with an appropriate agent, and reassembled. Thorough inspections are conducted to ensure that everything meets exacting standards of cleanliness before the plant is allowed to operate.

U.S. government guidelines called the Current Good Manufacturing Practices (CGMP) and the Sanitation Standard Operating Procedures (SSOP) are useful in maintaining adequate sanitation in processing plants. CGMPs are overseen by the U.S. Food and Drug Administration (FDA), while SSOPs are guidelines of the USDA. The latter are required for all meat plants. Each company must develop specific procedures that are in compliance with these regulations from the receipt of raw materials, to the cleanup and sanitizing of all processing equipment, to the monitoring of chlorine usage, to management of wastewater generated by the plant, and many other activities.

Food processing plants generate large amounts of waste. Governmental regulations impose certain requirements on the quality of the water and air leaving the plant. The quality department of a company is responsible for ensuring that these standards are met. Whenever possible, waste from food processing operations is turned into usable products (by-products), such as exotic items for foreign markets (chicken feet), animal feed (fruits and vegetables not fit for human consumption), or fertilizer.

Sanitation is not limited to the food processing plant. It is also critical in every operation of a restaurant. Strict guidelines are provided for all food service operations from locally owned and operated restaurants to major food chains. Restaurant chains develop extensive guidelines and have company inspectors who check each unit to make sure it is operating appropriately. In many places the local health board inspects restaurants within its community. The types of things inspectors check that are considered critical violations are described in Chapter 1 (see Insert 1.7). Restaurants are required to post their health rating in a prominent place. Do you check those ratings when you eat out, before you order?

Consumer acceptability

The ultimate goal in ensuring product quality is customer satisfaction. Although quality focuses on the characteristics of the product, acceptability must consider the attitudes of the consumer. *Consumer acceptability* is defined as the willingness to buy and eat a product.

While quality characteristics can be measured with accuracy and precision, it is much more difficult to measure consumer attitudes. Food processors using QC or QA systems don't become overly concerned about measuring consumer acceptability. They produce a consistent product, make it available to the public, and advertise. If consumers buy a product and like it, they are likely to buy it again. As long as sales meet or exceed the company's expectations for purchases and profits, the item will continue to be produced. Companies who take a QM approach will attempt to learn what it is that current and potential consumers want and need.

To determine acceptability of a product, a food company must determine market segments. For example, a cheddar cheese manufacturer knows that some consumers like it mild and others like it sharp. They may choose to cater to both mild and sharp segments, specialize in only one segment, or produce a medium and extra sharp for additional segments. With regard to peanut butter, some like it creamy and others like it crunchy. Within the creamy peanut butter category, there are subcategories. In my classes, we have run blind peanut butter consumer tests and there is a particular brand (which also happens to be the leading brand) that wins every time. The leading brand has developed their product to appeal to a segment constituting a majority of peanut butter lovers. It is hard to imagine that any company will develop a better peanut butter that will match or exceed that brand. There are obviously some consumers who do not like that brand. Developers of a new peanut butter might be better trying to appeal to a segment that does not prefer the leading brand. Being the leading brand, however, doesn't always mean that consumers like it best. In other blind tests we've done in classes, most students cannot tell the difference between the two leading cola beverages, and a less popular brand frequently wins preference tests. In this case, marketing is apparently a better determinant of success than actual preferences.

When determining market segments, it is important to look at potential consumers of a product as well as current consumers. If we only look at current peanut butter consumers or cola drinkers, we may miss an opportunity to get more people to consume our product. For example, the person who first cut big carrots into small ones, put them in a package, and sold them, opened the product to many people not willing to take the time to clean, peel, and cut the big ones.

Measuring acceptability is a difficult task. First, we must determine the segment or segments of the population interested in that product.

Acceptability	Willingness to purchase	Hedonic	Smiley Face
3 – Tastes great	5 – Definitely would purchase	9 – Like extremely 8 – Like very much	A
	4 – Probably would purchase	7 – Like moderately 6 – Like slightly	
2 – Acceptable	3 – Might or might not purchase	5 – Neither like nor dislike	B
	2 – Probably would not purchase	4 –Dislike slightly 3 – Dislike moderately 2 – Dislike very much	
1 – Unacceptable	1 – Definitelywould not purchase	1 – Dislike extremely	C

Insert 7.6 Scales used in consumer tests.

Then we need to develop a test that presents consumers with one or more similar products and asks them if they like it or not. We must be careful not to get too complicated with the test questions, as normal consumers are not trained to measure specific quality characteristics. Frequently, this type of information can be inaccurate. However, consumers are the only ones who can let us know if they like it or not. When conducting research with consumers, I like to present them with a sample and ask them if it "tastes great," is "acceptable," or is "unacceptable." Other researchers like to use a nine-point scale that ranges from "like extremely" to "dislike extremely." See Insert 7.6 for some examples of scales used in consumer tests.

Remember this!

- Although quality focuses on the characteristics of the product, acceptability considers the attitudes of the consumer.
- The ultimate goal in ensuring product quality is to satisfy the consumer.
- Food processors must be careful to prevent contamination, beginning with the raw materials and on through every unit operation, using sanitation as the key to preventing contamination.
- The Hazard Analysis and Critical Control Point system is a means of ensuring microbial safety in a product.
- Statistical process control involves the measurement of the effectiveness of a process or unit operation based on the idea that by doing everything right in each operation of the process, the final product will be good.
- Quality management defines quality in terms of consumer acceptability.
- Quality assurance considers the whole process when evaluating quality with the emphasis on "getting it right the first time."

- Quality control involves inspecting the product to make sure it is of good quality.
- As food production becomes more mechanized and less personal, it is necessary to develop ways of making sure that each product is of good quality.
- For a quality measurement to be useful it must be accurate, precise, sensitive, and relevant.
- Quality includes sensory characteristics, which can be readily detected by the five senses, and hidden characteristics, which are not readily detectable by consumers.
- Product quality is defined by the properties of the food that can be measured by food scientists.

Looking ahead

This chapter introduced us to how food scientists define quality, measure it, and ensure that foods are safe and of consistent quality. Chapter 8 describes how new food products are developed for the marketplace. Chapter 9 emphasizes the role of food scientists in basic research and in developing and enforcing regulations. In the final section, the fundamentals of food science are presented. Chapter 10 addresses the chemical basis of food quality. Chapter 11 describes the nutritional basis of food quality, and Chapter 12 presents biological aspects of quality with a particular emphasis on microbiology. Food engineering is the subject of Chapter 13, and in the final chapter, sensory studies (actual tasting of products) are covered.

References

Deming, W. E. 2000. *Out of the crisis.* Cambridge, MA: MIT Press.
Hurst, W. C. 2002. Safety aspects of fresh-cut fruits and vegetables. In *Fresh-cut fruits and vegetables*, ed. O. Lamikanra. Boca Raton, FL: CRC Press.
Pirsig, R. 1999. *Zen and the art of motorcycle maintenance: An inquiry into values.* New York: Morrow.

Further reading

Alli, I. 2004. *Food quality assurance: Principles and practices.* Boca Raton, FL: CRC Press LLC.
Carpenter, R. P., D. H. Lyon, and T. A. Hasdell. 2000. *Guidelines for sensory analysis in food product development and quality control*, 2nd ed. Gaithersburg, MD: Aspen Publishers.
Florkowski, W. J., R. L. Shewfelt, B. Breuckner, and S. E. Prussia. 2009. *Postharvest handling: A systems approach*, 2nd ed. San Diego, CA: Academic Press/Elsevier.

Hubbard, M. R. 2003. *Statistical quality control for the food industry*, 3rd ed. New York: Kluwer Academic/Plenum Press.

Man, D., and A. Jones. 2000. *Shelf-life evaluation of foods*. Gaithersburg, MD: Aspen Publications, Inc.

Marriot, N. G., and R. B. Gravani. 2006. *Principles of food sanitation*, 5th ed. New York: Springer Science.

Vasconcellos, J. A. 2007. *Quality assurance for the food industry: A practical approach*. Boca Raton, FL: CRC Press, Taylor & Francis Group.

chapter eight

Product and process development

For her group's assignment in her introductory food science class, Jennifer's team decided to develop a one-calorie beer. The first part of the assignment was to come up with a fantasy food product not currently available but that would plug a hole in the market. The second part of the assignment was to create a marketing plan. Jennifer's team finally decided to develop a one-calorie beer. The students designed a nice-looking package and a clever marketing campaign aimed at NASCAR fans. The slogan was, "The Buzz without the Calories!" They made sure to include a public service campaign to discourage drinking and driving. Jennifer and her team were very disappointed to receive a C minus. Written in green ink next to the grade they read, "Technically impossible." One of the group observed, "Hey, if they can make a one-calorie can of cola, why not beer? A whole lot more people have a beer gut than a cola gut!"

As part of a national food product development competition sponsored by the Institute of Food Technologists (IFT), Martin led a project to design an edible soup bowl. In some restaurants we can order chowder in a bread bowl, eat the soup, and then eat the bowl. While this idea is not exactly new, it is one thing to bake the bread fresh and serve it in a restaurant and quite another to put it in a package as a shelf-stable food. His group designed the product, the package, the process operations, a HACCP (Hazard Analysis and Critical Control Point) plan, a nutritional label, and a marketing plan for the bowl. One of the toughest problems to solve was how to package the soup and bread together. In putting the project report together, they found that they had to use information from every food science course they had taken and some of their other courses as well. They learned that it is difficult for one person to know enough to completely develop a product and a process. Different team members had different skills. By working together they were able to produce a much better report than they would have working separately. While their project did not win the IFT prize, Martin considered his experience to be one of the most satisfying and rewarding ones during his college days. He now works with a large food company developing new sports drinks.

Because she was disgusted that so many foods were filled with artificial colors, sugars, fats, preservatives, and stabilizers, Hannah's idea was to develop a nutritious food product. Her particular interest was in soy. The problem with soy was that once we get past tofu, there are not many products available and most of them taste terrible. After working on her project for a while, she began to understand the difficult challenge soy posed to a product developer. Soy lipids easily and rapidly oxidize during storage. The oxidation produced strong beany and cardboard flavors. Hannah learned that developing nutritious food is much more difficult than developing junk food, and that sugar and fat are used to overcome many undesirable characteristics. One of her biggest dilemmas was figuring out how to balance the good with the not-so-good ingredients to design a product that was both nutritious and appealing.

These three students tasted both the thrill and the disappointment of new product development. Product and process development allow food scientists the opportunity to use both creative and technical skills. Creativity is important in designing a product or process that results in something many consumers will enjoy eating and come back for more. Technical knowledge is essential in separating the ideas that are possible from those that are not and in extending development beyond trial-and-error to the level of skill and logic. This chapter provides some insight into how food scientists work with others in the food industry to create new processes and products that feed our insatiable desires for variety in the supermarket. Product development is perhaps the most important, interesting, and challenging activity that a food scientist performs as a professional.

Looking back

Previous chapters focused on food issues we deal with in our daily lives and with the types of products we encounter in the marketplace. Some key points that were covered in those chapters help prepare us for understanding product and process development.

- Quality includes sensory characteristics, which can be readily detected by the five senses, and hidden characteristics, which are not readily detectable by consumers.
- Quality of a product is defined by the properties of the food that can be measured by food scientists.
- Formulated foods are products that involve the mixing of ingredients.
- Individual fat replacers cannot perform all functions of lipids, as lipids are important functional ingredients in foods.
- Preservatives are food additives that prevent or slow spoilage.
- Food processing and preservation increases the shelf stability of a raw material but usually decreases affordability, nutrition, and quality of the product.
- Nutraceuticals are foods specifically designed to act as drugs.
- Nutrition labels can help us keep up with the quality of our diets.

- Packaging protects food from microbial and chemical contamination, physical damage, and invasion of insects and rodents.
- Processed foods should be designed to spoil before they become unsafe.

The proliferation of food products

Next time you are in the supermarket, take a look around at the wide variety of products that are available to us. Each one of these products was developed by a food scientist, probably as part of a research team. Some of these products are truly new products, something very different from anything we've ever seen before. Others may be a "new and improved" product or a line extension, such as a new flavor. Sometimes one company brings out an alternative to a best-selling brand. All of these items provide us with a number of choices. For example, Jennifer was in her local supermarket the other day and counted 149 different brands, flavors, sizes, and types of ice-cream products. Despite all of her choices, they didn't have her favorite "no sugar added" flavor—chocolate marshmallow.

It has been estimated that less than 10% of the products developed each year make it to supermarket shelves. Of those, most do not last a year before they are pulled by the manufacturer. The ones that remain are survivors in a very competitive marketplace. That is why we may find a great new product and enjoy it for a short time only to be disappointed when it is no longer available. In the competitive marketplace there is such a rapid push to get new products on the shelves that there isn't enough time to really identify the factors that make the product a success or failure. One thing is obvious; not enough consumers were buying the product for it to survive. Perhaps others didn't like it as well as we did, or maybe they just didn't find out about it in time. Many other explanations are offered, but no one seems to really know. If food companies knew the answer to this question, there would be more successes and fewer products developed. See Insert 8.1 for the story of an initial product developer.

Supermarkets are not the only place we can find new products. Fast-food and casual-dining restaurants also introduce them to their customers. While we might think that these items are made from scratch in the back of the restaurant, it is usually not the case. Most fast-food products are processed or formulated in a food manufacturing plant and shipped to a central warehouse and then to the individual location where the products are prepared and assembled for sale. Even those delicious entrées and desserts at those upscale chains are frequently prepared elsewhere, divided into individual servings, frozen, and shipped to the restaurant for our dining pleasure. These products may be developed by food scientists working for the restaurant chain, a supplier company (like a chicken

---

Insert 8.1 Chocolate science.

Early in his life Milt, like his father Henry, failed at everything he did. When he was out of money after another failure, he was working in Denver for a caramel maker. He learned that adding milk instead of paraffin to caramels improved the flavor and texture of the candy. He returned to Pennsylvania, added more milk to his caramels to make them easier to chew and more buttery, found a wholesale distributor, and made a small fortune. Instead of retiring on his fortune, he plowed his profits into a new idea—milk chocolate.

Milt found that it was much more difficult to incorporate milk into chocolate than it was to incorporate it into caramels. He moved to Pennsylvania close to dairy farms, and worked through trial and error to come up with the right combination of ingredients. He found that the water of the milk and the fats in the chocolate did not mix well, so he needed to boil most of the water from the milk. He encountered many other difficulties before he was able to succeed, making the milk chocolate that Americans love.

Milt was not well educated. He never completed grade school, but he was part of a generation of entrepreneurs who designed their own products and manufactured them. He is actually credited with introducing mass production just a few years before Henry Ford. Ironically, Milt was never very interested in the business side of chocolate. Fortunately, he developed a partnership with William Murrie, who handled all of the business aspects while Milt focused on product design and manufacturing. Although his chocolate was a success, many of his other inventions, such as onion-flavored ice cream, were not. It is thought that his cigar-smoking habit may have dulled his ability to perceive flavor. Milton Hershey gave his name to the chocolate Americans love.

Today, people who design products require a sound technical education. Products are not usually developed by a single individual now. Rather, they are designed by a team generally led by a food scientist and composed of several other employees such as those with a background in plant operations, quality and safety testing, package engineering, graphic design, product distribution, accounting, marketing, and sales. Some products may require a nutritionist, process engineer, lawyer, or other expert on the team. For more on the life of Milton Hershey, read *The Emperors of Chocolate* (Brenner, 1999) and *Hershey* (D'Antonio, 2006).

---

processor for chicken sandwiches), or a team composed of scientists from both companies.

Generating new food product ideas

Before a product is developed, someone must come up with an idea. A concept then develops from the idea (like the one-calorie beer Jennifer's team developed). Maybe somebody likes a product but has an idea to make it better. Maybe someone has experimented at home and combines two or more favorite products into a dynamite eating sensation. Have you tried peanut butter and vanilla ice cream together? Sounds awful, but it's really good! Maybe someone has a wild dream that turns into a novel product. There are many different routes to new ideas.

Many new product ideas come in raw form from marketing departments. Marketers are free thinking, creative people who are always looking at what is and how it can be different or more interesting, but their ideas are not necessarily practical. Marketing is about getting people to try new products. Ideas can also come from the sales department. Salespeople collect ideas from a variety of sources:

- Rumors of new products even before they hit the shelves
- Products that are currently available (this includes competitors)
- What is working and what is not (shelf stability)
- How good products could be made better
- New products consumers would like

These ideas tend to be less original and more practical than those coming from the marketing department because salespeople are on the frontlines dealing with customers daily and must see their products from the customers' viewpoint.

Another source of product ideas is the consumer. As mentioned earlier, it is the consumer who makes or breaks a new product by either buying it or not buying it. Marketing departments spend much time and money using surveys and focus groups to systematically collect information. During a survey, consumers make known their likes and dislikes if asked about a new idea or if given the opportunity to try a new product. Surveys may also be conducted from 1-800 calls or e-mails to the company's Web site. A focus group consists of a few customers and a skilled moderator with a carefully scripted set of questions designed to get new ideas and test product concepts. Although some product ideas are generated from consumers, it is difficult to separate those that will appeal to the general market and those that are "unique" and will appeal to only a few.

Finally, ideas come from product developers themselves. They work with new ideas all the time. Sometimes a failed product will give them an idea for a new product that may work. Other times, a product not previously envisioned happens quite by accident with the addition of the wrong ingredient or the right ingredient at the wrong level, which produces unexpected results. Although food scientists bring a systematic approach to the development process, product development still includes some art as well as science. A successful developer will be able to bring some of both to bear.

Improving existing products

We are all familiar with the phrase, "new and improved." It may be just a more consumer-friendly package or a product that doesn't spoil as quickly. Sometimes change in a product can make it less desirable to some people, eliciting a comment like, "I wish they wouldn't mess with a good thing!" Other times it may actually make something good even better. As times change so do consumer tastes and preferences. A developer must diligently keep products current with changing times while simultaneously not making changes that will cause the loss of loyal customers. It can be a delicate balance.

Product line extensions are defined as changes in a product feature, such as flavor, content, size, name, or packaging. For example, a new papaya flavor for ice cream, milk, or juice mix adds a new product to an existing line. Adding cherry, lemon, or vanilla to colas or jalapeno peppers to cheeses, represents a more dramatic line extension. Creating a line of vegetable yogurt was an even more revolutionary change that did not work in the American marketplace. To appeal to a different market such as children or seniors, products may be packaged in different sizes or under different names or labels. Any modification to product ingredients requires a corresponding change in the labeling.

As mentioned earlier, improvements are made for many different reasons and come from suggestions made by a variety of sources, including marketing and sales departments and consumers. Improvement of the flavor, color, or texture of a product generally happens because of consumer complaints. The sensory quality of a product may be fine when it leaves the manufacturing plant but is lost during distribution. The improvement needed here is to extend the shelf life by increasing the product's stability. This change can be accomplished with a new formulation (change in ingredients) or a modified process at a specific unit operation (change in temperature). Packaging can also be a source of product improvement simply by making it easier to open. Easier preparation, such as microwavable maple syrup containers, is a common type of product improvement.

Some new products are merely different twists on an old theme. These twists could be innovations in an existing line but which go beyond the typical line extension. For example, many years ago a mustard with onion chunks was introduced so that the consumer wouldn't have to deal with chopping onions for hot dogs. A more recent example is colored ketchup. The green ketchup, in a new container, was guaranteed to deliver ketchup more quickly for impatient children. It was probably reformulated to flow more quickly than the old style product and the green novelty color added to get the attention of children.

Another reason to modify an existing product might be the need to substitute ingredients. Occasionally an ingredient becomes unavailable, like the loss of pistachios when we went to war with Iraq (now we can get them from California). At other times the ingredient is available but the quality is too variable, making it impossible to provide product consistency. Perhaps the ingredient is available but it has become too expensive. Then again, maybe the developer has found an ingredient that is simply better than the one presently being used. Many products are now being formulated to produce low-calorie versions using fat and sugar substitutes. Substituting an ingredient is not as simple as it might seem. The product developer must conduct many experiments to find the right balance. It may take two or more ingredients to perform the same function as a single ingredient in the original formula, or one ingredient may substitute for two or more. Substitute ingredients may interact with other ingredients to provide quality or stability problems not encountered with the original formulation. The new ingredient(s) may induce slightly different sensory properties, but it is not clear whether these new properties are great enough to affect consumer acceptability. Another concern when ingredients are substituted is that the statement on the label must be changed. Companies want to be certain that the ingredient change is a definite improvement before changing their label and tossing a large stock of unused labels. Cost is always a concern when reformulating a product.

Product improvements could also result from process modifications. Production of truly new products (like the 3-D corn chips) requires new equipment or at least major modifications to existing equipment. Major modifications in equipment or processes are rare because they are very expensive. Thus, new product sales projections must be high for management to make any changes. Sometimes minor modifications are made to equipment in a food manufacturing plant, which results in a more consistent quality product, but in most cases the developer will modify the product to fit the existing equipment rather than modify the equipment to fit the product.

Brand new products

New products are developed for a variety of reasons, such as a new idea, a new trend, a problem confronting consumers, a new technology for food production, a new type of material for packaging, or the introduction of a foreign dish. Some new products represent major changes and are truly innovative. These products may change the way consumers eat or even think about eating.

A few years ago a major dog-food manufacturer was faced with a problem. Dogs liked the flavor of canned dog food, but their human servants preferred the convenience and cleanliness of dry food. A product development team at General Foods explored the possibility of producing a dog food that would satisfy both the consumer and the purchaser. After some initial challenges they came up with a revolutionary product—Gaines-Burgers®. It had the flavor the dogs loved without the mess that bothered the dog owners. For more details on how this product was developed, see Insert 8.2. This product and others like it are part of an entirely new category of foods called *intermediate moisture foods*. These foods have low water activity, meaning they are less likely to spoil by microbial decay. Some "natural foods" like dates and pepperoni can be classified as intermediate moisture foods, but GainesBurgers® was the first food to be specifically designed to take advantage of spoilage-resistant properties of low water activity.

Another source of truly innovative products is a new idea of what a food is or how to eat it. One of the more famous is the hot dog. There are many stories about how it was developed, but this one stuck. Apparently a vendor decided to put a German sausage into a French roll. The sausage, being long and round, prompted him to call it, dachshund. The name didn't work, but a modification––hot dog––did. The product had appeal because it could be eaten without utensils, a trait that is important in fast-food restaurants of today. This innovation led to the development of specific frankfurters for hot dogs, including the substitution of turkey for pork or beef. Buns were designed to fit the wieners, which then took on different lengths. More wieners and buns could be sold if there were different quantities of buns and wieners in their respective packages. Toppings like mustard, ketchup, coleslaw, sauerkraut, and chili have increased in sales and consumption thanks to the development of the hot dog.

Consumers are sometimes considered fickle by product developers, but maybe they just like something different. When Martin was younger, he used to put blue food coloring in his drink so his little sister wouldn't touch it. Blue was considered gross and people wouldn't eat blue foods with the exception of blueberries. Now many sports drinks and other products are bright blue. Developers were able to identify a new trend among consumers and tapped into it. Sometimes trends are brief, like the clear colas (remember that most colors are flavorless and most flavors are colorless), but others stick like the bright blue sports drinks.

The big trend these days is health foods. Developers like Hannah have been trying to incorporate trendy ingredients into acceptable products. They have found that many of these ingredients have unpleasant flavors or odors. For example, garlic has been touted as being a healthy ingredient, proclaiming to help prevent cancer and heart disease, but garlic is also a major cause of halitosis (bad breath), a socially unacceptable condition. Some dispensers of garlic tablets provide the best of both worlds— odorless garlic. If, however, the physiologically active components are also the ones that provide its odor, any health benefits will be lost when the odor is lost.

Insert 8.2 The development of GainesBurgers®.

My story about a new food product deals with the introduction in the early 1960s of GainesBurgers, a soft, moist, shelf-stable dog food. This is a particularly interesting one because in this case the purchaser is not the consumer. The consumer is the dog. The purchaser is the dog owner. So when you evaluate the keys to success—consumer need, consumer delivery, and consumer acceptance—you have to think in terms of the real consumer *and* the purchaser (the surrogate consumer, if you will).

At the time of introduction there were two kinds of products on the market for dogs. One was a dry, homogenized dog food, and more recently, extruded and expanded dry dog food. As dry products, these clearly had a limit of palatability with the pets, but they were cheap, convenient, and stable, so they satisfied the purchaser, even if they didn't do a very good job of satisfying the consumer. On the other end of the spectrum were canned dog foods, which were highly palatable to the animals, but they were highly unacceptable to the purchaser because they were odiferous, messy to handle, and unaesthetic-looking mixtures. Further, the leftovers had to be stored in the refrigerator providing unsavory connotations of having pet foods in the same area as human foods.

So, consumer need number one was pretty clear. There was a need for a better product for both the purchaser and the consumer—a product with the positive attributes of both dry and canned dog food, but with none or few of the negatives of either. I remember that the specification for this product was drawn up before anybody had a real idea of what technology was required to deliver it, or for that matter, even *if* the technology existed to deliver it. Experimental products were developed, ranging all the way from the most obvious, which was to take canned dog food and put it in a sausage casing, to various forms of very expensive pressed jerky meat forms. The dogs indicated that we were on the right track by showing high acceptance for these prototype products, but giving us little clue as to how to preserve and deliver them in an affordable way that would satisfy the purchasers.

Invention was made in a relatively short time by a talented group of food technologists keying off of other preservation systems employing bound moisture similar to that in jams and jellies and partially dehydrated fruit. There was eventually further research into so-called *water activity preservation* (bound water or intermediate moisture) technology by General Foods technologists. This technology was the basis for GainesBurgers formulation and preservation. The dogs loved the product! It appears that the three keys to success—consumer need, consumer delivery, and consumer satisfaction—were now being satisfied. However, after one year of test marketing, the product was on the verge of being discontinued because of low sales volume. In fact, I believe that it would have been discontinued if it wasn't for the overriding evidence of extremely high acceptance by the dogs in the kennels where we could observe them on a controlled basis.

What had happened was that the marketing story did not explain clearly to the *purchaser* what this product was by simply calling it GainesBurgers, "the burger for dogs." It sounded more like a novelty item rather than a dog *food*. This name was just too strange and unclear to the human purchaser. At that time, as I've said, people knew that dogs loved canned dog foods and that bulk, dry dog food was convenient. What did the new product provide other than a fancy name and a concept that was not understood? The problem was in the message, not in the delivery. A new marketing position was developed identifying the product as "the canned dog food without the can," and that did it. It immediately told the purchasers why they should buy this strange new concept in dog foods. It communicated the message that this product delivered the palatability of canned with the convenience of dry. The story from there on out is well known—an entirely new product category in pet foods was created that is still growing, spreading from dog foods to cat foods and eventually back to human foods where the intermediate-moisture principle is being used today. This is a case where consumer need and consumer acceptance were satisfied, but the product was almost killed in marketing by inadequate consumer delivery.

Al Clausi (Retired General Foods research chemist.)

New technologies become another source for truly innovative products. Extruders permit a whole range of new foods. By changing the die on the end of the extruder, new sizes and shapes can be created while still using the same piece of processing equipment. In the late 1960s, using extrusion technology, two companies collaborated to develop a new bite-size egg-roll without the traditional fold. The companies decided it would be too expensive to retool the plant. A new company was started by some of the employees from each of the former two companies. The new manufacturing plant with new equipment produced a much better quality product. The two original companies sued the new one for breaking confidentiality rules, and the new company went out of business. Many years passed before the newer, better, bite-size egg-rolls came back into the market.

New technologies can also be introduced in the home. An important trend in the past forty years has been to reduce meal preparation time. What used to take hours now takes minutes. In the 1950s the TV dinner was a convenient item that took 45 minutes to prepare. The quality of the food was poor, but it was reasonably priced and required little effort. However, consumers wanted even more convenience, and they wanted it now. Microwave ovens heat foods quickly. Aluminum TV dinner trays gave way to microwavable trays and pouches. In the 1970s the microwave oven revolutionized the way Americans prepare meals and how developers design new products New packaging was needed and was designed to meet this requirement, and the food products were designed to meet the needs and wants of consumers. Initially the quality of microwave products and packaging was as poor as their TV dinner predecessors, but improved packaging, products, and microwave ovens can deliver premium quality products.

Innovation can also be inspired by a new type of packaging. An overlooked health food on the market, liquid milk, is not as popular as it once was. Milk, the most available source of calcium, does raise some health concerns like most of the other health products that were discussed in Chapter 3. For consumers who do not get enough calcium, more milk can be desirable. New products that contain reduced lactose, reduced fat, improved flavor, and have a richer texture have helped skim milk hang on. Traditional milk packages don't fit very well with today's lifestyle. They have been redesigned to open and pour more easily, which is particularly important to elderly consumers. Milk chugs are designed to fit into car cup holders and are primarily marketed at one-stop gas stations as an option for students to replace colas and other nonnutritious beverages.

Traditional foreign dishes provide new products in new environments. Almost all Americans ate good old American foods in the 1950s and 1960s (meat, potatoes, two vegetables, and dessert). Now "pasta" and other catchy sounding foreign names are the primary products available in low-calorie, flavorful, microwavable entrées in the frozen food section of the supermarket. Pizza and yogurt entered the United States because American soldiers returning from the World Wars created a demand. Both products have become so Americanized in the intervening years that Americans in Europe are frequently disappointed with Italian pizza and French yogurt.

In this chapter, we will learn about the projects and experiences of two teams in a food product development class at the University of Georgia from the perspective of these food science students. Many of the remaining illustrations reflect the steps taken in the development process.

Reality check

Not all new product ideas are practical. Not all product concepts are feasible. Marketing may go to the research and development (R&D) unit indicating that they want a fresh broccoli, zucchini, bean, and pasta product with no preservatives, full flavor, and a shelf life of seventy-five days under refrigerated storage. By now we know that this idea is not practical. It becomes the task of R&D to either design the product or let the marketing group know the constraints and provide an alternate product that meets some, but not all, of the guidelines. Food scientists must understand the capabilities and limitations of ingredients, processes, and distribution when designing new products. They rely on their technical training in chemistry, microbiology, processing, engineering, nutrition, sensory evaluation, packaging, and quality control to modify an existing product or develop a new one. Usually this effort, led by a food scientist, is conducted by a research team, which contains members with a wide range of skills including marketing. Food scientists must be particularly concerned about ingredient interactions and limitations, sensory quality, product safety, process constraints, shelf stability, and economics.

Sometimes an idea sounds good, even to a food scientist, but it just doesn't work out. With the mustard and onion product described above, it was learned in the development process that the onions became soggy in the mustard, which then became watery. Despite these limitations, the product did find its way to supermarket shelves, but it never was a big hit. Experienced food scientists can anticipate many of these problems and solve some of them through knowledge of the functionality of ingredients and processing techniques. They can also identify technical problems that are not likely to be solved and develop alternative solutions that meet some, if not all, of the desired characteristics.

Ingredients have many limitations, but a developer with a good knowledge of different ingredients can usually find one that will perform the function needed. When substituting low-calorie sweeteners for sugar or replacing fats, sometimes a blend is needed to produce the desired effect. College students seem to be very sensitive to some low-calorie sweeteners, and many reject any product containing them. It is very difficult to develop a fat-free product that delivers both flavor and mouthfeel, but food scientists have become very good at producing high-quality products with much of the fat removed. Many dieters would do well to find the low-fat, low-sugar alternatives that combine low calories with good sensory characteristics. Of course there are some components that cannot be changed. For example, Jennifer's one-calorie beer is technically impossible because ethanol (drinking alcohol) contains seven calories per gram. To remove the calories we must first remove the alcohol. Jennifer may be drinking beer for its flavor, but others may be looking for something else. No-alcohol malt beverages that taste like beer are available, but they are not very popular. They do have some calories, however, because they contain carbohydrates.

Product developers must be very sensitive to the quality of the product. Remember that quality doesn't have to be synonymous with perfection. It is more important that products meet or exceed consumer expectations. Developers would like to use the finest of ingredients for each product, but they must take cost into consideration. As stated previously, they may develop the finest of products in the laboratory, but conventional processing, storage, and distribution can lead to loss of quality. Other factors must be considered. For example, with microwavable syrup, developers must consider how repeated heating and cooling will affect flavor color and viscosity.

Product developers must always be concerned with safety. As we have learned, a properly canned product is about as safe a product as we can eat. Contrary to popular opinion, fresh foods are more likely to be unsafe than processed ones. Consumers demand foods with full, fresh flavor, and food scientists want to deliver. When a newer, fresher product is designed, the product developer must build safety measures into the process, package design, distribution and storage strategy, and preparation instructions to ensure that the product will be safe to eat under normal conditions. For perishable products, the food scientist also wants to make sure that the product will spoil before it becomes unsafe to eat.

Process constraints are another factor a developer must consider when designing new products. For example, any heat process, even mild heat treatments like blanching and pasteurization, can diminish flavor and nutritional quality. Trade-offs must be anticipated and the developer or development team must consider all aspects from production to consumption in its design. The fewer changes needed to existing equipment, conditions, or operations, the greater the chance the company will adopt the product for production.

Distribution constraints also limit the development process. A product must reach the consumer at an acceptable level of quality or it will not last long in the marketplace. Developers must test their new inventions under standard market conditions. Tests evaluating the quality changes of a refrigerated product at 35°F will not give the food scientist a clue as to how it will behave in supermarkets where it is stored at 45 to 50°F. Shelf-life studies are critical in determining its likely success or failure in the marketplace. The sales department frequently wants a longer shelf life than is realistic or necessary. It is important to determine realistic shelf-life requirements and the necessary trade-offs to meet these requirements. Food scientists must also consider economic constraints as described in Insert 8.3.

The assignment for the product development class was to develop a product that used organic chocolate from Ecuador as a raw material. The new product was part of an environmentally friendly project to help protect the rain forests. One team chose to develop organic syrup designed to appeal to the up-scale market, and the other chose to design a dipping sauce to appeal to children. Each package of dipping sauce would have an educational message to help children better appreciate the importance of the rain forests in global ecology.

Food formulation

As we learned in Chapter 5, formulated foods involve mixtures of ingredients. Most of the food products many of us eat daily are formulated. Among the key responsibilities of a product developer is to know and understand

Insert 8.3 Economic constraints associated with product development.

When I was working for an egg-roll company mentioned above, my boss had a great idea that we would develop a ham-and-egg-roll, the same outside but ham and eggs on the inside. We worked on our formulation, using the freshest of ingredients, and these egg-rolls were great! I took one of our early versions around to the front office, and everybody loved them. Then my boss suggested that I do cost estimation based on the best price we could obtain for the ingredients. The cost of ingredients alone was greater than what we would be able to charge for the product. We started looking for less expensive ingredients, and we found them. We developed a formulation that was reasonable cost-wise. Then we produced our modified version. They were terrible! News of the disaster spread faster than I could move from office to office. No matter how hard we tried we couldn't get a decent flavored product for a reasonable cost. I suspect many other great ideas meet a similar fate.

Robert L. Shewfelt

---

the function of each ingredient in a product, how the ingredients interact with each other, what ingredients are essential, what ingredients can be replaced, and what the potential effects ingredient substitution will have on quality, nutrition, stability, and safety of that product. Some ingredients have their own identity and cannot be replaced without changing the product. Pistachios are critical components in pistachio ice cream, just as strawberries are in strawberry jam. Others, such as starches and gums, vary widely in their functional properties. For a frozen custard pie we might wish to use carrageenan for freeze–thaw stability. To maintain the thickness of a salad dressing at room and refrigerated temperatures we could use carboxymethylcellulose.

Product matches are something that developers are asked to perform. The boss brings in a sample of a competitor's product her mother bought in the supermarket over the weekend, and wonders why our company can't make a comparable product. The ingredient statement on the label provides a place to start looking. Ingredients must be stated in order of amount added with the first being present in the largest amount. Some ingredients, like citric acid, are easy to match, while others, like natural flavors, are obscure and part of trade secrets. Once we have a list of potential ingredients, we must start guessing at the levels of the ingredients. Most developers will have a reasonable clue as to how much is needed from personal experience. An attempt at duplication may not produce a similar product if production details are unknown. Usually the first new product on the market will be the market leader if it meets the expectations of the consumer. Copycat products tend to be successful only if they come attached to a well-known brand that has loyal followers, have a noticeably lower price, or possess clearly superior quality.

Knowledge of the functional properties of ingredients separates the good developers from the superior ones. Functional properties affect not only the flavor, color, and texture of a product, but they also affect its convenience, nutritional quality, safety, and stability. Formulations of some of our favorite products today started out as old family recipes. While some of these recipes remain intact, most required major changes in ingredients to become commercially viable. As we have seen, freshly prepared products are not likely to maintain their quality during conventional processing, distribution, and storage. A home recipe might not withstand freezing and thawing if it is produced as a frozen product. Fresh-squeezed juice blends will not retain their delicate flavors upon pasteurization to keep them safe. The cost of typical ingredients may be too high to produce a product that enough consumers are willing to purchase at a price necessary to generate a profit.

In developing a new formulation, the food scientist must determine which ingredients to use and how much. If it is a line extension or minor modification, the basic formulation is already determined and ingredients are substituted. It may be as simple as substituting papayas for strawberries in an ice cream product line, but most development projects are not as simple as they might seem. There may be ingredient interactions with the papayas that we did not experience with the strawberries, similar to those we saw above with the onions in mustard. If we use an amount of papaya equivalent to the amount of strawberries in the original, it may be too overpowering or may be too weak. Papayas may give a strange color that would repel the consumer or the color may change from highly desirable to unacceptable during processing, handling, or storage. Papayas may become sticky and gooey in the product, resulting in a nauseating texture in the mouth. The nutritional level for selected nutrients may change greatly, resulting in a major modification of the product label. The microbial safety of papaya pureé may not be as good as that for strawberries, requiring a more rigorous heat treatment. Changes in processing may lengthen the time of production, thus increasing the cost of the product and lowering nutritional and sensory properties. The developer must consider all of these factors and more for a seemingly simple change.

Another consideration that a product developer must make is the order of the addition of ingredients. It is not just as simple as adding them in the order as they appear on the ingredient statement. Anyone who has baked a cake knows that we can't dump all the flour into all the water and expect them to mix properly. In candy making, certain ingredients must be mixed in carefully to get them to distribute properly. Texture is the sensory property most often affected when ingredients are not added in the right order.

One factor that developers of low-fat and sugar-free products must consider is how to make up for the loss of fat or sugar. Assuming that a good sensory match can be made with the sugared, fat-filled product, the developer still has a problem. All sugar substitutes and many fat replacers

Full Fat Peanut Butter

Nutrition Facts	
Serving Size	32g
Servings Per Container	16
Amount Per Serving	
Calories 190	Calories
from fat 130	
	% Daily Value
Total fat 16g	25%
Saturated Fat 3g	16%
Trans Fat 0g	0%
Cholesterol 0mg	0%
Sodium 150mg	6%
Total Carbohydrate 7g	2%
Dietary Fiber 2g	9%
Sugars 3g	
Protein 8g	
Niacin 20% • Riboflavin 2%	
Vitamin E 10% • Iron 4%	

Reduced Fat Peanut Butter

Nutrition Facts	
Serving Size	36g
Servings Per Container	14
Amount Per Serving	
Calories 190	Calories
from fat 110	
	% Daily Value
Total fat 12g	18%
Saturated Fat 2.5g	12%
Trans Fat 0g	0%
Cholesterol 0mg	0%
Sodium 250mg	10%
Total Carbohydrate 15g	5%
Dietary Fiber 2g	9%
Sugars 4g	
Protein 8g	
Vitamin B6 6% • Folic acid 6%	
Niacin 25% • Iron 4%	
Copper 10% • Zinc 6%	
Magnesium 15% •	

SoyNut Butter Peanut Free

Nutrition Facts	
Serving Size	32g
Servings Per Container	13
Amount Per Serving	
Calories 170	
Calories from fat 110	
	% Daily Value
Total fat 12g	18%
Saturated Fat 1.5g	8%
Trans Fat 0g	0%
Cholesterol 0mg	0%
Sodium 160mg	7%
Total Carbohydrate 5g	2%
Dietary Fiber 5g	20%
Sugars 1g	
Protein 10g	
Calcium 6% • Iron 7%	

Insert 8.4 Nutritional labels for three types of peanut butter.

are required at much lower levels than the ingredients they are replacing. Consumers looking for low-calorie substitutes are not likely to be amused when given less product for the same price. Bulking agents must then be added to provide the same serving size at a lower calorie level, and most of the time these agents perform their function well. Sometimes the consumer just eats a larger quantity, consuming as many calories as they would have had with the original product. Occasionally overconsumption of bulking agents can lead to excess gas production and other evidence of gastric distress. In the case of reduced-fat peanut butter, the fat replacer also has calories and the reduced-fat version actually has more calories than the full-fat product (see Insert 8.4).

Initial product formulations are produced in a laboratory to achieve the desired sensory characteristics. The time it takes to develop a new formulation depends on the technical skill of the developer, differences in the new formulation from existing ones, and difficulties posed in modification. A successful formulation is called a *prototype*, which can then be evaluated for processing effects. An initial formulation for the chocolate syrup is shown in Insert 8.5.

Process operations

Even if the formulation is perfect and the ingredients are added in the right order, the product still may encounter problems during processing. The product developer must anticipate the potential effects of each step in the manufacturing process. Many of the processes that protect a food from

Insert 8.5 Initial formulation for Choco-Andes organic chocolate syrup. Developed as part of a classroom project at the University of Georgia by Emily Smith, Nolan Morris, Anna Ellington, Brandi Haynes, Maretta Jankowski, and Boone Curtis in a class taught by Dr. Yaowen Huang.

Ingredient	Amount (g)	% by weight
Water	150	35.1
Organic Honey	150	35.1
Cocoa Powder (22/24)	75	17.5
Brown Sugar	50	11.7
Xanthan Gum	2.5	0.59

spoilage rob it of its freshness and diminish its quality. Other preservation processes, such as fermentation and roasting, develop characteristics that we think enhance its acceptability. Heat can produce desirable browning, aromas, and textures, but it can also destroy delicate flavors and nutrients. Oxygen that is incorporated into the product speeds up reactions that generally decrease desirable characteristics, but it can provide light fluffy textures in bread and ice cream. Product developers work closely with process developers to make sure the correct process is matched to the new product.

Process developers design new processes or modify existing ones, much like product developers. Process developers must have a strong understanding of engineering concepts. They must know the effects of each unit operation on microbial activity, ingredient functionality, ingredient interaction, and product quality. In completely new processes, the process developer or development team may design entirely new pieces of equipment to achieve the desired result. More frequently, they are interested in the order of operations and the specific conditions, such as time and temperature, for each operation using existing equipment. Process developers don't go straight to the manufacturing plant to test their ideas. The equipment in the manufacturing plant is too big, and the time it takes to run process development experiments is too valuable. Instead, the process developer uses a pilot plant, consisting of miniaturized equipment that mimics those in the manufacturing plant.

Products from a pilot plant are much more realistic than those made in the laboratory. The process developer can evaluate different orders and conditions of operations in a set of experiments. Like the product development process, the length of time to develop an optimal *process* for a new product depends on the technical skill of the process developer, differences in the new process from existing ones, and difficulties posed in modification. The product and process developers (in a small company they may be the same person, while in a large company there may be several of each on the development team) must work closely to ensure

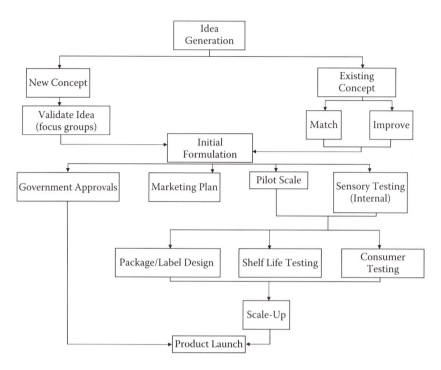

Insert 8.6 Schematic of product and process development. (Diagram by Carlos Margaria.)

that the desired texture, color, and flavor are obtained. See Insert 8.6 for a schematic of the overall development of a new product.

Even minor modifications in a process can have major implications in the microbiology and nutrition of a product. One result of a slight decrease in temperature in an operation could be inadequate killing of microbes, which then contributes to premature spoilage or, worse yet, a safety hazard. Heat and light also destroy vitamins, and use of too much heat to ensure safety can lead to lower nutritional quality.

Quality evaluation

Early in product and process development a standard or specification for the product should be developed. A standard or specification helps guide the development of a new product. Evaluation characteristics include the desired sensory properties, acceptable limits of microbes, and nutritional value. Usually the initial standard and specifications are high and difficult to accomplish. As development continues, it may be necessary to relax some of the restrictions to achieve a viable product. Occasionally, standards are raised. Each time the standard is relaxed a determination must be made as to whether to proceed with the new standard, reformulate or redesign the process to more closely achieve initial goals, or concede failure and discontinue the project.

Informal sensory testing by the developer and a few trusted colleagues is usually conducted for the initial formulations. Sensory quality is of major importance at this stage of development. More formal sensory testing is then performed on the most promising prototypes. Microbial and nutritional testing may also be performed at this stage if the developer is concerned that the differing formulations might affect nutrient value or microbial load. A preferred prototype formulation emerges from this testing, which then goes to the pilot plant for process development. The process developer then evaluates all of the sensory characteristics that are important to the product developer. The effects on microbes and nutrients are more important in process development than in formulation. Initial processing trials look at the effect on final product quality. Close evaluation of the effect of individual operations or the interactions of two or more operations may be necessary to decide on the best process for the new product. The process developer may need to send the prototype back to the product developer for minor or even major reformulation.

Formal sensory evaluation is a complex and time-consuming task. Product and process developers are unwilling to take every formulation or process-varied prototype to a formal panel. Remember from the last chapter that formal sensory testing does not guarantee consumer acceptability. Usually a few physical and chemical measures are sufficient to guide the development process. Any prototype that shows good promise should be presented to a formal panel to make sure that it is in line with standards. In some tests, such as product matching, the competitor's product may serve as a "gold standard" or the best possible product to be achieved. A gold standard can be very useful, but it can also eliminate formulations and process-tested prototypes that are superior. Storage stability is also a factor that must be considered when comparing our newly developed product to a gold standard. Insert 8.7 shows the HACCP plan for the chocolate syrup.

Storage stability

A superior formulation and process scheme does not guarantee success for our product. We must appreciate the effects of handling and storage on the quality of our best prototype. For example, the cookies on the end of a manufacturing line have much more fresh-cooked flavor than the same cookies in a vending machine. The "gold standard" product, which we were trying to match, is probably not as good as it was when first manufactured, and our freshly prepared product is better than it will be when it reaches the supermarket. Shelf-life testing is necessary to determine if

Insert 8.7 Summary HACCP Plan Worksheet for organic chocolate syrup.

Product: Choco-Andes Organic Chocolate Syrup	Plant: 1	Prepared by: Brandi
UPC Codes: 0123456789	Est. No.: 1	Date: 4/08/05 Rev. No.: 1

Process step	Hazard description	CCP	Critical limits	Monitoring procedures/frequency/person(s) responsible	Corrective actions/person(s) responsible	Verification procedures/person(s) responsible	HACCP records
Cooking	Biological: *Listeria monocytogenes, Zygosaccharomyces rouxii* (fermenter yeasts), *Bacillus cereus, Zalophus* species	1B	Cook to a minimum temperature of 100°C for 15 seconds.	Steam kettle operator will monitor the temperature of the syrup during cooking with a manual thermometer during each batch.	Steam kettle operator will adjust time/temperature if required.	Line supervisor will verify accuracy of the steam kettle log once per shift and observe plant employee performing monitoring. QA will check all thermometers used for monitoring devices for their accuracy and verify to within 2°C on a daily basis.	Steam kettle log—Steam kettle operator Thermometer calibration log- QA
Metal Detector	Physical: Foreign materials	2P	All chocolate syrup bottles pass through functioning metal detector.	Packaging operator will visually observe to ensure detector is on and bottles are going through.	Packaging line supervisor will check the metal detector using a passing wand every two hours to determine that limits are not exceeded.	QA will verify that the metal detector is functioning by running the passing wand through the metal detector twice per shift (once AM, once PM). All contaminated product is removed from system by functioning rejection mechanism.	Metal detector calibration log Corrective action log

the new product meets minimum requirements, establish an expiration date, and assess the quality of the product under conditions likely to be seen by the consumer.

The same types of physical, chemical, sensory, microbial, and nutritional tests used to evaluate initial formulations and processing prototypes are also used in shelf-life testing. During shelf-life testing, it is important to understand which characteristics deteriorate most rapidly and thus limit product stability. Changes do not occur at the same rate. Shelf-life tests usually are conducted at three temperatures. One of those temperatures should be the most likely storage temperature. During testing, higher temperatures are used to accelerate quality losses to provide a quicker, but sometimes less accurate, estimate of shelf life.

Package development

As a product is being developed, so is a compatible package. This task can be simple if it is a line extension or complex if it is a truly new product. In a large company, a packaging engineer will be on the development team, while in a smaller company, one person may serve as product developer, process developer, and packaging engineer. A packaging engineer from the company that supplies packaging materials to the company developing the product may also be involved in package development. Among other things, the package must contain the product, protect the product, provide important information to the consumer, and sell the product. The packaging engineer is responsible for choosing the right packaging material(s), making sure the package and the product are compatible and that the package serves all of the technical requirements of the new product, as described in Chapter 4.

Before selecting the material and designing a new package, we must list the functions a package must serve. Remember that the main function of a food package is to safely protect the processed food product from contamination. The package must also provide protection by keeping moisture, odors, and gases in or keeping oxygen, odors, light, and water out. *Permeability* is the property of a package that permits substances to cross the package from the food to the environment or vice versa. We must consider not only the properties of the primary package (the one directly in touch with the food product), but the secondary and tertiary packages too. We must also consider at what point during processing the product will be packaged. The packaging material must be able to withstand the conditions of processing and distribution. Canning usually occurs inside the package, but some frozen foods are packaged after freezing. If our product is processed by heat while in the package, it must be able to withstand the heat (and usually water associated with the process). If the package will be exposed to water, in either the processing plant or elsewhere, it must be water resistant. Our package must be able to withstand normal abuse experienced during its journey from the processing plant to the consumer. Most packages are designed with tamper-evident seals to notify the consumer of product contamination by unscrupulous humans. All packaging that is in touch with food material must be nontoxic and compatible with the food. Components of the food might interact with the package. Thus, we must study possible interactions to ensure that the package is not responsible for a safety hazard or premature spoilage. Any component of the package that will migrate into the product must be nontoxic and should not contribute to an off-flavor or odor. Some flavors, odors, or nutrients may also migrate from the food to the package causing loss of product quality.

Appropriate packaging materials for foods include glass, metal, fiberboard, paper, and plastics. Each material has its advantages and disadvantages. Glass containers hold a product nicely and prevent leakage of liquids and gases in or out of the product. Unfortunately, glass is heavy, which can increase shipping costs. It also makes a mess when it is broken. Metal also provides a nice barrier for liquids and gases and will not break under normal conditions. Most metal containers, particularly those for traditionally canned foods, are heavy. Thin aluminum cans used for popular beverages offer less weight, but at a greater risk for puncture. Some product ingredients such as acids can interact with the metal requiring an enamel coating to protect the can. Fiberboard is usually used for secondary containers. Many of us call these containers cardboard, but they are much more rigid than the flimsy cardboard we encounter in packages of other consumer products. Fiberboard is strong, flexible, and for the most part, recyclable. Fiberboard is susceptible to water damage and can collapse if placed under too great a load. Waxed fiberboard cartons provide protection from water damage, particularly with fresh fruits and vegetables, but waxed fiberboard is not recyclable. Paper and cardboard are relatively inexpensive and very good for displaying graphics. Paper and cardboard are not effective at keeping out gases or liquids and are easily penetrated by insects, pets, and vermin. Plastics encompass a wide variety of characteristics that can range from flexible to rigid, permeable to impermeable, clear to opaque, reactive with ingredients to inert, and have both graphic-friendly and graphic-hostile surfaces. Plastics have become the most widely used packaging material. A good packaging engineer understands the properties of each type of plastic and can select the most appropriate material or blend of materials at the appropriate thickness to meet all of the packaging requirements.

Graphics sell food products, and a great product will never be successful if it doesn't sell. A graphic artist may be part of the development team to design an appropriate label to merge the technical side of the product with the marketing campaign. Obviously, bright colors will help the product stand out, but may not be appropriate for a staple product such as rice or beans. Consumers have specific expectations about colors for specific foods and their labels. Meeting these expectations is critical in designing the graphics for the product. Focus groups may be used in testing types of designs that would be acceptable to current and potential consumers of the product. Package graphics must also be sensitive to the cultural impact of certain colors. In the United States, white stands for purity and black for death. Thus, we may see many white and few black food packages. In other countries, different meanings are attached to these colors, so package graphics may need modification for these markets. Size and shape of a container may also vary to meet the needs of the graphic artist as well as the technical aspects of the product.

In many cases the package may serve in food preparation or consumption. Enclosed plastic utensils, easy-to-open caps, the ability to place a beverage container in a standard cup holder, and spill-proof containers for coffee all provide the convenience so attractive to students on the run. While many packages are designed to be heated in a microwave oven, they are not convenient to those who want to use a conventional oven. One solution has been to produce dual-ovenable containers that can be heated in either type of oven. The ones Jennifer saw were made from sturdy paperboard that looked likely to burn up in a conventional oven. The first time she watched her turkey pot pie very carefully. Other innovations for microwave cooking include a metal sheet that helps to brown our pizza or the syrup bottle that can provide hot syrup for waffles or pancakes.

Package labels

Consumer information present on the label must include nutritional facts, product weight or size, and the ingredient statement. Contact information, such as the address of the manufacturer or distributor, a toll-free telephone number, a Web site address, and marketing information (such as how to get a space ring, T-shirt, or other valuable stuff) must also be included on the label. If a product requires special storage conditions or further preparation, there are generally specific instructions. For large products, package labels are not a problem; however, on small products the label can easily become cluttered, or the lettering is so small nobody over the age of forty can read it, even with glasses.

Comprehensive nutritional analysis is needed for the Nutrition Facts statement. Larger companies have the facilities to routinely perform these nutritional, formulation, and prototype tests for the development team. Smaller companies may not have a good idea of the nutrient content of the product until late in the development process as such analyses can be very expensive. Strict rules govern which nutrients must be tested and what claims can be made for a particular product. Since the nutrients on the label are listed in relation to the serving size, it is important that the appropriate serving size be determined. Some serving sizes are tricky. Many beverage containers, usually consumed whole, may actually contain two or more servings. The nutrient facts for the organic chocolate sauce can be found in Insert 8.8. As in other aspects of product development, it is not sufficient to perform nutrient analysis only on the freshly manufactured product. Rather, we must understand that the nutrients may change during handling, storage, and distribution. Generally speaking, the more rigorous the preservation technique (like canning) the more nutrients are lost during processing and the more stable they are during handling and storage. Over time, losses can also occur by interaction of nutrients with other components in the product or the package.

The ingredient statement is a very important, often ignored, part of the label. Much of this chapter emphasizes the importance of understanding the properties of ingredients in developing new products, processes, and packages. The statement warns allergic consumers to ingredients that may cause a reaction. It also alerts careful readers to unwanted preservatives or other chemical-sounding ingredients. Ingredient statements can provide important clues to developers making product matches as described earlier in the chapter. The ingredients must be stated in order of quantity added with the first being present at the highest amount. Simple products may have few ingredients while others, like beef stew, may have many ingredients, which in turn have ingredients of their own (designated

Nutrition Facts	
Serving Size	
Servings Per Container	
Amount Per Serving	
Calories 80	Calories from Fat 15
	% Daily Value*
Total Fat 1.5g	2%
Saturated Fat 0g	0%
Cholesterol 0mg	0%
Sodium 10mg	0%
Total Carbohydrate 19g	6%
Dietary Fiber 2g	9%
Sugars 16g	
Protein 1g	
Vitamin A 0% •	Vitamin C 0%
Calcium 2% •	Iron 30%

*Percent Daily Values are based on a 2,000 calorie diet. Your daily values may be higher or lower depending on your calorie needs:

	Calories:	2,000	2,500
Total Fat	Less than	65g	80g
Saturated Fat	Less than	20g	25g
Cholesterol	Less than	300mg	300mg
Sodium	Less than	2,400mg	2,400mg
Total Carbohydrate		300g	375g
Dietary Fiber		25g	30g

Calories per gram:
   Fat 9 • Carbohydrate 4 • Protein 4

Insert 8.8 Nutrient Facts for Choco-Andes organic chocolate syrup.

Insert 8.9 Ingredient statement for Choco-Andes organic chocolate syrup.

Choco-Andes organic chocolate syrup
INGREDIENTS: Water, organic honey, organic cocoa powder, brown sugar, xanthan gum.

in parentheses). See the labels shown in Chapter 5 for examples or Insert 8.9 for organic chocolate sauce.

Scaling up and consumer testing

Earlier in the chapter we discussed how a promising formulation developed in a laboratory becomes a prototype. The prototype is then used to develop the processing technique in a pilot plant. After the process has been developed, some reformulation may be necessary to help match the product and the process. Consumer testing is then needed to assess consumer acceptability of the best prototype. If a competitor's product is on the market, our product may be tested against it. If it is a "new and improved" product, it may be tested against the existing product. If it is a completely new product, it may be tested on its own. At this point, marketing or brand information are probably kept secret in order to determine the reaction of consumers to the product's sensory characteristics rather than its image characteristics.

Some pilot plants are large enough to provide adequate material for a limited consumer test. Other tests require more product than can be made in a small pilot plant and also require a limited run in the regular processing plant. Scaling up a product from the pilot plant to the manufacturing plant is not as easy as it might seem. While 2 to 10 pounds of a product might be produced at a laboratory bench and 50 to 200 pounds in a typical pilot plant, many manufacturing plants can produce tons of product in an hour. It may take hundreds of pounds of ingredients just to get a piece of equipment running effectively enough to produce useful samples, and then hours to clean the equipment after the test run prior to resuming production of the regular product. What works easily in the pilot plant with careful controls may not work as well in the manufacturing plant. Thus scaling up production runs must be carefully planned by the development team in conjunction with the plant production staff. Frequently these are done at the end of a normal production run prior to shutting down the operation for cleaning. Important lessons can be learned from running a test batch in a manufacturing plant. A complete microbiological, sensory, and nutritional analysis and shelf-life study should be conducted on product from this batch. The results might suggest reformulation or modification of process conditions prior to a full market test.

Market testing

At some point in the development process, a marketing plan for the new product is developed. A product name is selected and a target market identified. Advertising themes and venues along with promotional materials are considered. Merchandising efforts and product placement are determined. Focus groups are employed to test these marketing concepts. If some of the prototype is available it might even be presented at the end of the session to get consumer reaction. Collaboration between marketing and development personnel is necessary to match the product to the marketing campaign (e.g., the sensory and image characteristics).

Market testing begins when the best formulation has been prepared using the best process and the best package has been matched to the best marketing scheme. Control of the new product then passes from the R&D department to the marketing department. Consumer sessions, usually an intercept method at a shopping mall, are used to determine whether the marketing plan is reaching the right target market and if the product is acceptable. Image characteristics are more closely evaluated now than sensory characteristics. These tests are closely monitored. The results may suggest a need for modifications in the market campaign, package design, or product formulation. Major modifications may require backtracking to an earlier stage in development and repeating many steps. Complete rejection of the product or its marketing campaign by consumers may lead to its discontinuation. Minor modifications are feasible without further consumer testing.

Upon successful consumer testing, the product is sent for market testing. Items from one or more full production runs are selected. Many members of the development team will be present during the initial run(s) to help troubleshoot and be prepared for minor modifications. A certain region of the marketing territory is selected to present consumers with the new product in a typical setting. With full promotional effort, a fast-food chain will offer the new product for sale in small, medium, and large markets. A packaged food will be sold in a supermarket, possibly being promoted by someone offering free, prepared samples to passing shoppers. Based on market test results, an economic analysis is done to determine the probability of success. Full acceptance here sends the product to its launch. Problems with the economic analysis could lead to either further development or discontinuation. Many products disappear due to poor acceptance during market testing.

Product launch

The new product faces many go, modify, or no-go steps along the way. To make it to the launch, a product must have passed the challenges of concept generation, formulation, process design, package design, scale up, consumer testing, and market testing. Careful coordination is needed at this point to ensure that the product launch is successful. The news media is contacted to gain free publicity that enhances the marketing effort. All members of the development and marketing teams collectively hold their breath hoping that the launch is successful. Probably at no other point in the product's life will it receive more attention and advertising. This period is make-or-break for the product. Some follow-up testing is done to track its performance. On rare occasions, modifications may be made after the launch.

Success or failure

The first few weeks after launch are critical to the product's success. Marketers know that regardless of the marketing campaign, not everyone will try a new product. One consumer classification is the early adapter, like Isaac in Chapter 3. These consumers are usually the first in their neighborhood to try something new. Their opinion is very important as they will pass on their judgment to friends. If the early adapter recommends it, then every time a friend is reminded of the product the temptation to try it will increase. If the early adapter complains, the friend is unlikely to be tempted. As the second wave of friends make their personal judgments the product will take off, build steadily, or die. Milestones are set to see if the product is meeting company expectations.

The most important measure of product success is how much money it makes for the company. Investment costs of a nationally released product are such that it may take 1 to 1.5 years to break even. Sales projections are made as part of the economic analysis to see if the product is outperforming or underperforming. Serious underperformance could result in withdrawal of the product. With this many chances to fail, it is not difficult to understand why so few products are successful.

Some products are very successful early, and it may be difficult to produce enough to meet market demand. As we learned in Chapter 6, consumers are not familiar with the challenges in providing sufficient quantities of product. Inability to meet market demand can do serious damage to the success of a product when frustrated consumers finally give up trying to find it on the shelves. Examples of dramatic increases in new product demand are when catfish was introduced into a fast-food chain in the southeastern United States, and when colorless colas were introduced nationally. Possible reasons for unexpected early demand include the fact that the new product represents an unfulfilled consumer want or need or that it is seen as a novelty that becomes a fad. Fast-food catfish apparently met an unfulfilled need in southern consumers, but the colorless colas were a novelty that turned into a fad that faded as quickly as it started.

The goal of every developer is for the product to have long-term success with worldwide recognition and be a moneymaker for the company for several years. Such products exist, but for every one there are many that appear in the market for a short time only to fade away, and many more that never even enter the marketplace. Some of the successful products may cycle back to the original developer(s) or their successor(s) for consideration of a "new and improved" version or a line extension. Product and process development can be a very rewarding and challenging job for a creative person with strong technical training. Jobs always seem to be available for successful product developers.

Remember this!

- The most important measure of product success is how much money it makes for the company.
- During market testing, image characteristics are more closely evaluated than sensory characteristics.
- The ingredients in a product must be listed in order from the highest to lowest amount present in the product.
- The packaging engineer is responsible for choosing the right material(s), making sure the package and product are compatible, and ensuring that the package serves all of technical requirements of the new product.
- Shelf-life testing is necessary to determine if the new product meets minimum requirements, establish an expiration date, and assess the quality of the product under conditions likely to be seen by the consumer.
- A process developer uses a pilot plant, consisting of miniaturized pieces of equipment that mimic those in the manufacturing plant, to optimize process operations.
- In developing a new formulation, the food scientist must determine which ingredients to use, how much of each ingredient, and the proper order of addition.
- Food scientists must understand the capabilities and limitations of ingredients, processes, and distribution in order to design new products.
- A developer must be careful to keep the product updated while not making changes that will cause the loss of loyal customers.
- Line extensions are new flavors or slight modifications to a product that require changes on the label.
- Every product on supermarket shelves or in fast-food restaurants was developed by a food scientist, probably as part of a research team.

Looking ahead

This chapter introduced us to how food scientists develop new products and processes. Chapter 9 will emphasize the role of food scientists in developing and enforcing regulations as well as in basic research. In the final five chapters, the fundamentals of food science are presented. Chapter 10 presents the chemical basis of food quality. Chapter 11 covers nutritional aspects of quality, and Chapter 12 describes food microbiology. Food engineering is the subject of Chapter 13, and sensory studies (actual tasting of products) are covered in the final chapter.

References

Brenner, J. G. 1999. *The emperors of chocolate: Inside the secret world of Hershey and Mars*. New York: Random House.

D'Antonio, M.. 2006. *Hershey*. New York: Simon & Schuster.

Further reading

Brody, A. L., and J. B. Lord. 2008. *Developing new products for a changing marketplace*, 2nd ed. Boca Raton, FL: CRC Press, Taylor & Francis Group.

Carpenter, R. P., D. H. Lyon, and T. A. Hasdell. 2000. *Guidelines for sensory analysis in food product development and quality control*, 2nd ed. Gaithersburg, MD: Aspen Publishers.

MacFie, H. 2007. *Consumer-led food product development*. Cambridge, U.K.: CRC Press, Woodhead Publishing Limited.

Moskowitz, H. R., S. Porretta, and M. Silcher. 2005. *Concept research in food product design and development*. Ames, IA: Blackwell Publishing.

chapter nine

Government regulation and basic research

Olivia is fortunate enough to have landed a summer internship with the Food Safety and Inspection Service (FSIS) of the U.S. Department of Agriculture (USDA). Every morning she receives a shipment of raw chicken products, kept cool in sealed containers with dry ice, purchased from supermarkets in many parts of the country. Her job is to carefully record the purchase information and company names, perform routine swab tests on the surface of the birds, and transfer those swab samples to special media. These samples are then allowed to incubate at a specific time and temperature to determine if the samples are within specified tolerances for *Salmonella* species. If the bacterial count of the samples from a particular company is out of tolerance, more sampling is done on products from that company and store. Evidence of a continuing problem will result in a visit to the offending plant and a possible voluntary recall of product.

Greg's graduate research project focuses on double emulsions. Emulsions like mayonnaise and milk chocolate are dispersions of one phase in another phase (e.g., oil-in-water or water-in-oil). Double emulsions (e.g., water-in-oil-in-water) provide the ability for controlled release of ingredients. Unfortunately they tend to be highly unstable. Greg is particularly interested in delivering ingredients, such as tannins, which are beneficial to health but have objectionable flavors. He is looking at ways to increase the levels of phenolic compounds in energy drinks without the bitter flavors that accompany them. Traditionally the bitter flavors are masked with sugar or artificial sweeteners, but Greg would like to lower the level of sweeteners. Developing a stable double emulsion is a challenge that excites him.

Olivia and Greg are food scientists in training who are learning some of the aspects of the profession. Although most food scientists work in the food industry, many are employed by regulatory agencies of the government or by universities conducting research. In this chapter we will explore food science opportunities outside the food industry.

Looking back

Previous chapters focused on food issues we deal with in our daily lives and with food products we encounter in the marketplace. Following are some key points that were covered in those chapters that help prepare us for understanding governmental regulation and basic research.

- The ingredients in a product must be listed in order from highest to lowest level present.
- The Hazard Analysis and Critical Control Point (HACCP) system is a means of ensuring microbial safety in a product.
- Food is preserved to make it safer by reducing or eliminating harmful microbes.
- Nutrition labels can help us keep up with the quality of our diets.
- Many governmental agencies are responsible for maintaining the safety of our foods.
- A key to preventing future outbreaks is to understand how each occurs.

Most food scientists work in the food industry. They primarily work in quality management or product or process development, but there are many other functions they can perform. They may be involved in marketing, technical sales, manufacturing, regulatory issues, and packaging. Many food scientists work directly for manufacturers of food products, but others work for ingredient, packaging, and equipment suppliers. Other food scientists work for government agencies and universities. Many government agencies develop and enforce regulations to keep the food supply safe and prevent fraud. Both applied and basic research projects are conducted by food scientists in government, universities, the food industry, and trade organizations. This chapter will focus on governmental regulation and basic research.

Government regulation

We expect the food we eat to be safe. We expect the food at restaurants to be handled under sanitary conditions. We expect the labels on the food products we buy to be accurate. We expect the air we breathe and the water we drink to be free of noxious pollutants. Not all food companies or restaurants have their customers' best interests at heart. Governmental agencies in countries around the world police companies to ensure a safe food supply, that proper guidelines are being practiced, that we are not being defrauded, and that companies are not polluting. These agencies regulate all aspects of the food distribution system from the farm to the table. This section will emphasize agencies in the United States, but similar

Insert 9.1 Web sites that describe some of the current regulations of the food industry.

Federal Trade Commission: www.ftc.gov

Food and Drug Administration: www.fda.gov

Environmental Protection Agency: www.epa.gov

United States Department of Agriculture: www.usda.gov

Occupational Safety and Health Administration: www.osha.gov

activities occur in similar organizations around the globe. Since our food supply is very complex, the laws and regulations governing them are also complex. Foods are regulated by numerous agencies. It may be impossible to describe the regulatory process without using many acronyms for the agencies that regulate the food supply and the laws that govern these actions, so get ready for numerous abbreviations. At the time of this writing, the description in this chapter is the best assessment of the agencies involved in regulating the food industry. For a description of those regulations currently being considered, go to www.regulations.gov or to the Web sites shown in Insert 9.1.

On the farm

Contamination of many whole foods and raw materials starts on the farm. Sources of contamination include the air, soil, fecal material in the soil, insects, birds, humans, other animals, and rainwater. Sources of contamination come in the form of undesirable microorganisms as well as chemicals such as pesticides and pollutants. Although completely preventing environmental contamination is unrealistic, there are guidelines and regulations provided to decrease the chances for the development of safety hazards.

The United States Department of Agricultural (USDA) has developed a series of Good Agricultural Practices (GAPs) and Good Handling Practices (GHPs) for fresh fruits and vegetables. The Food Safety and Inspection Service (FSIS) conducts third-party audits to see if a grower or shipper is complying with these practices and provides proof of compliance to their customers. They perform a comprehensive inspection of facilities: field sanitation, manure and biosolids, packinghouse sanitation, sanitary facilities, ability to trace product back to the growing location, transportation, water supply, and worker health and hygiene. Since many fruits and vegetables are eaten raw, it is important that microbial contamination be minimized because simple washing may not be effective in removing human pathogens.

The U.S. Environmental Protection Agency (EPA) regulates pesticide application and residues through the Food Quality Protection Act (FQPA) and the Federal Insecticide, Fungicide, and Rodenticide Act (FIFRA). The EPA registers all pesticides that are used in the United States and determines the label requirements for each registered pesticide including application conditions on specific crops. Pesticides cannot be applied legally to unauthorized crops or under conditions other than those specified on the label without written consent from the EPA. Until 1996, there were many agencies responsible for pesticide regulation in the United States, sometimes resulting in contradictory requirements, but passage of the FQPA helped clear up inconsistencies and placed the responsibility for setting tolerances with the EPA and the responsibility for enforcement with the U.S. Food and Drug Administration (FDA).

Food manufacturing plants

Manufacturing plants are subject to inspections by many federal agencies. These agencies must be allowed access to the plant without prior notice, and appropriate records must be made available. Federal agencies are charged with the responsibility of ensuring a safe, wholesome food supply, enforcing quality standards, and making sure that the workplace is a safe environment for plant employees. Food manufacturers must keep up with the myriad of rules and regulations.

The FDA of the United States Department of Health and Human Services (HHS) is the primary agency responsible for maintaining a safe, unadulterated food supply in the United States. The key legislation that governs FDA regulations is the Food, Drug, and Cosmetic Act of 1936 as amended. The FDA does not permit any poisonous substances to be added to foods. It carefully regulates food additives, permitting only those that are generally recognized as safe (GRAS) or those that have demonstrated their safety through rigorous animal testing. The Delaney Clause of the Food Additives Amendment to the Food, Drug and Cosmetic Act states that any additive that can induce cancer at any level is not permitted in the food supply. *Adulteration* is the addition of ingredients to a product in an attempt to fool the consumer. Some examples of adulteration are dilution of beverages with water, the use of sugar instead of juice when juice is indicated on the label, or the use of pumpkin to give the impression of eggs present in a product.

The FDA has developed Good Manufacturing Practices (GMPs), which cover the proper operation of food manufacturing plants: sanitary procedures, minimizing safety risks, maintaining microbiological standards, design of facilities, production controls, and proper handling of ingredients and raw materials. For an indication of some of the products regulated by the FDA, see Insert 9.2. Types of things that FDA inspectors look for on a plant inspection of a candy factory include the cleanliness and safety of the raw materials, the cleanliness and condition of molding trays, the condition of scrap candy, as well as the proper use and levels of food and color additives. Inspectors collect samples of molding starch, scrap candy, and possible dangerous items (glass, stones, or pins) that could find their way into the candy. They also sample the finished products for misbranding, deceptive packaging, or filth.

The USDA is responsible for regulation of all meat products that are distributed in interstate commerce. Every carcass of chicken, pork, beef, and other edible animals is visually examined by a USDA-trained and certified inspector. No diseased animals are allowed to be processed or distributed for human food. Any meat processing operation must have at least one certified inspector present any time meat is handled and processed. While the inspectors are employed by the government, the meat processor must pay the USDA for the service. While not mandatory, the USDA will also provide grade inspection for fresh fruits and vegetables for a fee. Many distributors of fresh fruits and vegetables will not purchase them if they have not been graded by USDA-certified inspectors. The USDA is also the agency that certifies meat, poultry, fruits, vegetables, and other items as *organic* using clearly developed standards as to what is organic. The USDA has developed quality standards for many raw items such as corn, oil, soybeans, and wheat. The USDA is also responsible for approving HACCP plans for meat and poultry processors.

The EPA regulates the control of pests such as insects and rodents in processing plants. Insects and rodents are attracted to the plentiful food supply provided by inventories of raw materials. We obviously don't want filth such as droppings, insect parts, or rodent hairs in our finished food products. Likewise, we don't want pesticides or other control agents in them either. Food manufacturers must develop a strategy that ensures a safe, wholesome product.

The Occupational Safety and Health Administration (OSHA) of the United States Department of Labor is responsible for the health and safety of workers. Food manufacturing plants must ensure that the air quality in the plant is healthy, that workers are not exposed to undue physical and chemical risks, that protective equipment is available, that fire regulations are being followed, and that other guidelines are being followed.

The Bureau of Alcohol, Tobacco, and Firearms (BATF) of the United States Department of the Treasury is responsible for regulating any operation that produces alcoholic beverages including beer, wine, and distilled spirits. Although safety is a concern of the BATF, their main responsibility is to ensure that all alcoholic beverages produced are accounted for and that the appropriate taxes are paid to the federal government because the alcohol tax is an important source of government income.

Product labels

The information available on a product label is regulated by the FDA under the Nutrition Labeling and Education Act (NLEA). Each label must have an ingredient statement, a net weight, an address for the processor or distributor, and the name of the product that can be clearly understood by the consumer. An example of a product label is shown in Insert 9.3. Ingredients are listed in descending order of the level (amount) in the product's formulation. They must be identified using a common name. Note that when an ingredient in the product is composed of other ingredients, the other ingredients are enclosed in parentheses. The address must be provided so that the FDA or the consumer can contact the company about a complaint or problem with the product. Many products also have a processing date or expiration date.

Most products require a statement of *Nutrition Facts*. The Nutrition Facts statement must include the serving size and the number of servings per container. All other nutritional information on the label is presented in the amount per serving. The label must provide the amount of total calories and the calories from fat. In addition, the total amount of fat, saturated fat, trans fat, cholesterol, sodium, potassium, total carbohydrate, dietary fiber, sugars, and protein per serving are provided in grams. The %DV (daily value) for the fats, cholesterol, sodium, total carbohydrate, and dietary fiber are based on a 2,000-calorie diet. The %DV for vitamins A and C, calcium, and iron must be presented. Other vitamins and minerals providing at least 5% of the daily value may also be posted in the Nutrition Facts statement.

Insert 9.3 Example of a food label. Can you identify all of the required elements on the label?

Dietary supplements are regulated by the FDA under the Dietary Supplement Health and Education Act (DSHEA). Although producers of food additives must demonstrate the safety and efficacy (ability to perform the specific purpose for which it was added) of the additive, dietary supplements can be added without meeting safety and efficacy requirements. The FDA regulates claims of nutrient content and health benefits based on strict guidelines they have developed.

Packaging

As we learned in Chapter 8, some of the packaging components migrate from the package to the food product. Governmental agencies recognize that incidental contaminants enter foods unavoidably, such as packaging materials, processing aids, and low levels of pesticides. These substances must be demonstrated to be safe in the products at the levels at which they would be consumed. Before a new package is approved, extensive testing is conducted to demonstrate that no toxic compounds migrate into the product from the package. Packaging for processed, formulated, and prepared foods is regulated by the FDA, whereas packaging for raw meat and poultry products is regulated by the USDA.

Product recalls

Although neither the FDA nor the USDA has the authority to recall unsafe foods, when an unsafe situation arises, the appropriate agency will request a voluntary recall of the product by the manufacturing company. A Class I recall is one in which a clear danger to the public has been identified. A Class II recall involves less serious health hazards in which people are likely to get sick and to recover. A Class III recall involves a violation of agency rules, but the public is not endangered. Recalls frequently involve the presence of dangerous levels of pathogenic microorganisms, the presence of a toxin such as a cleaning fluid in a product, or the presence of a potent allergen like peanuts in a product that is not properly labeled. When a company is judged to be out of compliance, the FDA sends it a warning letter advising the company of the seriousness of the violation. If that does not work, the FDA can issue a press release informing the public of the problem. If the bad publicity does not work, the FDA can refer the matter to the Department of Justice, which can seize the product in question without prior notice. The FDA can then seek an injunction to prevent the manufacturer from continuing to produce the product. The Department of Justice can then file criminal charges against the violators.

Transportation, distribution, purchase, and consumption

Once a raw material or finished food product is being transported, it is subject to more regulation. Any truck, train, or ship that transports raw agricultural crops or livestock must be licensed and insured through the Federal Motor Carrier Safety Administration (FMCSA) of the U.S. Department of Transportation (DOT). The USDA certifies the weight measurements of grains and the safety of shipping containers.

Although we think of supermarkets mainly as places where we buy food, many modern supermarkets also prepare foods. In the deli, bakery, sushi bar, meat counter, produce department, and in-store restaurant, fresh foods are being handled and prepared. State agencies regulate the safety of most food items in a supermarket and other retail distribution points. Restaurants and other food service institutions are inspected by local health boards. A supermarket generally is not considered a restaurant until it installs seats for on-premise consumption of its food.

Regulation of the food industry doesn't stop with the product. The Federal Trade Commission (FTC) is responsible for making sure the advertisements for food products are not misleading. If products are compared to a competitor's product, there must be clear proof of the claim.

At this point you may be bewildered at the seemingly endless string of letters for agencies, legislative acts, and recommended practices. If you are up to it, try to see how many acronyms you can identify in Insert 9.4.

Basic research

Another function of a food scientist is basic research. The terms *basic* and *applied* research are relative. A fundamental chemist or microbiologist might consider basic food science to be applied chemistry or microbiology. In the pure sciences, basic research is generally regarded as work that has no direct application to practical problems, whereas applied research focuses on solving specific problems. In food science, applied research is directed at specific problems; basic research focuses on general problems related to food. Both basic and applied research are conducted at universities, in governmental laboratories, and in large food companies.

Insert 9.4 Now that you have mastered the regulatory process, can you identify these acronyms?

BATF	DOT	DSHEA	EPA
FDA	FIFRA	FMSCA	FQPA
FSIS	FTC	GAPs	GHPs
GMPs	GRAS	HACCP	HHS
	NLEA	OSHA	USDA

University and government researchers have an obligation to publish their results in scientific journals. Industry scientists, however, may keep their results secret to provide long-term economic advantage to their company. Food science research might be considered "use-inspired" as described by Louis Pasteur (Stokes, 1997). He looked for solutions to common problems using basic science. While food science was not a recognized area of research at the time, Pasteur's discoveries, particularly his contributions to the germ theory and pasteurization, make him one of the most important food scientists who ever lived.

As science moves forward, however, fewer breakthroughs are the result of a single scientist focusing on a single research problem. As the problems facing the world become more complex, scientists are working together to approach problems from different disciplinary perspectives. For example, food microbiologists work with food engineers to develop more effective ways of killing microbes to better preserve foods. Food chemists and sensory specialists may also be needed to assess the effects of such processes on the quality of a finished product. Food scientists work with nutritionists, biochemists, molecular biologists, physicists, and many other types of scientists to help better understand the science behind our food. The remainder of this chapter is highly technical, as food scientists are at the cutting edge of some very exciting scientific and technological breakthroughs.

Fundamental physical and chemical properties

In food chemistry, structure–function relationships are of critical interest. This subject will be detailed in the next chapter, In some cases, molecules with similar structures have similar properties, but other compounds with similar structures function quite differently. Food chemists are trying to find ways to relate chemical structure to functional properties by the following progression:

- chemical structure determines physical properties,
- physical properties influence chemical properties,
- chemical properties relate to kinetics, which involves how likely it is that two molecules will react with each other,
- kinetics determine functional properties of molecules,
- functional properties of molecules in an ingredient determine the effectiveness of that ingredient in a food product, and
- effectiveness of ingredients to perform their function affects final product quality.

The physical state of a food product plays a critical role in whether a chemical reaction can take place and how fast it can occur. For molecules to react, they must be able to touch each other. Molecules within most foods are not stationary; they vibrate and move around in what we term a *phase*. They can also be transported from one phase to another. Examples of phases are the states of matter (gas, liquid, and solid), as well as water-based and oil-based phases. In living organisms, membranes separate compartments within a cell providing limited transfer of molecules from one compartment to another. As the temperature increases within a food product, molecular motion speeds up, more molecules bump into each other, and chemical reactions are more likely to occur. As a product cools, molecular motion slows and so do molecular interactions and chemical reactions.

Chemical reactions, such as development of succulent flavors during cooking or the golden brown color of toast, can be beneficial. Other chemical reactions, such as the separation of fat from chocolate, leaving a white coating on the surface, or the loss of vitamins during long-term storage of a fresh vegetable, are detrimental. Molecules in a gaseous state are more able to move and interact than in a liquid. Likewise, molecular motion is more restricted in a thick liquid than in a watery one, and in a solid than in a liquid. Some solid foods are light and fluffy like bread; others are rubbery like a steak, and still others are glassy like ice cream, which is frozen solid. When in the glassy state, all molecular motion ceases and so do all chemical reactions. Food scientists are developing a better understanding of the transitions to rubbery and glassy states and the implications of these transitions for preserving the highest quality of foods.

Long used in the drug industry, microencapsulation has some interesting potential uses for food products. Microencapsulation allows the controlled release of desired compounds. Bacterial cultures can be encapsulated and used to induce more efficient fermentations. Acidulants can be released to alter the pH and thus the properties of the food. Flavorants can be encapsulated to provide a prolonged flavor experience during eating or they may extend the life of the flavor, and thus the shelf life, of a formulated product. Much of the technology is currently available for microencapsulation, but the expense of the technique decreases its value (Anal, 2007).

Cyclodextrins and hydrocolloids are food ingredients that can be used in encapsulation. Cyclodextrins have an interesting molecular structure (see Insert 9.5, Szente, 2004). By binding other molecules, they can slow molecular interactions and provide controlled release. They can be used to stabilize ingredients in foods such as colors, flavors, unsaturated fatty acids, and vitamins. They can protect compounds from oxidation and heat degradation. Hydrocolloids are carbohydrates that are not absorbed by the body but are useful ingredients as edible coatings, microencapsulators, thickeners, flavor-delivery systems, and texture modifiers. The small variations in molecular structure and the effect the variations have on the physical properties of these molecules are being studied. Physical properties are also important in maintaining surface tension at interfaces

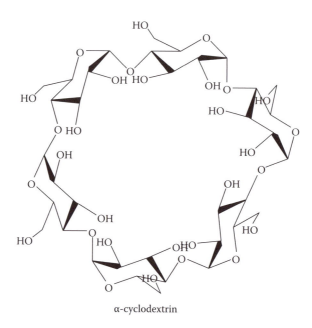

α-cyclodextrin

Insert 9.5 Chemical structure of α-cyclodextrin, a multipurpose compound that can serve as a bulking agent and a stabilizer.

in dispersed systems such as emulsions and foams. Sophisticated techniques such as *atomic force microscopy* and *scanning near-field optical microscopy* are used to determine the interrelationships among the structure of proteins, the physical structure of the food matrix, and the stability of food dispersions. These techniques can also be used to study starches, how they bind to form granules, and how they interact with the enzymes that break down these granules (Zuniga, 2008).

Another method of slow release is the formation of double emulsions. Double emulsions can be in the form of oil-water-in-oil or water-in-oil-in-water. Although double emulsions were described over eighty years ago, it has not been until recently that they have been applied to food products. They are tricky to form as Greg is finding out, but the emulsification of membranes has simplified the process. A double emulsion is a two-step process: forming a single emulsion first and then converting it to a double emulsion. Surfactants are needed to lower surface tension. A double emulsion requires more than one surfactant to stabilize the dispersion. They are primarily being used for drug delivery, but the applications for food systems include the release of flavorants or delivery of nutrients (vander Graff, 2005).

Microbiology

Food microbiology involves the isolation, identification, and sometimes classification of microorganisms in foods, which we will explore further in Chapter 12. Basic food microbiology explores the molecular mechanisms of pathogenesis, which include questions such as the following:

- Why are some microorganisms pathogenic, but most are not?
- Why are some microorganisms pathogenic to some animal species, like humans, and not to other species?
- What is the molecular basis of pathogenicity?
- How can we rapidly determine the presence or absence of food-borne pathogens in a food product?

Probiotics are microorganisms that when digested, have beneficial aspects in the body. *Prebiotics* are nondigestible substances that stimulate the growth of beneficial microorganisms in our intestines (Grootaert, 2007). Probiotics offer many benefits: improved digestion, competition with harmful bacteria, reduced gas production, stimulated immune response, and increased vitamin synthesis. Research in this area is directed at determining which species of microorganisms can serve as probiotics, which compounds can serve as prebiotics, and how they survive the digestive process. The extent of beneficial effects and the potential detrimental effects also must be determined.

Because many consumers wish to see less or no preservatives in their foods, many food scientists have been looking at ways of preserving foods with fermentations. Many fermented foods are consumed around the world, but have not been consumed in large quantities in North America. Research is being conducted to look at the microbial ecology of novel fermentation microbes. Cereals and grains provide the starting material for many promising fermented foods. Knowing and understanding the natural microflora associated with a grain and how these fermentations can be enhanced can lead to better control of these processes (Hammes, 2005).

Other exciting areas in food microbiology include developing a mechanistic understanding of the effects of food processes that do control microorganisms in foods without heat; developing a better understanding of the antimicrobial action of preservatives; identifying new pathogens and spoilage microorganisms and methods for their control, particularly parasites and viruses; investigating the implications of antibiotic resistance in pathogens; understanding microbial attachment to foods and equipment, particularly in biofilms; and protecting our crops from agroterrorism.

Molecular biology

Advances have been made in genetic modification of food materials from both plants and animals. As more genomes are mapped and organisms transformed, there will be more opportunities for better and safer food products. Genetically modified organisms (GMOs) can help reduce our reliance on pesticides, produce oilseeds with more unsaturated fats, and slow food spoilage. Research has been quite successful when a characteristic can be manipulated by a single gene. Many characteristics, however, are tied to multiple genes. Future research will look at how to improve foods requiring modification of multiple genes and unexpected changes to the whole organism. Research is also focusing on the detection of GMO ingredients in foods. Although most food scientists believe GMOs show great promise for improving safety, nutritional value, and sensory quality of foods, some consumer groups consider foods with GMO ingredients to be contaminated and unworthy of consumption (Evenson, 2004).

Proteomics involves identifying proteins expressed by a genome. The field also encompasses everything involved in understanding the structure, location, and function of protein in the cell. Proteomics helps the food scientist understand interactions of proteins in foods with other proteins, as well as other components in the food structure. It provides an understanding of how proteins were formed in the raw products and allows prediction of how they will be affected by food processes. Structural changes of proteins can be used as markers of the degree of processing to make sure that a product has been adequately processed to ensure product safety. Proteomics may also be useful in understanding the basis of food allergens and how to decrease allergenicity of food ingredients (Han, 2008).

Bioinformatics applies mathematical and statistical techniques to understand biological processes. Advances in bioinformatics are aimed at the areas of molecular biology with particular emphasis on gene expression and protein structure. Studies are focused on the structure–function relationships of proteins in foods and food-associated pathogens. Bioinformatics can also be useful in understanding protein interactions with other food components, allergenic proteins, and genetic marking for GMOs.

Nutritional properties

New functional foods hit the supermarket weekly to treat a wide range of diseases: obesity, diabetes, cancer, heart disease, and other diseases of civilization. Research is being conducted to determine how molecules that act as antioxidants and anticarcinogens work to slow or prevent diseases. Phenolic compounds, like anthocyanin pigments and tannins, are of particular interest. Scientists are studying the molecular mechanisms of these diseases and compounds, the doses needed to produce a positive outcome, whether higher levels can lead to toxicity or harmful interactions with other beneficial compounds such as vitamins, and how food processing affects the amount and potency of these compounds. Eventually, the sensory properties of these molecules must be studied to make sure that they are palatable at the effective dose and will be consumed. By understanding the specific nutritional needs and sensory preferences of the consumer, foods will be designed to meet those needs (Ronteltap, 2007).

As we become more aware of the nutritional properties of bioactive compounds and develop better profiles of the health and nutritional status of individual consumers, we will have the ability to custom design food products to meet the needs of an individual consumer. Diagnostic techniques could also be expanded to determine the preferences of each consumer so these nutritionally perfect foods could be tailored to match personal preference. Although such a world sounds ideal, it tends to ignore the social aspects of food choice, as described in Chapter 3, and could lead to other problems.

Postharvest physiology

Fruits and vegetables are living organisms. Food scientists and postharvest physiologists monitor the physiological reactions that fresh fruits and vegetables undergo during handling and storage (Hertog, 2007). *Hormesis* is the application of low doses of stress to induce a protection response. The exposure of fruit to low doses of ultraviolet light delays ripening and stimulates production of compounds that slow the growth of molds. Likewise, short-term exposure to heat protects susceptible fruits to chilling injury. Such processes are being studied to determine if they can be used to prevent waste by slowing senescence and decay during storage of fresh fruits to extend shelf life. Quality changes must be carefully observed, however, to make sure that the protective effect is not associated with unexpected losses in flavor or texture.

Fresh-cut fruits and vegetables are a convenient way to obtain dietary fiber. One of the problems with fresh-cut fruits is that they lose aroma rapidly after cutting. One way to maintain the fresh flavor of these cut fruits is to add higher levels of the natural, chemical precursors for the aromatic compounds in the atmosphere of the item. The enzymes in the fruit convert these precursors into the flavor compounds we associate with that fruit. Such a process is difficult because if too much of the compounds are added to some fruits, it may cause the flavor to become objectionable if they accumulate at higher levels (Lanciotti, 2004). Also, some consumers might object to adding compounds to cut fruits, an action that is not really "natural."

Food processing

Some of the innovative new processes to preserve foods that are being studied by food engineers were described in Chapter 4. Heat exchange has been the primary means of preserving food in such processes as canning, freezing, drying, and cooking. Heat is very effective in killing microbes and inactivating enzymes, but it also damages vitamins and flavor. Freezing is gentler on vitamins and flavor but requires energy to maintain the product in the frozen state. New techniques are being evaluated that are as effective as heat on microbes and enzymes but not as damaging to vitamins and flavor. One such preservation technology is the use of ultrasound. Ultrasound is currently being used in many processes, including those to clean the surfaces of foods, tenderize meats, and form emulsions. Other applications include the inactivation of spoilage and pathogenic microbes, improving the flow of liquids during filtration, and evaluating the sensory quality of food products. Unfortunately, ultrasound techniques are energy intensive, making them very expensive to use.

Current food processes usually require large amounts of water and chemical solvents resulting in large amounts of pollutants. Research is underway to develop processing methods that require less water without reducing the safety or quality of the product. Food processing also consumes energy and produces greenhouse gases. For a perspective of energy use in the food industry, see Insert 9.6. Note that the emphasis on reducing food miles will probably not make a big difference in either energy use or greenhouse gas production relative to other aspects of the food distribution system. Use of enzymes and "greener technology" shows promise for decreasing the amount of solvents used in current processing methods. An exciting development is the use of enzymes to remove toxic

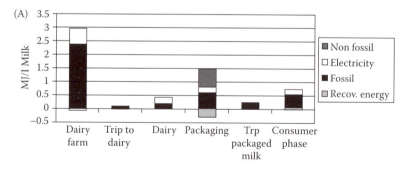

Insert 9.6 Use of energy (A) and production of greenhouse gases (B) during the life cycle of milk from farm to the consumer. (From C. Cederberg, "Life Cycle Assessment of Animal Products," in *Environmentally-Friendly Food Processing*, ed. B. Mattsson and U. Sonesson (Cambridge, U.K.: Woodhead Publishing Limited, 2003), chap. 5. With permission.)

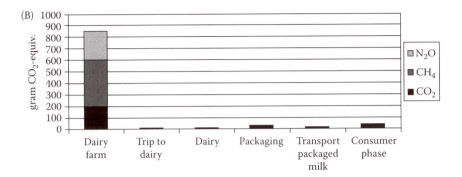

Figure 9.6 (Continued.)

compounds, such as reactive oxygen species, which are formed during processing, storage, and cooking of food products.

Toxicology

Acrylamide has been identified as a carcinogen that occurs in many food products, particularly ones that have been heated to high temperatures. It is not clear how dangerous this compound is, how it is formed, or how it can be minimized in foods. Research is progressing in all these areas as well as in identifying other dangerous compounds present in our foods (Skog, 2006).

Food allergies range from minor irritants to those that can lead to anaphylactic shock and death. Research is being conducted to develop a better understanding of specific chemicals within a food that lead to a true allergic reaction and possible mechanisms of these responses in the human body. When a specific chemical has been identified with an allergic response, foods can be formulated to replace the offending food or ingredient containing the allergen. Wheat-gluten intolerance can cause allergen responses leading to celiac disease. Development of gluten-free products is challenging because wheat gluten is responsible for the light textures of bread and bread products. Combinations of hydrocolloids and nonwheat starches can result in promising replacements. Another approach to this problem is to determine the portion of the gluten molecule that is responsible for the allergenic response and modify the gluten molecule, either chemically or genetically, to eliminate allergenicity while maintaining functionality.

Nanotechnology and other frontiers

The new field of nanotechnology suggests that food could be constructed one molecule at a time. While food nanotechnology is currently in the

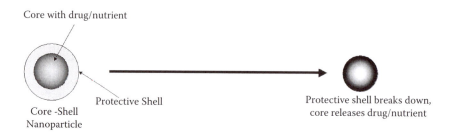

Core with drug/nutrient

Protective Shell

Core -Shell
Nanoparticle

Protective shell breaks down,
core releases drug/nutrient

Insert 9.7 Diagram of how nanoparticles can be used to provide controlled release of drugs or nutrients. (Diagram provided by Dr. Jake Mulligan.)

realm of science fiction, food proteomics or food bioinformatics would have been considered impractical just a few years ago. Nanoparticles can be tailored for specific ingredient applications such as nutrient delivery (see Insert 9.7). Nanoscale reactions may revolutionize food processing operations, and nanobased sensors and tracers could be used to help protect foods from bioterrorists. This field is so new that it will be important to develop appropriate regulations that will both protect consumers and allow us to take advantage of a new technology (Chan, 2007).

As other basic sciences advance, food scientists will be paying attention. Detailed statistical analysis of images such as those provided by x-rays, CAT (computed axial tomography) scans, and colonoscopies have given medical doctors important tools in assessing health and diagnosing diseases. Likewise, similar types of images can give food scientists a peak inside foods to determine molecular interactions and hidden quality defects.

Remember this!

- New processing techniques are being evaluated that are as effective as heat on microbes and enzymes, but not as damaging to vitamins and flavor.
- As we become more aware of the nutritional properties of bio-active compounds and develop better profiles of the health and nutritional status of individual consumers, we will have the ability to design specific food products to meet the needs of an individual consumer.
- Genetically modified organisms can help reduce our reliance on pesticides, produce oilseeds with more unsaturated fats, and slow food spoilage.
- The physical state of a food product plays a critical role in whether a chemical reaction can take place and how fast it can occur.
- In food science, applied research is directed at specific problems; basic research focuses on general problems related to food.

- Governmental agencies in countries around the world police companies to ensure a safe food supply, that proper guidelines are being practiced, that we are not being defrauded, and that companies are not polluting.
- Contamination of many whole foods and raw materials starts on the farm.
- Manufacturing plants are subject to inspections by many federal agencies.
- The information available on a product label is regulated by the FDA under the Nutrition Labeling and Education Act.
- Governmental agencies recognize that incidental contaminants, such as packaging materials, processing aids, and low levels of pesticides, may enter foods unavoidably.
- When an unsafe situation arises, the appropriate agency will request a voluntary recall of the product by the manufacturing company.
- Restaurants and other food service institutions are inspected by local health boards.

Looking ahead

This chapter focused on the roles of food scientists as governmental regulators and researchers. The final five chapters of the book provide more depth in the primary subdisciplines of food science, which form the basis for further advances in the field.

Answers to chapter questions

Insert 9.3

An ingredient statement, a net weight, and an address of the processor or distributor.

References

Anal, A. K., and H. Singh. 2007. Recent advances in microencapsulation of probiotics for industrial applications and targeted delivery. *Trends in Food Science and Technology* 18: 240.

Cederberg, C. 2003. Life cycle assessment of animal products. In *Environmentally-friendly food processing*, ed. B. Mattsson and U. Sonesson. Cambridge, U.K.: Woodhead Publishing Limited, chap. 5.

Chau, C-F., S-H. Wu, and G-C. Yen. 2007. The development of regulations for food nanotechnology. *Trends in Food Science and Technology* 18: 269.

Evenson, R. E., and V. Santaniello. 2004. *Consumer acceptance of genetically modified foods*. Cambridge, MA: CABI Publishing.

Grootaert, C., J. A. Delcour, C. M. Courtin, W. F. Broekaert, W. Verstraete, and T. van de Wiele. 2007. Microbial metabolism and prebiotic potency of arabinoxylan oligosaccharides in the human intestine. *Trends in Food Science and Technology* 18: 64.

Hammes, W. P., M. J. Brandt, K. L. Francis, J. Rosenheim, M. F. H. Seitter, and S. A. Vogelmann. 2005. Microbial ecology of cereal fermentations. *Trends in Food Science and Technology* 16: 4.

Han, J-Z. and Y-B. Wang. 2008. Proteomics: Present and future in food science and technology. *Trends in Food Science and Technology* 19: 26.

Hertog, M. L. A. T. M., J. Lammertyn, B. DeKetelaere, N. Scheerlinck, and B. Nicolai. 2007. Managing quality variance in the postharvest food chain. *Trends in Food Science and Technology* 18: 320.

Lanciotti, R., A. Gianotti, F. Patrignani, N. Belletti, M. E. Guerzoni, and F. Gardini. 2004. Use of natural aroma compounds to improve shelf-life and safety of minimally processed fruits. *Trends in Food Science and Technology* 15: 201.

Ronteltap, A., J. C. M. van Trijp, and R. J. Renes. 2007. Expert views on critical success and failure factors for nutrigenomics. *Trends in Food Science and Technology* 18: 189.

Skog, K., and J. Alexander. 2006. *Acrylamide and other hazardous compounds in heat-treated foods.* Cambridge, U.K.: Woodhead Publishing Limited.

Stokes, D. E. 1997. *Pasteur's quadrant: Basic science and technological innovation.* Harrisburg, VA: R. R. Donneley and Sons.

Szente, L., and J. Szejtli. 2004. Cyclodextrins and food ingredients. *Trends in Food Science and Technology* 15: 137.

van der Graaf, S., C. G. P. H. Schroën, and R. M. Boom. 2005. Preparation of double emulsions by membrane emulsification: A review. *Journal of Membrane Science* 251: 7–15.

Zuniga, R. N., and J. M. Aguilera. 2008. Aerated food gels: Fabrication and potential applications. *Trends in Food Science and Technology* 19: 176.

Further reading

Corredig, M., and M. Alexander. 2008. Food emulsions studied by DWS: Recent advances. *Trends in Food Science and Technology* 19: 67.

Doyle, M. P., and L. R. Beuchat. 2007. *Food microbiology: Fundamentals and frontiers,* 3rd ed. Washington, DC: ASM Press.

Eskin, N. A. M., and S. Tamir. 2007. *Dictionary of nutraceuticals and functional foods.* Boca Raton, FL: CRC Press, Taylor & Francis Group.

Goodman, R. E., S. L. Hefle, S. L. Taylor, and R. van Ree. 2005. Assessing genetically modified crops to minimize risk of food allergy: A review. *International Archives in Allergy and Immunology* 137: 153.

Gowan, A. A., and C. P. O'Donnell. 2007. Hyperspectral imaging—An emerging process analytical tool for food safety and quality control. *Trends in Food Science and Technology* 18: 590.

Katz, L. B., ed. 2005. *Agroterrorism: Another domino.* New York: Nova Science Publishers, Inc.

Maleki. S. J., A. W. Burks, and R. M. Helm. 2006. *Food allergy.* Washington, DC: ASM Press.

Mine, Y., and F. Shahidi. 2006. *Nutraceutical proteins and peptides in health and disease.* Boca Raton, FL: CRC Press, Taylor & Francis Group.

Nussinovitch, A. 2003. *Water-soluble polymer applications in foods.* Oxford, U.K.: Blackwell Publishing.

Pina, K. R., and W. L. Pines. 1998. *A practical guide to food and drug law and regulation.* Washington, DC: Food and Drug Law Institute.

Rodríguez-Lázaro, D., B. Lombard, H. Smith, A. Rzezutka, M. D'Agostino, R. Helmuth, A. Schroeter, B. Malorny, A. Miko, B. Guerra, J. Davison, A. Kobilinsky, M. Hernández, Y. Bertheau, and N. Cook. 2007. Trends in analytical methodology in food safety and quality: Monitoring microorganisms and genetically modified organisms. *Trends in Food Science and Technology* 18: 306.

Scarafoni, A., C. Magni, and M. Duranti. 2007. Molecular nutraceutics as a mean to investigate the positive effects of legume seed proteins on human health. *Trends in Food Science and Technology* 18: 454.

Shetty, K., G. Paliyath, A. Pometto, and R. E. Levin. 2006. *Food biotechnology.* Boca Raton, FL: CRC Press, Taylor & Francis Group.

Shi, J. 2006. *Functional ingredients and nutraceuticals: Processing technologies.* Boca Raton, FL: CRC Press, Taylor & Francis Group.

Sun, D-W. 2006. *Thermal food processing: New technologies and quality issues.* Boca Raton, FL: CRC Press, Taylor & Francis Group.

Vetter, J. L. 1996. *Food laws and regulations.* Manhattan, KS: American Baking Institute.

Weiss, J., P. Takhistov, and D. J. McClements. 2006. Functional materials in food nanotechnology. *Food Technology* 71(9): R107.

Scientific principles

chapter ten

Food chemistry

Ever since his fateful ride back from New Orleans with Brian, Carlos became interested in what caused people to get sick. As a chemistry major, Carlos was more interested in what toxic chemicals the microbes made than in the microbes themselves. He was also interested in all the chemicals that were added to foods, like preservatives, but he wasn't so sure that he wanted all those chemicals in his mouth. While he was looking for an elective course online during open registration, he came across one called Food Chemistry. It sounded interesting, like it might be a course where he could apply his basic knowledge in chemistry to the real world. He took the course, loved it, and found out that he had much more to learn about the chemicals in his food. Much of what he learned changed his understanding of and appreciation for chemistry in everyday life.

Hannah, by now a committed vegetarian, decided to major in nutrition because she wanted to learn how to eat healthier. We can imagine her shock when she learned that nutrition is mainly about chemicals and how they act in the body. It really doesn't have much to do with food at all. There was in-depth study of digestion and all the metabolic pathways that occur in food components once they are broken down, but healthy eating wasn't exactly the main topic of discussion. When her professors talked about a vegetarian diet, they had some good things to say but also a lot of negative things as well. Mostly they talked about chemicals. They talked about large chemical compounds like carbs (only they called them carbohydrates), fats (lipids), and proteins. She learned that vitamins are actually organic chemicals, and that minerals are elements straight from the Periodic Table she had memorized (and quickly forgotten) in first-year chemistry. She wanted to know more about nutrients, but she wanted to know about the nutrients in the types of foods she ate regularly.

Kyle is a food science major and is learning more about all the chemicals in his foods. He learned about the toxins and preservatives Carlos was into, as well as about the nutrients that Hannah was studying. He was fascinated by the many other chemicals in foods and how they all work together to produce the quality of a food. He found out that foods are made of ingredients and ingredients are made of chemicals. Each ingredient has a function in the food, but it is actually the chemical components that give an ingredient its function. The more he learned about pigments and stabilizers, the more he wanted to learn about texturizers and humectants. He thought that the taste of a food was simple, but now he knows that flavor combines taste and aroma. He learned that tens and even hundreds of chemical compounds could be contributing to the aroma of foods. Upon completion of his degree, Kyle plans to go to graduate school to learn more about food chemistry and then to work in the food industry to develop new food products.

Kyle, Hannah, and Carlos are learning that the basic components of foods are chemicals. Chemicals are not just nasty-smelling substances we encounter in chemistry labs and then spend the rest of our lives avoiding. Chemicals are the basis of life, and food is the delivery system that provides the chemicals we need to our bodies. This chapter introduces us to the types of chemicals in our foods, why they are there, and what they do.

Looking back

Previous chapters focused on food issues we deal with in our daily lives, with the types of food products we encounter in the marketplace, and with activities that food scientists perform. The most basic science we need to know as it relates to foods is chemistry. The following key points were covered in previous chapters and help prepare us for a basic understanding of the chemical aspects of foods.

- In developing a new formulation, the food scientist must determine which ingredients to use, how much of each ingredient, and the proper order of addition.
- Individual fat replacers cannot perform all functions of lipids.
- Formulated foods are products that involve the mixing of ingredients.
- Food processing and preservation increases the shelf stability of a raw material but usually decreases affordability, nutrition, and quality of the product.
- The main function of a package is to prevent microbial or chemical contamination of a processed food product.
- Nutrition labels provide information to assist us in making good decisions to improve the quality of our diets.
- Prolonged fasting can result in severe nutritional deficiencies.
- Nutraceuticals are foods specifically designed to act as drugs.
- Preservatives are food additives that prevent or slow spoilage.
- Natural chemicals and products are not necessarily superior to artificial chemicals and processed products.

Food chemistry

By now it should be apparent that everything we put into our mouths is composed of chemicals. Everything good in our foods is composed of chemicals, everything bad in our foods is composed of chemicals, and everything else is composed of chemicals. For people not educated in nutrition, physiology, or food science, the idea of chemicals and food being used together seems inappropriate, as Hannah discovered. Our bodies, however, are complex organisms that are composed of chemicals and need chemical replacements to maintain health. Food scientists believe that chemicals are not to be feared but rather to be understood.

Chemicals in different types of food

Our first thought about chemicals in our food usually involves preservatives and additives. The two most widely used chemical preservatives, sugar (sucrose) and salt (sodium chloride), are not usually considered chemicals. Other common preservatives, such as potassium sorbate or citric acid, are present in foods in much lower quantities. Food additives perform a wide variety of functions, including natural and synthetic colorants and flavorants. Other additives function as acidulants, antibiotics, anticaking agents, antioxidants, bulking agents, emulsifiers, fat replacers, sequestrants, stabilizers, sweeteners, texturizers, and thickeners. Many additives fit into more than one of these categories. Each of these categories will be discussed later in the chapter.

All nutrients are chemicals. Proteins, carbohydrates, and fats are chemicals that provide calories. Vitamins and minerals are chemicals that are needed for proper body metabolism. Dietary fiber is a complex mix of chemicals that aids digestion and binds toxins. Even the most important nutrients of all, water and oxygen, are chemicals. Nutrients also represent a food additive category that was not mentioned in the previous paragraph. Next time we see a weird chemical-sounding name like ferric orthophosphate or pyridoxine hydrochloride, we can rest easier knowing that these are simply iron and vitamin B_6.

All toxins in foods are chemicals. Most toxins in foods are natural and are made by microbes. Toxins from bacteria are large protein molecules. Mycotoxins from molds are much smaller (see Insert 10.1). Many natural components in foods are toxic chemicals. Fortunately, most of these are present at such low levels in foods that they pose no threat to most humans. Unfortunately, other toxic compounds, such as pesticides, find their way into foods. These compounds are strictly regulated by the government to protect the consumer.

When we eat food from plants like fruits or vegetables or from animals like chicken or beef, we are actually eating food tissues. The chemicals that were instrumental in maintaining the life of that plant or animal are being recycled in our bodies as we consume these tissues.

Aflatoxin B1   Patulin

Insert 10.1 Chemical structures of mycotoxins in foods. Mycotoxins are produced by molds.

Plant tissues

The chemicals in some foods we derive from plants go far beyond the pesticides that normally come to mind. Solanine in green potatoes and amygdalin (a compound that releases cyanide) in apricot and peach pits are among the many naturally occurring toxins. Organic acids like citric acid in citrus fruits and benzoic acid in cranberry products are effective natural preservatives. Caffeine is a bitter chemical extracted from the coffee bean and cola nut to stimulate our senses. Plant pigments are chemicals. Pigments like chlorophyll make lettuce green, lycopene makes tomatoes red, and anthocyanins make grapes purple. Kyle was surprised to learn that over 200 chemical compounds contribute to the flavor of fresh-squeezed orange juice. Vitamins A and C are important nutrients found in fruits and vegetables. Grains provide carbohydrates; soybeans and peanuts are good sources of protein and oil. Papain is an enzyme that can be extracted from papayas and used as a natural meat tenderizer. The most abundant chemical in plants, however, is water, which accounts for up to 96 percent of volume in lettuce and cucumber.

We don't eat whole plants, but we do eat raw plant parts, processed plant parts, or ingredients derived from plant parts. Harvested plant products like fruits, vegetables, grains, and oilseeds are living tissues. As a result, these tissues are undergoing chemical and biochemical changes due to the metabolic activities driven primarily by enzymes. The genetic differences between species (wheat versus rye) or variety (Red Delicious versus Gala apples) will also affect these changes. Chemical reactions can also be affected by the plant part (organ) like a broccoli head (flower), celery stalk (stem), corn (seed), and carrot (root). Microbes on the plant can lead to decay by breaking down the chemical components of cell walls and membranes.

Physiological changes in metabolic activities occur in a fruit or vegetable from initial development to refrigerated storage. During growth and development of plants, rapid changes are occurring as parts of the plant change from inedible to edible. One of the most dramatic changes occurs during fruit ripening. After pollination of the flower, a tiny fruit forms. This fruit grows in size as cells divide rapidly and minerals are taken in through the roots and stems. Then a series of complex metabolic reactions occur to produce

- the pigments that provide its color,
- volatile compounds for the aromatic component of flavor,
- conversion of starch to sugar to provide sweetness,
- breakdown of cell walls to induce softening,
- synthesis of important vitamins, and
- the production of many other compounds as well.

Another dramatic physiological event is the separation of the organ from the parent plant. Chemical signals from the detached plant organ, particularly from the simple gas ethylene, cause major changes in enzymatic activity, leading to a cascade of chemical reactions. The mechanical damage caused by cutting, impact, and vibration increases these signals, as well as breaking apart the delicate membranes around and inside each cell that maintain the normal metabolic reactions. Too much mechanical damage can cause visible bruising and lower the quality of the produce.

Physiological and microbial changes in the plant organ are affected by environmental conditions like temperature and relative humidity (RH). Generally speaking, the lower the temperature the slower the chemical reactions in the cell and the longer the shelf life. Most fruits and vegetables have a range of RH that is best for that item. If the RH is too low, it will wilt or shrivel; if the RH is too high, microbes can grow more rapidly and the item will deteriorate more quickly. Each day that a fresh fruit or vegetable is stored, there will be a loss in sensory and nutritional quality.

During the processing of plant organs there are several unit operations that cause changes in the chemical composition of that fruit or vegetable. Washing or soaking operations remove microbes; however, the water can remove soluble minerals and vitamins reducing the nutritional value of the product. This type of nutrient loss is called *leaching*. Cutting and slicing operations can increase enzymatic activity, causing unacceptable texture, off-flavors, product discoloration, and a loss of nutrients.

Heating bleaches out pigments, softens cell walls, and changes flavor compounds. If we compare a raw green bean with a canned green bean in our minds, we can see the difference between the two. Heating is also very destructive to vitamins and minerals. Heating inactivates the enzymes in the product that cause it to deteriorate, and therefore canned products are more stable than fresh products. Drying is usually a more gentle process than heating, although some heating may be involved. Some of the same chemical reactions may occur in drying as in heating, but the biggest change is in the loss of water. As water is lost, other chemicals become more concentrated. Water loss causes an increase in some reactions, like browning of sugars and oxidation of lipids, but a decrease in others such as enzymatic reactions because these reactions require water to function effectively. Freezing is a more gentle food process than heating and drying, leading to much slower chemical reactions during frozen storage. If a frozen food is held for a long time, however, even small changes can add up to quality losses.

Animal tissues

Food from animals is also composed of chemicals. Most meat is muscle tissue. Muscle tissue contains fat, protein, vitamins, minerals, and hormones, but very little carbohydrates. *Heme* is the red pigment of red meats and is attached to a protein called *hemoglobin*. Heme is rich in iron. Meat flavor is produced by the chemical reactions that are induced by heating. Heat changes chemicals already present (called *precursors*) into volatile compounds that we associate with the flavor of cooked meat. The type of cooking (roasting, frying, baking) affects the temperature and the types of reactions and reaction products that occur, thus affecting the flavor. The most abundant chemical in muscle is water. The chemical composition of muscle tissue is similar to the chemical composition of our bodies, making them much more complete sources of essential nutrients than plant products. Other foods from animals that are not muscle include milk, eggs, and organs (liver, kidney etc.). These foods are very good sources of vitamins and minerals.

Unlike fresh plant tissues, animal products are not living. Physiological changes in muscle tissue are associated with the physiology of death. After slaughter, two major proteins in muscle tissue, actin and myosin, react with each other to form a thick tough mass called *actomyosin* during the development of rigor mortis. The presence of actomyosin causes toughness in meat, particularly beef and pork. The larger the animal, the more severe the toughening effect. Rigor is resolved by conditioning the meat at low temperatures to allow natural enzymes, called *proteases*, to break down the actomyosin at sites away from the main point of attachment of the actin and myosin.

Meats, milk, and eggs are rich sources of nutrients for consumers and for microbes. Meats are perishable because the microbes use their enzymes to convert the nutritious chemicals (known as *substrates*) in the meat to slime, off-colors, off-odors, off-flavors, and other reaction products, which make the meat less desirable. Some metabolic disorders in the meat itself can result in poor quality meat. One example is PSE (pale, soft, exudative) pork. The pork is soft and mushy because of a rapid decrease in the pH of the meat right after slaughter.

Processing is very important in meat products because it prevents spoilage and the growth of harmful microbes. Processing causes changes in flavor, color, and texture. Since we are used to eating most of our meats cooked, we associate many of these changes with improvements in quality. The red color of wieners and bologna is due to nitrates and nitrites, which are added to inhibit the growth of microbes as well as to stabilize meat color. Drying removes water and provides unique flavor sensations by encouraging specific chemical reactions in intermediate-moisture products such as pepperoni and moist pet foods. Freezing of meats slows the growth of microbes and preserves the flavor, color, and texture of fresh meats. Many meat products, like chicken nuggets and *surimi* (ground fish) use high-quality ground meat and rely on proteins to form gels to help hold the muscle fibers together. Pasteurization of milk and eggs kills harmful microbes but does not kill all the spoilage microbes. This process changes the flavor of the milk or egg and also decreases the nutritional quality. In most processes, particularly those involving heat, food scientists are willing to trade off some loss in nutritional quality to keep the food safe. Although processed meats, milk, and eggs are much more stable than their fresh counterparts, the fats in these products can become oxidized and cause off-flavors and toxic compounds.

Formulated foods

With foods derived from plant or animal products, food scientists like Kyle must work with the chemical compounds that are present in the plant or animal tissue. In formulated foods, however, food scientists can take advantage of specific chemicals in ingredients to produce the desired result. In addition to the physical and chemical properties of compounds, food chemists are also interested in functional properties of chemicals, which translate into the functionality of ingredients. The functionality of ingredients is defined as the physicochemical properties that affect their performance in a food product during processing, storage, and preparation.

Product developers must know about the functionality of ingredients when designing a new food product or redesigning an existing one. Natural pigments or synthetic food colorants give a food or beverage its color. Food colorants are added to color products like multicolored cereals, blue sports beverages, green and purple ketchup, and pink butter. Flavorants (natural or synthetic) are added to products to achieve a desired flavor or to enhance an existing flavor. These compounds produce fruit flavors in beverages and candies. Creative uses of gums, celluloses, starches, pectins, and other carbohydrates, as well as proteins and fats, produce a wide range of textural sensations.

Types of chemicals in food

The building blocks of foods are chemicals. Chemicals in foods can be natural components of the food like the complex carbohydrates in plant cell walls or the stored lipids in muscle tissue. They can be added to perform a specific function like sugar to sweeten or salt to preserve a food. Food scientists seek to understand the functions of chemicals in our foods to make them safe, wholesome, and desirable.

Toxic compounds

As we learned in Chapter 1, some foods can make us sick. Toxic compounds in foods come in many forms. The most likely chemicals in foods that can make us sick are the toxins produced by microbes. Bacterial toxins are large proteins. Mycotoxins are produced from molds and are much smaller molecules. Pesticides in foods are also toxic. Fortunately, pesticides permitted in foods are chosen for their ability to break down rapidly into nontoxic reaction products prior to purchase and consumption. Most pesticides are very toxic to pests in the field, but if handled properly, they are not toxic to consumers. Some compounds in food tissues are also toxic, like solanine, amygdalin, or ciguatera, a potent seafood toxin. Some of these toxins can be converted to nontoxic products by heating, but others are heat resistant.

The toxicity of a chemical is related to its dose. Most of us consume minute amounts of toxic compounds daily without feeling their effects because the dose is less than what is needed to make us sick. Some compounds, like the botulinum toxin, are so powerful that very small doses can make us ill and cause death. Even a potent toxin like botulinum (botox) can be diluted to a small level and injected to help reduce wrinkles! On the other hand, some compounds, like vitamin A, which are necessary for life, can be toxic at higher concentrations. Thus, food scientists like Carlos must not only know which compounds are toxic, they must also determine the toxic dosage level. Currently, any compound that is a known carcinogen cannot be used in foods. Toxic compounds are carefully monitored and a 100-fold safety factor is built into any product.

Preservatives and other food additives

Chemical preservatives provide one of the oldest and most effective defenses against microbial activity in foods. As we learned in the first chapter, the most common preservatives in foods are sucrose (table sugar) and sodium chloride (table salt). Preservatives work by slowing the growth of a microbe, stopping it, or even killing it. One way that preservatives work is by binding water and making it inaccessible to microbes.

Sugar and salt are very effective at decreasing water activity and thus making it harder for microbes to grow. Other preservatives, like organic acids such as acetic acid (vinegar) and benzoic acid (a primary component of cranberry juice), lower the pH in the food, which produces an unfavorable environment for most microbes. Still others, like potassium sorbate, interact with compounds in offending microbes to prevent them from growing. As mentioned earlier in the chapter, preservatives are chemical compounds and ingredients. Acidulants, like citric acid (found naturally in orange juice and many other fruits and their products), are added to citrus-flavored soft drinks to impart a tart flavor sensation as well as to act as a preservative (see Insert 10.2).

Carlos and Kyle have learned that food additives are incorporated into food products for many reasons other than as preservatives. Antibiotics show up in milk as residues of treatments given to prevent diseases in dairy cows. Anticaking agents are added to powders to prevent clumping, such as the addition of calcium silicate to sodium chloride so it will flow smoothly when the relative humidity is high. Antioxidants prevent fats from becoming rancid because rancid fats smell bad and can become toxic. Butylated hydroxytoluene (BHT) is a very effective synthetic antioxidant used in formulated foods, and α-tocopherol (vitamin E) is an excellent antioxidant present in tissue foods (see Insert 10.3). Lecithin is an emulsifier that allows oil and water to coexist in a product and keeps them from separating in salad dressings and peanut butter. Sequesterants or chelating agents like sodium tripolyphosphate bind metal ions, preventing them from oxidizing fats and causing flavor problems. Stabilizers are compounds that help keep food components from separating. An emulsifier is a type of stabilizer. Gums are a class of chemicals that serve as texturizers helping to make ice creams and yogurts smooth.

As consumers like Hannah become more health conscious, they want less sugar and fat in their products, but they still want full flavor. Food scientists use many classes of additives to help achieve these objectives. Bulking agents like starches and maltodextrin replace the fats that are removed

Citric acid

Acetic acid

Insert 10.2 Chemical structures of acids found in natural and formulated foods. Citric acid is the primary acid in citrus fruits and many carbonated beverages. We most commonly consume acetic acid in vinegar.

Butylated hydroxytoluene

α-tocopherol

Insert 10.3 Chemical structures of two potent antioxidants. Butylated hydroxytoluene (BHT) is an artificial antioxidant that is found in many formulated foods.
The α-tocopherol molecule is vitamin E and is the most effective antioxidant in
cellular membranes.

in making low-fat foods. If these agents weren't added, we would just get
less food. Fat replacers like sucrose polyesters (e.g., Olestra® in the fat-free
potato chips) must be able to replace many functions of fats, including flavor, mouthfeel, and cooking properties. Few compounds can perform all of
these functions. Thickeners are added to foods to provide a pleasant mouthfeel. They have been particularly effective in making skim milk products
more appealing. Artificial sweeteners like aspartame (NutraSweet®) and
sucralose (Splenda®) replace sugars, permitting dieters and diabetics the
opportunity to eat sweet foods without guilt (see Insert 10.4).

Colors and flavors

Natural and synthetic colorants differ in structure and stability. Nature
provides a wide range of colors, including the reds of lycopene in tomatoes and heme in uncooked steaks, to the oranges of β-carotene in carrots and astaxanthin in salmon, to the yellows of xanthophylls in summer
squash and in prepared mustard, to the greens of chlorophyll in broccoli,
to the blues of anthocyanins in blue corn and blueberries, to the brown pigments created enzymatically in a cut apple or nonenzymatically in golden
brown pancakes. Unfortunately, many of these compounds are unstable

Insert 10.4 Chemical structures of artificial sweeteners in food products.

under most processing conditions and even during storage of fresh, whole foods. Many times the color is lost when the flavor, texture, safety, and nutrition is still fine. Consumers may throw away perfectly good food because it does not look right. Because of the instability of many natural pigments, food scientists frequently rely on synthetic colorants to produce the bright colors so important in beverages, candies, and breakfast cereals. These synthetic colorants are listed in the ingredient statement (e.g., Yellow #6 and Red #40). They provide a bright color at very low concentrations. These synthetic dyes are derived from petroleum products and are carefully regulated by the FDA (see Insert 10.5).

Flavor can be narrowly defined as the combination of the senses of taste and smell or broadly defined to encompass all sensations encountered in the eating experience. Taste occurs when specific compounds come in contact with receptors on the tongue. Sugars and artificial sweeteners interact with sweet receptors, whereas acids interact with sour receptors. The perception of bitter taste is stimulated by many compounds, including caffeine, theobromine (responsible for bitterness in chocolate), quinine, and humulone (a component of hops that contributes to bitterness in beer). Sodium chloride is the most common chemical affecting saltiness perception. Aromatic compounds contribute to a full flavor sensation. Character-impact compounds are chemicals associated with a particular product such as benzaldehyde in cherries and almonds, isoamyl acetate in bananas, and methyl anthranilate in grapes. Many foods are complex mixtures of many aromatic compounds, none of which qualify as character-impact compounds. Limonene is a compound that contributes to the flavor of many fruit products but is not distinctive for any fruit.

Most meat flavors develop during the cooking process. Among these heat-generated compounds are the pyrazine compounds associated with Maillard browning. Although most flavors are colorless and most colors are flavorless, the browning-reaction products also contribute to flavor. In addition to taste and aroma compounds, other components are responsible for the sensations we experience as spicy heat, coolness, pungency, and

Yellow # 6

Red # 40

Insert 10.5 Chemical structures of two of the most popular artificial colors, FD&C Yellow # 6 and FD&C Red # 40. Look for them in the ingredient statements in your favorite colored foods.

astringency. The heat of hot peppers is attributed to capsaicin, and menthol provides the cooling sensation of mints. Onion pungency is attributed to sulfur compounds, and tannins are responsible for the astringency of tea and unripe persimmons (see Insert 10.6).

Numerous chemical compounds contribute to the flavor of fresh and cooked whole foods. Extracted natural flavors or added artificial flavors are present in formulated foods. Thus individual compounds serve as flavor additives in these foods. Vanillin is a synthetic compound that is added to foods to provide vanilla flavor. Many fruit-flavored beverages, soft drinks, and frozen desserts use naturally extracted or synthesized chemicals to convey a particular flavor. Character-impact compounds can be used for these applications. The addition of specific compounds that are not normally found together can produce unique flavor sensations. Other

Capsaicin

Menthol

Insert 10.6 Chemical structures of compounds responsible for chemical feeling factors. Capsaicin is responsible for the heat of hot peppers, and menthol provides a cool sensation in many mints, candies, and cough drops.

compounds, like monosodium glutamate and maltol are flavor enhancers. At high concentrations, monosodium glutamate has a unique taste (called umami), but at lower concentrations it brings out flavors, particularly meaty flavors. Maltol, a component of sweet potatoes and the character-impact compound for cotton candy, is a product of nonenzymic browning and can enhance flavors in sweet products. Flavor compounds are delicate and can be easily modified by enzymes during storage of fresh items or by heat processing. Interaction with oxygen or light can also change flavor compounds.

Vitamins and minerals

Vitamins are organic chemicals that are essential to health. Fat-soluble vitamins are more difficult to incorporate into formulated foods than water-soluble vitamins. Fat-soluble vitamins are susceptible to oxidation. Vitamins A, C, and E are antioxidant vitamins because they help protect fats from becoming oxidized (see Insert 10.7). Processing, β-carotene

Ascorbic acid

Insert 10.7 Chemical structures of two compounds with vitamin activity. The molecule β-carotene is provitamin A and ascorbic acid is vitamin C.

particularly heat processing, can be very damaging to vitamins. Canning or home cooking can destroy over 50% of the vitamins, but further losses during subsequent storage are minimal. Most consumers don't realize that vitamins in fresh foods degrade during storage, either through normal metabolic processes or in response to external stress. Thus, major losses in vitamins, particularly water-soluble vitamins, can occur in whole foods stored in the refrigerator. Unlike vitamins, minerals are elements, and at least twenty-five are essential for life (see Insert 10.8).

Insert 10.8 List of essential minerals and their chemical symbols.

Calcium	Ca	Nickel	Ni
Chlorine	Cl	Phosphorous	P
Chromium	Cr	Potassium	K
Copper	Cu	Selenium	Se
Fluorine	F	Silicon	Si
Iodine	I	Sodium	Na
Iron	Fe	Sulfur	S
Magnesium	Mg	Tin	Sn
Manganese	Mn	Vanadium	V
Molybdenum	Mo	Zinc	Zn

Carbohydrates

We derive energy from macronutrients—proteins, lipids, and carbo-hydrates. Most people around the world obtain more than half of their calories from carbohydrates (see Insert 10.9). Monosaccharides, the basic units of carbohydrates, consist of carbon, hydrogen, and oxygen, and are known as simple sugars. The most common monosaccharides are glucose (blood sugar) and fructose (the primary sweetener in honey). Common

Glucose

Sucrose

Trehalose

Insert 10.9 Chemical structures of three sugars. Glucose is a monosaccharide that is also known as blood sugar. Sucrose is a disaccharide commonly called table sugar. Trehalose is also a disaccharide found primarily in yeasts, mushrooms, and many insects.

disaccharides include sucrose (table sugar) and lactose (milk sugar). Sucrose is composed of a molecule of glucose linked to a molecule of fructose, and lactose is a molecule of glucose linked to a molecule of galactose (another monosaccharide). Starches are complex carbohydrates that consist of long chains of monosaccharides (primarily glucose). Cellulose is another long molecule consisting of glucose molecules linked together. The molecular linkage in starch is different from that of cellulose. Our digestive systems have enzymes that can break down the bonds in starch into single glucose units, which in turn can be used by our bodies, but we do not have the enzymes that can break the cellulose linkages. Cattle and other ruminants do have the enzymes that can break down the bonds in cellulose allowing them to eat grass and derive energy from it. Dietary fiber, found in fruits, vegetables, whole grains, and their products, is also composed of complex carbohydrates. These compounds (hemicellulose, pectin, and lignin) are more complex in structure than starch and cellulose and are not broken down during digestion.

Food scientists are interested in carbohydrates for more than their nutritional value. They are interested in sugars because they are sweet and add to the flavor of many food products, and in complex carbohydrates because they contribute to the texture of food products. Starches can be chemically modified to improve ingredient functionality. Modified starches have desirable properties that make it possible to produce instant foods (instant puddings or quick grits), maintain creamy textures upon thawing (frozen custard pies or other frozen desserts), and cause thickening (gravies). Pectins, celluloses, and gums are also effective at providing desirable textural properties. Food scientists must be aware of the properties of the carbohydrate, the properties of the food, and how both will change with typical processing operations. For example, the usefulness of a gum as a thickener depends on its solubility, stability to changes in temperature and pH, compatibility with other gums, and its freeze–thaw stability.

Starch is the major component of bread and other bakery products. One of the most common quality problems associated with bakery products is *staling*. Staling results from the retrogradation of starch. *Retrogradation* of starch involves the crystallization of the starch molecules such that they begin to separate from the other molecules present in the product. This separation leads to loss of acceptable color, flavor, and texture of the product. Anyone who has used starch to prepare a food product knows that starch can't just be dumped in water because it will not dissolve. The hydroxyl (–OH) groups on the starch molecule interact with the water. Using just a small amount of water with the starch permits hydration of these hydroxyl groups by hydrogen bonding. Once these are hydrated, the starch will more easily dissolve in the water.

Lipids

The chemical category that combines fats and oils is lipids. Fats are solid at room temperature and oils are liquid. Lipids are defined as chemical compounds that are soluble in organic solvents but not soluble in water. The three primary groups of lipids in foods are triacylglycerols, phospholipids, and sterols (see Insert 10.10). Triacylglycerols contain three fatty acids and are found in bulk lipids such as the visible fats we cut off of a steak, lard used in baking, and vegetable oils that come in bottles. Phospholipids contain two fatty acids with a polar head group and are found in membranes of plants and animals. Phospholipids are not visible in tissue foods, but they can be used as emulsifiers. Lecithin, the most common phospholipid emulsifier, can be extracted from egg yolks and soybeans. Cholesterol is the most well-known sterol, but there are many other sterols in plants and animals. Sterols are multiringed structures without fatty acids that help stabilize cellular membranes in living tissue. They are also important precursors of hormones in our bodies. Since many of us consume too many lipids, it is sometimes hard to appreciate that we all need small amounts of lipids.

In earlier chapters we learned about saturated, unsaturated, polyunsaturated, and trans fats. It is the fatty acid part of a lipid molecule that varies by degree of saturation. Fatty acids consist of chains of hydrocarbons

Phospholipid

Cholesterol

Insert 10.10 Chemical structures of a phospholipid, and a sterol.

(a carbon backbone with hydrogen atoms attached). A fatty acid is classified by the number of carbon atoms and the degree of saturation. A fatty
acid is completely saturated when it has the maximum number of hydrogen atoms attached and no double bonds. A monounsaturated fatty acid
contains a double bond and two less hydrogen atoms. A polyunsaturated
fatty acid contains at least two double bonds. As the number of double
bonds increases, the melting temperature decreases and the more likely
the lipid is to become an oil. Saturated lipids are solid fats and are associated with heart disease. Polyunsaturated lipids tend to be oils and are
not readily spread on bread or other products. These oils can be hydrogenated by adding hydrogen atoms across a double bond to make the fatty
acid less saturated. While this process provides a better spread, trans fatty
acids become an unwanted by-product. *Cis* fatty acids, which occur in fats
and oils that have not been modified, have the hydrogen atoms on the
same side of the double bond, while *trans* fatty acids have the hydrogen
atoms on opposite sides (see Insert 10.11). Our bodies cannot use *trans* fatty
acids. At best they affect us in the same way as saturated fatty acids. There
is growing scientific evidence that *trans* fatty acids provide more serious
threats to our health.

Lipids are considered unhealthy by many consumers. The health
risks associated with lipids include a doubling of calorie content for the
same weight of carbohydrates or proteins, and the accumulation of lipids

cis-oleic acid

Trans-elaidic acid

Insert 10.11 Chemical structures of a cis fatty acid and a trans fatty acid. The
hydrogen atoms are on the same side of the double bond in the cis fatty acid
resulting in a kink in the chain. The hydrogen atoms are on the opposite side of
the double bond in the trans fatty acid.

in our arteries leading to atherosclerosis and heart disease. Contrary to popular belief, we do need some fat in the diet. Linoleic (18:2, 18 carbon atoms and 2 double bonds) and linolenic (18:3) are essential fatty acids that we must get in our foods as we cannot produce them in our bodies. The fat-soluble vitamins require some lipid to allow our bodies to absorb them in the intestines. In addition, lipids contribute to many of the pleasant and unpleasant aromas in foods and provide lubrication of foods in the mouth. If we cut out too many lipids, our food becomes tasteless, and we might not eat enough to keep us healthy. Low-fat foods are becoming more popular, but if we overconsume them, we will still gain weight.

Lipids can be broken down in foods by hydrolysis and oxidation. Hydrolysis involves the separation of a fatty acid from the glycerol backbone of a triacylglycerol or phospholipid. If these fatty acids are short chain (4 carbons or less), the aromas can be objectionable, but hydrolysis usually does not cause quality problems. Lipid oxidation is a reaction of lipids with molecular oxygen that proceeds through a free-radical mechanism. Lipid oxidation can result in severe off-odors and flavors (known as *oxidative rancidity*). Too many of these oxidative products can be toxic to humans. The free radical consists of a carbon atom with an extra electron, resulting from the removal of a hydrogen atom. Oxygen can attach to the free electron resulting in a lipid hydroperoxide. Through a very complex series of chemical reactions, these compounds are formed, leading to oxidative rancidity. Antioxidants either inhibit or retard oxidation through one or more mechanisms. Antioxidants such as β-carotene (pro-vitamin A), α-tocopherol (vitamin E), lycopene (red pigment in tomatoes), or ascorbic acid (vitamin C) may occur in the food tissue itself and help slow or prevent oxidation. Other compounds such as BHA (butylated hydroxyanisole) and BHT (butylated hydroxytoluene) are synthetic antioxidants that are added to foods to prevent rancidity. The naturally occurring antioxidants tend to be more effective in tissue foods to prevent oxidation of phospholipids in cellular membranes, while the synthetic antioxidants tend to be more effective in the triacylglycerols in formulated foods.

Proteins

Proteins are also important macronutrients in our foods. Although they can be used for energy in the body, they are much more important as a source of essential amino acids. Proteins are large molecules composed of long, folded chains of amino acids, which are linked together by peptide bonds. Amino acids can act both as acids and as bases with the carboxyl (–COOH) end acting as the acid and the amino (–NH_2) group acting as a base. The peptide bond forms as the carboxyl group from one amino acid binds with the amino group from the next one, with the net loss of a water (H_2O) molecule. There are twenty-one amino acids that are typically found in food products, but only nine are essential to human health (phenyla-lanine, methionine, tryptophan, isoleucine, threonine, histidine, leucine, lysine, and valine). The remaining twelve amino acids can be synthesized by our bodies and thus do not need to be ingested in our foods. The nutritional quality of a food protein is based on the balance of the essential amino acids present. Animal proteins tend to be more balanced in proteins than plant proteins.

While proteins are an essential part of the diet, problems can develop when too much protein is consumed. Proteins are important components of muscle tissue (those of our own, which we use, and those we eat as meat). Amino acids are also components of enzymes in our bodies that are needed for metabolic processes. During digestion the proteins are broken down into individual amino acids in the intestines and absorbed across the intestinal mucosa into the blood stream, where they are then transported to the tissues and cells where they are needed. Athletes who want to build muscle mistakenly think that an increase of protein consumption and a decrease of carbohydrates will help build muscle. Actually, carbohydrates are needed to build muscle. Increased protein consumption relative to carbohydrates increases hunger and causes the body to use the amino acids (both essential and nonessential) for energy rather than to build proteins. The brain needs glucose present to function properly, but the body cannot produce enough glucose for the brain if there is insufficient carbohydrate present. The use of proteins as energy results in the production of ketone bodies leading to bad breath associated with ketoacidosis and the excretion of urea in the urine. The presence of ketone bodies on the breath and urea in the urine is unhealthy.

The structure of proteins is very complex. Primary structure refers to the chain of amino acids linked by the peptide bonds. Secondary structure relates to whether the chains tend to form α-helix or β-sheet shapes. *Tertiary structure* involves the interfolding of the primary chain such that amino acid number 42 may be close to amino acid number 16. *Quaternary structure* refers to the linkage of more than one peptide chain. Physiochemical properties (how they act in living systems) of proteins and peptides are influenced by the hydrophobicity and hydrophilicity of the constituent amino acids. Some amino acids are soluble in lipids and not in water (*hydrophobic*) while others are soluble in water and not in lipids (*hydrophilic*). In the tertiary structure of a protein, regions can form that are primarily composed of hydrophobic or hydrophilic amino acids. These regions then associate with like environments within a cell. Other factors that affect physiochemical properties of proteins include ionic charge, hydrogen bonding, and van der Waals forces.

Food scientists are also very interested in the functional properties of proteins in foods. The texture of omelets, meringues, yogurt, and many other products is due to the presence of denatured egg proteins.

Denaturation of proteins affects the secondary, tertiary, and quaternary structure of the protein by inactivating the physiochemical properties while changing the physicochemical properties (physical and chemical properties that in turn affect the functional properties) of the protein. Chemical modification of amino acid side chains alters functional properties of the proteins, which in turn alter the physicochemical properties of the food. During denaturation or chemical alteration of the proteins, some of the essential amino acids lose their nutritional value. Lysine is particularly susceptible to these changes.

Enzymes

Enzymes are specialized proteins that catalyze specific reactions converting substrates to products. Living systems need enzymes to speed up reactions in the cell. Enzymes break down foods in the intestine into small components that can be absorbed across the intestinal walls into the blood stream. They are also needed to synthesize new proteins and pigments, build and repair membranes and cell walls, replace worn-out DNA, and perform many other functions in cellular metabolism. Without enzymes, life as we know it would not be possible. Many compounds in the cell are *stereospecific* (the structure represents one side of a mirror image). The enzyme will recognize one side of the mirror image and not the other (e.g., L-amino acids and not D-amino acids). Enzyme reactions thus are stereospecific while normal chemical reactions produce equal amounts of the two mirror images.

Cellular compartmentation is an important factor in controlling enzyme reactions in plant and animal tissues. The plasma membrane, composed primarily of phospholipids and proteins, surrounds each cell. Within the cell are smaller structures called *organelles*, which are also surrounded or composed of membranes. Substrates can be protected from enzymes by cell membranes or they can be directly attached to a membrane. When these membranes are broken by a physical blow, bruising develops, as evidenced in a muscle or in a fresh peach. In the peach the enzyme polygalacturonase causes the flesh to become mushy, while another enzyme, polyphenoloxidase, is responsible for brown color development. During cutting, slicing, and other processing operations, these membranes can become damaged, and quality of the product suffers due to the effects on the enzymes.

Many enzymes are inactivated during processing, handling, and storage. Heat can denature an enzyme. Once an enzyme unfolds from its secondary and tertiary structure it cannot operate as an enzyme, but it can still contribute functional properties that affect the structure of the food product. Enzyme activity (the rate at which it acts) is also affected by the pH and moisture content of the food, the storage temperature, the presence of oxygen and carbon dioxide in the food environment, and light.

Genetic engineering alters genetic composition to modify a specific trait of a raw food product. Genetic engineering works on modifying the DNA in the cell's nucleus (*transformation*), which affects RNA synthesis (*transcription*) and placement of amino acids in sequence in the proteins (*translation*). Many of the proteins affected are enzymes which are responsible for modifying chemical composition in cells leading to changes in food quality.

Water

Water is the most abundant chemical in foods. Everyone knows its chemical formula, but most people do not think of it as a chemical. It is the only chemical commonly found as a solid, liquid, and gas under normal conditions. Water is both an amazingly simple compound to understand and a deceptively complex component of almost all food systems. Vegetable oils are among the few foods that contain no water. Water is known as the universal solvent as it will dissolve most solids and liquids. Some compounds, such as fats, oils, and organic solvents like hexane, are not soluble in water. The solubility of organic compounds in water is a function of hydrogen bonding. Hydrogen bonding is primarily a function of the presence of hydroxyl (–OH) groups with the H in the hydroxyl group forming a hydrogen bond with the O in water, and the O in the hydroxyl group forming a hydrogen bond with an H in water. The longer the chain of a hydrocarbon, the less soluble it will be.

Water is important in foods for many reasons. It provides a medium to support microbial growth. Generally speaking the more water that is present, the greater the opportunity for microbes to grow. More important than the amount of water is the amount of water available to the microbe. Water activity (a_w) is the term that relates to the ability of a microbe to grow in a particular environment. Removal of water by drying or concentrating will reduce its susceptibility to microbial attack, but has a profound effect on the texture of foods as well as color and flavor. Moisture sorption isotherms are plotted by food scientists to determine the relationship between water content and water activity (see Insert 10.12). Such plots help predict the behavior of a food during drying and rehydration. A large difference in the plots of the drying and rehydration isotherms (known as syneresis) of a breakfast cereal means that it will not become soggy in a bowl of milk.

Water content affects food quality. The presence of water can be perceived as juicy and succulent or watery and soggy, depending on the product. Likewise, the absence of water can be perceived as dry or crunchy. Water-holding capacity (WHC) results from physical

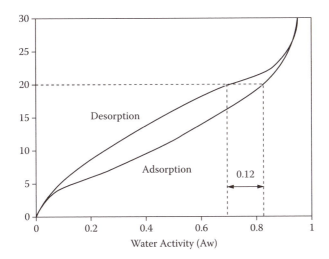

Insert 10.12 Moisture sorption isotherm, which relates moisture content to water activity. (From B. Ray, *Fundamental Food Microbiology*, 3rd ed. Boca Raton, FL: CRC Press, Taylor & Francis Group, 2004. With permission.)

entrapment, which restricts flow but not other physical properties of the water molecules. In packaged meats, lack of WHC leads to drip loss or unsightly microbe-laden water in the bottom of the package and a tougher cooked product. Water is also the medium for many enzyme reactions in food. Molecules must be able to move within a food product to react with each other. During food storage, most of these reactions are detrimental to food quality (aging of wine and beef as well as ripening of fruits are notable exceptions). *Molecular mobility* is the term used to describe this movement of molecules within a food product. During freezing of a product, the molecular mobility decreases as the water freezes. Not all water freezes at the same rate, however, so the water in a food does not go immediately from a liquid to a solid state. At a temperature below the freezing point of a food, a product reaches its *glass transition temperature*. Below this temperature, the food is like a glass, and there is no molecular mobility. With no molecular mobility, all reactions stop and quality is maintained. The glass transition temperature varies among foods depending on the other components present. It is too low for most foods to be of practical use for maintaining quality of food products.

Dispersions

Many foods and beverages that we consume are dispersions (a discontinuous phase dispersed in a continuous phase). For example, milk is a dispersion of lipids in water. A milk fat globule is composed of triacylglyc-erols layered by proteins and encapsulated by a plasma membrane. Milk, as it comes out of the mother mammal, is not a stable dispersion. Left out, the milk fat globules (the discontinuous phase), which are less dense than the water (the continuous phase), float to the top of the milk. This fat can be skimmed off the top of the milk with the fatty portion called the *cream* and the remaining watery product, the *serum* (or skimmed milk). To prevent this separation, milk processors homogenize the milk by disinte-grating the fat globules into tiny fat particles that are too small to coalesce (come together) and rise to the top.

Dispersions can consist of a phase of one state (gas, liquid, or solid) dispersed in the phase of another. For example, a solid dispersed in a liquid is called either a *solution* or a *suspension*. True solutions are homo-geneous with the individual molecules of the solute (solid, discontinuous phase) separated in the solvent (liquid continuous phase). Suspensions are less stable than solutions. *Colloidal suspensions* are formed when big-ger particles are dispersed in a liquid, but the particles are still not vis-ible to the naked eye. Liquid spreads are colloidal suspensions called *sols* with fat globules dispersed in liquid oil. Jelly is also a colloidal suspen-sion of pectin in a sugar solution with water as the solvent. A semisolid colloidal suspension like jelly is called a *gel*. Other interesting dispersions are *foams* like whipped cream (air dispersed in liquid fat), *aerosols* like breath spray (liquid flavorants dispersed in air), and *smoke* (solid particles dispersed in air).

The most common food dispersion is an emulsion that consists of a dispersion of two immiscible liquids (don't dissolve in each other). Emulsions can be subdivided into two primary types—oil-in-water and water-in-oil. The discontinuous phase (the lesser of the two liquids by volume) is dispersed in the continuous phase (the greater of the two liquids). Salad dressings tend to be oil-in-water emulsions, while may-onnaise is a water-in-oil emulsion. Chocolate products are also water-in-oil emulsions formed when both the oil and water phases were liquid and then cooled to form a solid product. Since the two liquids of an emulsion are immiscible, help is needed to form the emulsion. Physical methods can be used to form this emulsion, such as vigorous shaking of oil and vinegar to form a salad dressing. Unfortunately, these two liquid phases separate quickly. Food additives known as stabilizers are used to form and maintain dispersion. Emulsifiers are specific stabiliz-ers used to form and maintain an emulsion. Emulsifiers are classified according to their HLB (hydrophile–lipophile balance) index. A high HLB indicates that the emulsifier is more soluble in water than in oil and would be more suitable for use in an oil-in-water emulsion, while one with a low HLB would be more effective in water-in-oil emulsions.

Phospholipids like lecithin are particularly effective emulsifiers as there are both hydrophilic and lipophilic ends in the molecule.

Food chemistry as an integral part of food science

In this chapter, the emphasis has been on contributions of individual chemicals to nutrition and food functionality. One thing that makes food chemistry so interesting is that foods are complex mixtures of many chemical compounds. As a result there are many chemical reactions that can occur in foods—some of them are desirable and many are undesirable. It is not sufficient to know about individual chemical compounds in a food. Rather we must know how chemical compounds react with each other in the environment within a food product. Such reactions could lead to compromised safety, loss of nutritional quality, or loss of sensory quality.

Safety hazards develop from the production of a toxic compound, usually produced by a microbe, but also from the presence of a natural toxin or development of a toxin during storage. Solanine is a toxic alkaloid that can develop in potatoes exposed to light. Fortunately, toxic potatoes tend to turn green (production of chlorophyll) under the same conditions, providing a warning to observant consumers. Another protection for the consumer is that solanine is readily inactivated by heat such as baking, boiling, or frying. Toxic compounds can enter a food from package materials or the interaction of the food with the primary package. Food scientists carefully screen package materials and study food–package interactions to make sure that the products are safe.

Loss of sensory quality of a food is the result of an increase in an undesirable characteristic or a decrease in a desirable characteristic. Development of slime on wieners, yellowing of lettuce and broccoli, and objectionable aromas in milk are all undesirable characteristics that signal poor quality. More subtle changes, such as softening of bananas, loss of bright red color in strawberry jam, or decreased flavor of fresh tomatoes, are all examples of losses of desirable characteristics. Frequently an undesirable characteristic develops as a desirable characteristic is lost. All of these characteristics can be traced back to specific chemical reactions. Food chemists can measure the loss of a reactant or the accumulation of a product in a chemical reaction to predict the loss of quality. Food engineers use these data to calculate the rate of deterioration using chemical kinetics. Many reactions in foods can be classified as zero-order reactions (linear degradation, such as nonenzymatic browning) or first-order reactions (semilogarithmic, such as loss of vitamins or lipid oxidation) (see Insert 10.13). Food engineers can then use these equations to estimate the shelf life of a product under typical handling and storage conditions and develop an expiration date for the product.

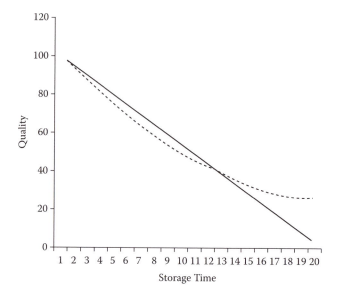

Insert 10.13 Quality degradation as affected by zero-order kinetics (solid line) and first-order kinetics (dotted line).

Remember this!

- Molecular mobility encompasses the rate movement of all molecules within a food product.
- The dispersion of two immiscible liquids is called an emulsion.
- Cellular compartmentation is an important factor in controlling enzyme reactions in plant and animal tissues.
- Chemical modification of amino acid side chains alters functional properties of the proteins, which in turn alters the physicochemical properties of the food.
- Lipid oxidation is a reaction of lipids with molecular oxygen, which proceeds through a free-radical mechanism.
- Starches are chemically modified to improve ingredient functionality.
- Flavor can be narrowly defined as the combination of the senses of taste and smell, or broadly defined to encompass all sensations encountered in the eating experience.
- Natural and synthetic colorants differ in structure and stability.
- Chemical preservatives provide one of the oldest and most effective defenses against microbial activity in foods.
- The toxicity of a chemical compound is related to its dose.

- Functionality of ingredients is defined as the physicochemical properties that affect its performance in a food product during processing, storage, and preparation.
- Everything we put into our mouths is chemical.

Looking ahead

This chapter introduced us to food chemistry. Chapter 11 introducees us to nutrition, Chapter 12 to food microbiology, and Chapter 13 covers food engineering, including the design of processes to protect us from disease and foods from spoilage due to unwanted microorganisms. The final chapter describes the science associated with the use of humans to evaluate the quality of food products.

References

Ray, B. 2004. *Fundamental food microbiology*, 3rd ed. Boca Raton, FL: CRC Press, Taylor & Francis Group.

Further reading

Belitz, H-D., W. Grosch, and P. Schieberle. 2004. *Food chemistry*, 3rd ed. New York: Springer Science.

Damodaran, S., K. L. Parkin, and O. R. Fennema. 2007. *Fennema's food chemistry*, 4th ed. Boca Raton, FL: CRC Press, Taylor & Francis Group.

deMan, J. M. 1999. *Principles of food chemistry*, 3rd ed. Gaithersburg, MD: Aspen Publishers.

Hui, Y. H. 2007. *Food chemistry: Principles and applications*, 2nd ed. West Sacramento, CA: Science Technology System.

Roday, S. 2007. *Food science & nutrition*. Oxford, U.K.: Oxford University Press.

Singh, R. P. 2000. Scientific principles of shelf-life evaluation. In *Shelf-life evaluation of foods*, 2nd ed., ed. D. Man and A. Jones. Gaithersburg, MD: Aspen Publishers, Inc.

chapter eleven

Nutrition

Tanya is very interested in the contribution of specific foods to the growing obesity problem in America. She has read many diet books and they seem to contradict one another. She wants to get a better perspective, based on true scientific data and not on propaganda. For her senior thesis project she decided to research the physiological mechanisms that lead to obesity. As she got into her library research she found that nutrition was only one part of the obesity problem.

Hannah is now interested in the health benefits of whole grains, fruits, and vegetables. When they changed the food pyramid, she noticed that plant products took on an added importance. She was also beginning to realize that as a vegetarian, you had to be careful or you might miss out on some essential nutrients. She was still a vegetarian and needed to be careful that she was eating a balanced diet and that she was getting enough calories to maintain her weight.

Martin learned about food products when he worked as part of a team that designed an edible soup bowl. One of the most interesting things they studied was nutrient loss during formulation and processing. He wanted to learn more about how nutrients are lost during manufacturing and storage and what could be done to prevent this from happening, so he decided to pursue graduate studies in this area.

Tanya, Hannah, and Martin are learning more about nutrition and how it interfaces with food science. Although food scientists aren't nutritionists, they must understand basic nutrition to properly evaluate food products and processes because many of the activities food scientists perform affect the nutritional quality of a product. Many times these activities require trade-offs between nutrition and some other factor like sensory quality or shelf stability.

Looking back

The previous chapter focused on the function of chemicals in foods. This chapter expands that idea to the function of food chemicals when they are consumed. The following key points were covered in previous chapters and will help prepare us to understand the nutritional aspects of foods.

- Functionality of an ingredient is defined as the physicochemical properties that affect its performance in a food product during processing, storage, and preparation.
- Chemical modification of amino acid side chains alters the functional properties of the proteins, which in turn alter the physicochemical properties of the food.
- The information on a product label is regulated by the U.S. Food and Drug Administration (FDA) under the Nutrition Labeling and Education Act.
- The conditions that affect nutritional value also affect sensory quality.
- The higher the temperature and shorter the duration of a heat process, the less vitamins are lost in the process.
- Food processing and preservation can increase shelf stability of a raw material, usually at the cost of affordability, nutrition, and quality.
- Although many of us try to eat healthy, there are many temptations and influences affecting our food choices that we do not consciously consider.
- Technology has transformed our food supply: more formulated foods are produced, leading to the consumption of less whole foods.
- Although there are individual dietary needs and restrictions for a select few, in general, our nutritional requirements tend to be quite similar.
- Good nutrition involves adequate consumption of required nutrients in the context of moderate energy intake.
- Healthy eating involves a balanced diet.

Nutrients in foods

As we learned in Chapter 2, the main function of food is to provide us with the energy we need. The energy in food comes in the form of calories (technically speaking, kilocalories). We can obtain kilocalories from four sources––carbohydrates, lipids, proteins, and ethanol. Despite claims to the contrary, the body does not differentiate between kilocalorie sources––a calorie is a calorie is a calorie. Nutritionists recommend that our kilocalories come from a balance of foods and not a single source. As we learned earlier, too many or too few kilocalories can cause severe health problems and death. Likewise, too much reliance on any one of the four sources of energy can cause disease, either from consuming too much of something, or more often, from not consuming enough of certain nutrients. The consequences of too much or too little of a nutrient are described later in this chapter. The most serious nutritional problem affecting Americans and Western Europeans is related to the consumption of too many kilocalories.

In many parts of the world, the most serious nutritional problem is consuming too few kilocalories. When we consume too few kilocalories, we are unlikely to consume adequate amounts of vitamins and minerals. Vitamins and minerals do not provide kilocalories, but rather they are key components of the enzyme cofactors that our bodies need to process the energy sources and perform the daily functions of life. Deficiency diseases are the result of a lack of certain vitamins and minerals.

Other substances have nutritional implications in our bodies. Dietary fiber does not provide energy nor does it function as an enzyme cofactor, but it helps prevent constipation and has been associated with the prevention of or decrease in the incidence of a number of diseases, from varicose veins to cancer. Oxygen and water are also essential to the maintenance of life. Many other components, such as antioxidants and dietary supplements, are also thought to contribute to health, but their role is not as clearly defined.

Proteins

As we learned in the previous chapter, proteins are composed of amino acids. Our bodies are constantly turning over our proteins (breaking down old proteins and building new ones). Our bodies can synthesize many, but not all, of these amino acids. Those that our bodies can't synthesize are called *essential amino acids* (shown in Insert 11.1). If an essential amino acid is not available, either from the recycling of the ones from old proteins or from dietary sources, protein synthesis cannot continue. During digestion the proteins are broken down into individual amino acids in the intestines and absorbed across the intestinal mucosa into the blood stream where they are then transported to the tissues and cells where they are needed. Proteins perform many functions in our bodies. Proteins are important components of muscle tissue. Enzymes are also proteins and are needed to catalyze metabolic processes in our bodies.

Insert 11.1 Chemical structures of the essential amino acids.

Although proteins are an essential part of the diet, problems can develop when too much protein is consumed. Athletes who want to build muscle mistakenly think that an increase of protein consumption and a decrease of carbohydrate consumption will help build muscle. Actually, carbohydrates are needed to build muscle. Increased protein consumption relative to carbohydrate consumption causes the body to use the amino acids for energy rather than to build proteins. The brain needs glucose to function properly, but the body cannot produce enough glucose for the brain if there are insufficient carbohydrates present. The use of proteins as an energy source results in the production of ketone bodies leading to bad breath associated with ketoacidosis and the excretion of urea in the urine. The presence of excess ketones in the body and urea in the urine is unhealthy.

Lipids

As we learned in the previous chapter, fats and oils are lipids. In the diet, we generally refer to all lipids as fats. Fats are critical components of our bodies; we could not function without them. As described in Chapter 2, fats provide twice the energy per gram as carbohydrates and proteins. They are compact sources of kilocalories, which can be quite convenient on long hikes where weight and volume are critical. Fats contribute to the overconsumption of energy, particularly among sedentary consumers. Fats are also associated with the accumulation of undesirable lipids in the arteries, contributing to heart disease. To cut kilocalories, food scientists have replaced fats with synthetic compounds that mimic fats but contain fewer kilocalories.

Lipids are in every cell of our bodies. The plasma membrane surrounding each cell and the membranes within each cell contain phospholipids, cholesterol, and specialized proteins. When we consume excess food, our bodies convert that food to lipids. This reserve lipid is stored for times when we do not have enough food, is burned to keep us warm (thermogenesis), and is used as a padding to protect vital organs. Prolonged overconsumption of kilocalories leads to excess padding. This excess padding is additional weight that stresses the heart, skeletal muscles, and other organs. These additional burdens can make exercise difficult, which in turn makes the problem worse.

We need to consume lipids for the essential fatty acids and sterols they provide (shown in Insert 11.2) and because lipid-soluble vitamins cannot be absorbed unless they are dissolved in lipids. Our bodies cannot synthesize these essential fatty acids, so they must be consumed in the diet. These fatty acids are needed to synthesize phospholipids, the major component of cell membranes (see Insert 11.3). Sterols are also very important in our bodies. Enzymes in our bodies convert plant sterols to cholesterol, which helps stabilize cell membranes. Consumption of plant sterols

Linolenic acid

Linoleic acid

Insert 11.2 Chemical structures of two essential fatty acids. Note that they are *cis* fatty acids.

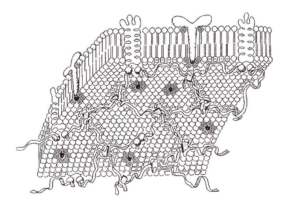

Insert 11.3 Schematic of a cellular membrane. A membrane is composed of a phospholipid (lollipop-like structures with a water-soluble head group on both surfaces of the membrane and two fatty acids in its interior), bilayer spanned by proteins (structures that cross from one side of the membrane to the other), and sugars (the ribbon-like structures coming out of the bottom of the membrane. (From J. A. Tuszynski and M. Kurzynski, *Introduction to Molecular Biophysics*, Boca Raton, FL: CRC Press, Taylor & Francis Group, 2003. With permission.)

is also associated with lowering cholesterol, particularly low-density lipo-proteins. Sterols are also needed for proper brain function and to produce key hormones such as progesterone and testosterone. Vitamins A, D, E, and K are the fat-soluble vitamins. They need lipids in the intestines during digestion to be properly absorbed. In an effort to reduce kilocalories by cutting our lipid consumption drastically (e.g., less than 10% of total kilocalories), we may be trading one problem (too many kilocalories) for a more serious one (a vitamin deficiency).

Too much lipid in the diet can be a problem. In addition to contributing to an overconsumption of kilocalories, lipids can also accumulate in our arteries and cause heart disease. Heart attacks are caused by either thrombosis (blood flow is blocked due to clotting of platelets) or athero-sclerosis (a buildup of plaque thickens arterial walls). A blood test can check the level of lipoproteins in the blood and indicate a consumer's risk of heart disease. The presence of a higher proportion of high-density lipo-proteins (HDL) is an indicator of a reduced risk of heart disease, whereas a higher proportion of low-density lipoproteins (LDL) and cholesterol are indicators of a greater risk of the disease. Evidence is growing that the presence of oxidized lipids in the arteries increases the risk of heart disease, suggesting that the consumption of antioxidant vitamins will help lower the risk of heart disease. Unfortunately, experiments relating increased consumption of antioxidant vitamins to heart disease have not been conclusive. Some of the most common contributors to heart disease are body weight, diabetes, gender, genetics, high blood pressure, and smoking (including secondhand smoke). In general, we have more control over the lipids we put into our mouths than any of the other risk factors.

As we have learned earlier, fat contributes to the flavor and enjoyment of our foods. For those who enjoy eating but don't want to consume too many kilocalories, food scientists have developed fat replacers. If fat replacers are used as part of an overall plan to cut kilocalories, they can be useful; however, many times the consumption of food made with fat replacers provides some of us with an excuse to eat more and not reduce kilocalories at all. Fat replacers may lead to other problems: sucrose esters can cause loose stools and other digestive difficulties when consumed in excess.

Carbohydrates

Carbohydrates should be our primary source of kilocalories. Carbohydrates include sugars, starches, and nondigestible cell wall materials. Starches are broken down into simple sugars prior to absorption. Our blood-stream circulates glucose, a simple sugar, which is essential to proper brain function. Any excess glucose is converted to glycogen, which is an animal starch. Glycogen is stored in the muscles as a source of immediate energy reserves. Since our muscles have limited capacity to store glycogen, additional excess glucose is converted to fat. Nondigestible cell wall materials are known as *dietary fiber.* Although they are not absorbed in the body, they have an important dietary function. Dietary fiber helps maintain peristalsis (keeping the material moving in our bowels). The slowing or stopping of peristalsis leads to constipation. Constipation, in addition to the discomfort it can bring, has been associated with a range of diseases and disorders, including varicose veins. Fiber is also an excellent binder of toxins, as well as some nutrients in our foods. Although the binding of toxins can protect our bodies, excess fiber consumption may limit the absorption of essential vitamins and minerals.

Unlike proteins and lipids, there are no essential carbohydrates that we need to consume. As mentioned above, we must have glucose circulating in our bodies. Insulin and glucagon are the hormones responsible for regulating sugar metabolism. Our bodies must properly regulate the levels of glucose in our blood stream. Too much glucose in the blood over long periods of time can cause blindness, or it can damage nerves and blood vessels, requiring amputation of the hands and feet. Too little glucose in the blood can impair brain function, cause fainting and possibly death. Amino acids can be converted to energy sources, but they cannot provide enough glucose to adequately maintain brain function. Since too much protein and too much lipid can cause dietary problems, that leaves carbohydrates as our most important energy source.

The consumption of too much carbohydrate is also a problem. Products high in sugar, such as pies, cakes, and candies are quick sources of energy that are very appealing. We can consume them rapidly without feeling full. There is strong evidence to suggest that the level of glucose in our blood helps regulate our feeling of fullness, but we can consume large amounts of sugar or starchy treats before we receive the signal to stop consumption, which over time can lead to obesity. Sugar replacers present the same advantages and disadvantages as fat replacers described in the previous section. Diabetics must be particularly careful of sugar consumption. Diabetes is a disease in which the body cannot properly regulate sugar degradation, causing elevated levels of sugar right after a meal and a sugar crash when the glucose is eventually broken down. Excess sugar on the teeth can cause dental caries (cavities). Although sugars and starches (since they are converted to sugars) have been associated with many other diseases, including hypoglycemia (low-sugar blues), most of this criticism is probably not valid.

Vitamins and minerals

In our bodies, vitamins are cofactors to enzymes. Without these cofactors, enzymes can't catalyze reactions and our metabolism is affected. Prolonged deficiency of these vitamins leads to the diseases discussed in earlier chapters. Many vitamins, including the B-vitamins and vitamin C are soluble in water. These vitamins are added to most foods but are susceptible to leaching during processing operations. Vitamins A, D, E, and K are lipid-soluble vitamins. Like vitamins, minerals also serve as cofactors in enzyme-catalyzed reactions in our bodies. They also become parts of important molecules, such as the calcium in our bones, phosphorous in our cell membranes, and iron in our blood. Usually they enter our bodies as parts of other chemicals (both organic and inorganic) in our foods.

The usefulness of a vitamin or a mineral depends on its bioavailability. *Bioavailability* refers to how much of a vitamin or mineral can actually be used by the body. First the vitamin or mineral must be absorbed in our intestinal tract. Although a food enters our body when it enters our mouths, its nutrients don't nourish the body unless they are absorbed. Some chemicals, like oxalic acid in spinach and rhubarb, inhibit absorption of nutrients; other chemicals, like ascorbic acid (vitamin C), enhance absorption. As indicated above, some lipids are necessary for the absorption of lipid-soluble vitamins. The chemical form of a nutrient also affects its absorption. Even when a nutrient is absorbed, there is no guarantee that it will be used by the body. Certain forms of nutrients are more likely to be used than others.

Although vitamins and minerals are essential to health, overconsumption can be dangerous. Lipid-soluble vitamins like A, D, and E can be toxic at levels not much higher than those required. For example, too much vitamin D can cause hypercalcemia, which is an irreversible accumulation of calcium in the kidneys, heart, and lungs. Excess lipid-soluble vitamins are not excreted but are deposited in body fat. Most water-soluble vitamins are excreted in the urine when consumed in excess and do not pose a health threat. Likewise, we can overconsume minerals. For example, too much iron can cause increased oxidation of lipids in the body, and can interfere with the absorption of zinc. Too much magnesium interferes with the absorption of calcium and iron.

Electrolytes

Electrolytes are ionic forms of minerals, generally found in beverages like sports drinks. The electrolyte that we need in our cells is potassium (K^+), whereas the one found in our bloodstream is sodium (Na^+). Chloride (Cl^-) is also an important electrolyte found in conjunction with both K^+ and Na^+. For every positive charge in the body, there must be a corresponding negative charge. When table salt (NaCl), the most abundant source of sodium in our diets, is dissolved in water it dissociates into Na^+ and Cl^-. As the primary electrolyte in intracellular fluids, potassium is a critical nutrient in our diets. Low levels of K$^+$ lead to high blood pressure since it is probably the balance of K$^+$ and Na$^+$ that determines the blood pressure. K$^+$ is particularly important in helping the body burn kilocalories, in ensuring a proper acid–base balance, and in helping the muscles relax after contraction. Na$^+$ is also an essential nutrient in the body, maintaining the osmotic pressure of the blood and other fluids located outside the cell. For most of us, getting enough Na$^+$ is not a problem; rather we tend to consume too much sodium, which can cause high blood pressure. Light salts have an equal ratio of NaCl and KCl to help prevent an electrolyte imbalance. When we work out or perform other strenuous exercise, we lose electrolytes through perspiration. Sports drinks and energy bars help replenish these electrolytes. The consumption of other fluids is also recommended since perspiration leads to a loss of water as well. In particularly dry environments, we may need even more fluids since the perspiration evaporates quickly. Dehydration initially causes reduced urination and thirst; if not corrected, it can cause rapid respiration, pulse elevation, muscle spasms, nausea, and an inability to perform physically.

Alcohol

The fourth source of kilocalories in the human diet is alcohol. Ethanol provides 7 kilocalories per gram (190 kilocalories per ounce), intermediate between the other sources. A moderate amount of alcohol—a glass of wine, 12 ounces of beer, or a single mixed drink—is recommended by many physicians for good health. Originally it was thought that it was primarily the antioxidants in red wine that provided the health benefits, but more recent evidence suggests that the ethanol itself is providing health benefits. One of the most highly regarded popular guides to healthy eating suggests a daily alcoholic drink (*Eat, Drink and Be Healthy*, Willett, 2001). All too often, however, we don't limit ourselves to a single drink. Excess alcohol puts the kilocalories on quickly (see Insert 11.4). An evening of binge drinking (5 drinks for a male at a sitting and 4 drinks for a female) can result in an increase in kilocalories of about 200 per drink (a 12-ounce beer, 4-ounce glass of wine, or a mixed drink with 1 ounce of liquor) with little or no vitamin or mineral intake. Heavy drinkers tend to either gain weight (the infamous beer gut) or, to keep from gaining weight, cut back on their food intake and deprive themselves of needed nutrients. In addition, excess ethanol impairs motor skills and judgment. As a drug and a weak toxin, ethanol probably causes more emotional problems, injury, and death than any other single chemical consumed. Responsible consumption of the chemical can be enjoyable and even health promoting, but too often it is misused and abused.

Insert 11.4 Kilocalories associated with popular items on college campuses. (Calculated from data at MyPyramid.gov (http://www.mypyramid.gov/) and Starbucks Coffee (http://www.starbucks.com/retail/nutrition_beverages.asp).)

Product	Calories per serving	Calories per 5 servings
Beer, 12-ounce bottle	145	725
Beer, 28-ounce oilcan	340	1,700
Bowl of ice cream, 2 scoops	265	1,325
Cola, 12-ounce can	150	750
Light beer, 12-ounce bottle	100	500
Milk chocolate, 1 ounce	145	725
Popcorn, buttered, 2 handfuls	55	275
Rum and cola	260	1,300
Salted peanuts, 1 handful	260	1,300
Starbucks® original Frappuccino®	240	1,220
Red wine, 4-ounce glass	85	425
Strawberry Daiquiri	175	875
Wine cooler, 12 ounces	250	1,250
White wine, 4-ounce glass	80	400

Nutrient composition of foods

Most of us consume our nutrients in the form of foods. Once they enter the body they are broken down into nutrients that we can use. It is easier to design diets around foods than around nutrients. Thus, it is important to understand the composition of various food groups.

Grains

Carbohydrates are the most important nutrient in grains. Grains are cereal crops that are grown for their seeds, which are then further processed into foods or feeds. Grains are the most prominent source of carbohydrates in the world and most people around the world get more than half of their kilocalories from carbohydrates. The primary carbohydrate in grains is starch. Starch molecules have one of two structures: *linear* or *branching*. The body has enzymes to break down both linear and branched starches. As we learned in the previous chapter, grains also have indigestible carbohydrates: cellulose, hemicellulose, lignin, and pectin. These indigestible carbohydrates form the dietary fiber we need to keep us regular and help bind toxins. Milling removes the coarser fractions of the grain resulting in more desirable, more functional, and whiter flour, but it also removes the dietary fiber. MyPyramid.gov recommends increasing consumption of whole grains. It is generally agreed that the average American does not get enough dietary fiber, whereas consumers in other areas of the world may consume too much fiber, binding minerals and vitamins and preventing their absorption into the body. Generally, fiber containing more pectin and less cellulose is better at achieving desired results. Oat fiber is considered to be one of the best food fibers.

Vegetables

Vegetables are defined as vegetative plant tissues that do not contain the reproductive parts of the plant. They can include leaves (lettuce, turnip greens), flowers (broccoli, cauliflower), stems (celery, asparagus), roots (carrots, parsnips), and tubers (potatoes, sweet potatoes). Tomatoes, squash, and cucumbers are classified as fruits botanically but are commonly considered vegetables by most consumers and in MyPyramid.gov. Although vegetables have minerals and vitamins, particularly water-soluble vitamins, the most important nutrient in vegetables may be fiber. Vegetable fiber tends to be higher in pectin and lower in cellulose than most whole grains. Another nutritional benefit of most vegetables is that they have high moisture content. Foods high in moisture and fiber and low in lipids and protein are more filling and less loaded with calories. The carbohydrates in vegetables tend to be indigestible fiber and starch, not sugars. Canned and frozen vegetables tend to be lower in vitamins than fresh vegetables, but the dietary fiber is usually more beneficial to the body in processed products than fresh products.

Fruits

A fruit is defined as a ripened ovary. Like vegetables, fruits contribute vitamins, minerals, and fiber to our diets. The nutritional benefits of fruits are similar to those of vegetables. In addition to the benefits listed above, fruits and vegetables are excellent sources of potassium. Fruits tend to have less starch and more sugars than vegetables.

The fruit group in MyPyramid.gov also includes processed fruit products, including juices. Juices tend to have much less dietary fiber than whole fruits. With the exception of avocado, most fruits and vegetables are low in lipids. Americans tend to eat less fruits and vegetables than consumers in other countries.

Milk

Milk, cheese, and yogurt are considered part of the milk group. Milk products are high in calcium, protein, potassium, and vitamin D. Low-fat alternatives are available for most milk products and are highly recommended. Many cheeses are high in fats, particularly saturated fats, so caution should be exercised in choosing cheese products. Ice cream is high in fat and sugar with much less calcium than other products and thus not considered part of this food group and should be eaten only occasionally. The calcium and vitamin D present in milk products is important in maintaining bone health and preventing osteoporosis. Most milk products are fortified with vitamin D. As a fat-soluble vitamin, vitamin D is susceptible to oxidation. Oxidized vitamin D is toxic. Since milk is highly perishable, the product is likely to spoil before vitamin D is oxidized, making it an excellent fortification vehicle. It is difficult to get sufficient calcium and vitamin D in American diets without consuming adequate amounts of milk products or the use of mineral and vitamin supplements. Lactose-intolerant consumers tend to be deficient in calcium and vitamin D because they avoid milk products. Fortunately, yogurt is low in lactose and is an excellent source of both nutrients.

Meat and beans

The meat and beans group contains dry beans, eggs, fish, poultry, nuts, and red meats. The primary nutrient that links all of these products is protein. Many of these items are also rich in the B vitamins, iron, magnesium, vitamin E, and zinc. The animal products in this group tend to be higher in fat, saturated fat, minerals, and balanced proteins than the dry bean and pea products. Lacto-ovo vegetarians, who consume milk and egg products, will have no trouble balancing proteins. Strict vegetarians, however, must ensure that their protein comes from diverse sources. By careful mixing of amino acids in plant sources such as grains, beans, and seeds, vegetarians like Hannah can balance their proteins. Some examples of how to balance proteins are shown in Insert 11.5. Although many vegetables contain iron, the iron in vegetables is not as bioavailable, meaning it is not absorbed as well in the intestines; rather, it is more likely to be bound by indigestible carbohydrates and excreted in the stool. Consumption of an iron supplement such as iron-fortified yeast can help overcome this deficiency. Vitamin B_{12} (cyanocobalamin) is readily available in animal, but not plant, products. Strict vegetarians can get their vitamin B_{12} from fortified breakfast cereals, soy beverages, veggie burgers, yeast, a supplement, or a shot.

Oils

Fats are solid at room temperature and have higher levels of saturated fats than oils, which are liquid at room temperature. Oilseeds are crops whose seeds are high in lipids. Primary examples of oilseeds are peanuts,

Insert 11.5 Balancing amino acids with grains and oilseeds. (From the U.S. Department of Agriculture National Nutrient Database, http://www.nal.usda. gov/fnic/foodcomp/cgi-bin/list_nut_edit.pl.)

Food	Limiting amino acid(s)	Good source for amino acid(s)
Almonds	Lysine	Histidine
Barley	Lysine	Methionine, tyrosine
Pinto beans	Methionine	Isoleucine
Buckwheat	Lysine, phenylalanine	Tryptophan, valine
Corn	Lysine, tyrosine	Leucine, valine
Millet	Lysine	Leucine, methionine
Oats	Lysine	Histidine, methionine
Peanut	Lysine, tyrosine	Histidine, phenylalanine
Quinoa	Methionine, tryptophan	Isoleucine
Rice	Lysine	Methionine, valine
Rye	Lysine	Methionine, valine
Soybean	Methionine	Isoleucine, phenylalanine, valine
Wheat	Threonine	Methionine

soybeans, and sunflowers. For examples of the saturation of fatty acids present in selected oilseeds and meats, see Insert 11.6. Oils pressed from seeds are used for cooking, frying, and salad dressings. It is important to have some oils in our diet because they are sources of essential fatty acids and omega-three and omega-six fatty acids, which are associated with maintaining proper HDL/LDL ratios. Oils are also essential for proper absorption of lipid-soluble vitamins. It is recommended that any oil product containing trans fats be avoided.

Processed, formulated, chilled, and prepared foods

The effect of food processing on the nutritional quality of food is described in Chapter 4. Many vitamins and minerals are lost during processing. To compensate for these losses, products are enriched with nutrients during processing operations. Products such as breakfast cereals may also be fortified by increasing the amount of vitamins and minerals above those found in the original grain. Some consumers mistakenly believe that vitamins that occur naturally are better than added ones. If the natural chemical form and the synthetic form of a nutrient are the same, the body cannot distinguish the difference between the two compounds. Processing can change the form of a nutrient to increase or decrease its bioavailability. For example, in many vegetables the form of dietary fiber components has been changed to increase the beneficial aspects of the fiber. Processing can also remove components that help in the absorption

Insert 11.6 Lipids from selected animal and plant sources. (Calculated from data at MyPyramid.gov, http://www.mypyramid.gov/.)

Food	Percent daily value	
	Total fat	Saturated fats
Fried, breaded chicken breast	57	50
French fries, 8 ounces	46	25
Grilled chicken, 4 ounces	9	10
Ground beef, one-quarter pound	31	45
Macaroni and cheese, 1 cup	34	20
Oatmeal, 1 cup	3	3
Pinto beans, 1 cup	0	0
Pork chops	18	23
Quarter-pound hamburger with cheese	45	65
Salted, oil-roasted peanuts, 1 handful	37	18
T-bone steak, 12 ounces	120	175
Taco salad	86	80

of key nutrients. For example, if vitamin C is removed, calcium and iron absorption will be restricted. Generally, however, the potency of natural and added vitamins and minerals is similar. The primary purpose of processing is to increase shelf life and reduce waste. In many places around the world, simple processes like canning prevent large losses of fruits and vegetables.

As we learned in Chapter 5, formulated foods are mixtures of ingredients. Among these ingredients are nutrients such as minerals, proteins, and vitamins. Other health-promoting compounds are also included. A careful reading of the nutrition label is necessary to understand the nutritional composition of formulated foods. Many formulated foods are primary sources of carbohydrates, proteins, and lipids with low levels of a few vitamins and minerals. Others, such as energy bars, breakfast cereals, and functional foods have high levels of many nutrients. Heavy reliance on products with high levels of nutrients and then supplementing with additional vitamins and minerals could lead to overdosing, causing metabolic imbalances.

Chilled foods are whole foods that have the nutrients typical of the food group described above. Any preparation that requires heating can destroy minerals and vitamins. Minerals are not destroyed by heat, but they leach out during cooking. Quick cooling of cooked products helps keep them safe and slows the loss of vitamins. Unlike nutrients in processed and formulated foods, nutrients in chilled and prepared foods are perishable. Refrigeration will slow spoilage and vitamin loss, but vitamins are still not as stable in fresh foods as they are in processed and formulated foods.

Digestion and intermediary metabolism

As we learned earlier, kilocalories are the unit of energy in our bodies. The dietary guidelines that provide the basis for nutritional labeling assume that the typical intake of an American is 2,000 kilocalories per day. MyPyramid.gov recommends that you consume about 1,735 kilocalories to obtain all the required nutrients, allowing approximately 265 kilocalories for fun (discretionary calories). See what you can eat with 265 kilocalories in Insert 11.7. That amount, unfortunately, is much less than most of us consume.

For our bodies to use kilocalories or any other nutrient, they must be absorbed into the body after digestion. Digestion is a metabolic process that extracts nutrients from food products for use by our body. Digestion begins as we put the food in our mouth. The saliva we secrete moistens the food and releases amylase, an enzyme that breaks down starch. Nutritionists call this glob of food mixed with saliva the *bolus*. The bolus leaves the mouth, proceeds through the esophagus on its way to the stomach. In the stomach it is mixed with mucous, acid, and pepsin (an enzyme that starts breaking down proteins into short amino acid chains). The bolus dissolves in the stomach and is released slowly into the small

Insert 11.7 What you can do with the 265 discretionary kilocalories given to you by MyPyramid.gov. (Calculated from data at MyPyramid.com, http://www.mypyramid.gov/.)

1 glazed doughnut (200)
2 slices white bread (240)
1 breaded, fried chicken thigh (240)
6 breaded onion rings (245)
1, 20-ounce bottle of cola (255)
1 Milky Way candy bar (255)
1 cup of cream of mushroom soup (260)
1 small fast-food hamburger (265)
4, 1-inch cubes of cheddar cheese (270)
1 scoop premium ice cream (270)
2 whole, large bananas (275)
4 chocolate chip cookies (275)
3 tablespoons of peanut butter (285)
1 can chili with beans (290)
2 slices white bread with 2 pats of butter (310)

intestine. In the approximately 25-foot-long journey through the small intestine, numerous enzymes break down the proteins, fats, and carbohydrates into smaller compounds like individual amino acids, fatty acids, and sugars. The process of degrading the large molecules to smaller ones is called *catabolism*. Catabolism is necessary for absorption across the intestinal mucosa. Digestive aids are secreted from the gallbladder, liver, and pancreas to enhance the absorption process as it releases small molecules into the bloodstream. If a nutrient is not absorbed across the intestine, it is useless to the body. Remember that dietary fiber is not a nutrient, but it exerts its positive effect by maintaining peristalsis and binding toxins. Any undigested parts of the bolus, including dietary fiber, proceeds to the large intestine where it is stored prior to propulsion out of the body. Excreted solid human waste (stool) provides the nutritionist and epidemiologist with a gold mine of information about nutrient utilization and potential causes of food-associated illness.

The small molecules absorbed into the bloodstream are either used immediately or transported to cells throughout the body. In the cells, these small molecules are reassembled into larger molecules in a process called *anabolism*. Anabolism is necessary because we need these larger molecules to maintain proper bodily functions. Proteins are needed for muscle tissue and to function as enzymes. Lipids are needed for energy storage, padding to protect organs from physical damage, and as key components of cell membranes. Carbohydrates are needed for quick energy, and glycogen (animal starch) serves as a short-term energy reserve. Sugars can also be converted to other compounds our body needs. All anabolic reactions require energy to assemble these larger molecules. Energy is derived from catabolic reactions occurring in cells as glycogen is converted to glucose, triglycerides break down into glycerol and fatty acids, and proteins break down into amino acids. Anabolism uses energy stored in the form of adenosine triphosphate (ATP) to build larger molecules. Catabolic reactions degrade larger molecules to produce ATP. With the exception of the molecules in the lens of the eye, all other molecules are subject to turnover or replacement by other molecules. The genetic code in our DNA is what keeps us who we are despite all the modifications at the cellular and molecular level.

There are numerous things that can go wrong during digestion, and digestive disorders can reduce the quality of life. Chances for having digestive difficulties increase as we age, so here are some things you can look forward to. A low-fiber diet, low consumption of liquids, and little exercise can cause constipation. Constipation alternating with explosive diarrhea produces the symptoms of irritable bowel syndrome, which is frequently brought on by anorexia and bulimia. Swallowing air during eating can cause belching, and fermentation reactions in our intestines can cause flatulence. Consuming large meals without sufficient liquids, lying down within two hours of a meal, smoking, or obesity can cause heartburn and acid reflux. Acid reflux allows the consumer to relive the dining experience over and over again, along with the stomach acid and other digestive components. Caffeine can increase the severity of acid reflux by relaxing the epiglottis, the structure that prevents food from reentering the mouth from the esophagus. Finally, contrary to popular opinion, ulcers are caused by a bacterium, *Helicobacter pylori,* and not by stress or fatty foods. Alcohol, aspirin, caffeine, and smoking can aggravate ulcers, but they do not cause them.

Nutritional deficiency diseases

Metabolic reactions through catabolism and anabolism require enzyme catalysis. For the enzyme to act, it must have its necessary cofactors, which are derived from vitamins and minerals. If we do not consume enough kilocalories, our bodies will not have sufficient energy to use for metabolism. We will not have the materials to replace the components designated for turnover. Starvation, anorexia, and bulimia all prevent us from leading active lives, causing listlessness and eventually death. Kwashiorkor is a disease that affects people who obtain adequate kilocalories but inadequate protein. Marasmus is a disease that affects people who do not obtain enough kilocalories. If we consume enough kilocalories and protein, but the proteins are incomplete sources of amino acids, the protein anabolic reactions can't continue and our body can't synthesize needed proteins. Assuming our proteins are balanced, if we do not consume enough of a vitamin or a mineral over a long period of time, we can develop a deficiency disease. Examples of some common deficiency diseases and their symptoms are shown in Insert 11.8.

These deficiency diseases are unlike contagious diseases or food poisoning, which occur within a few hours or days after exposure, result in intense symptoms, followed by full recovery. Nutritional deficiency diseases develop slowly over a long period of time. Early indications of illness may be difficult to distinguish from other disorders or diseases. Once a deficiency disease is diagnosed, it usually takes a long time to treat because the body can't absorb and assimilate large doses of nutrients.

Antioxidants, supplements, and antinutrients

As described in Chapter 10, vitamins A and E are antioxidants that can reduce oxidation in foods and tissues. All of the so-called *diseases of civilization,* such as cancer, diabetes, and heart disease have a component that oxidizes membrane lipids in the body. A form of vitamin E, α-tocopherol, is a particularly effective antioxidant in membrane lipids. The tocopheroxyl free

Insert 11.8 Deficiency diseases resulting from lack of selected, crucial minerals and vitamins.

Disease	Deficient nutrient	Disease symptoms
Anemia	Iron	Fatigue, listless, lethargy
Beriberi	Thiamin	Wasting of muscles leading to paralysis
Childhood blindness	Vitamin A	Night blindness degenerating to total blindness
Goiter	Iodine	Enlarged neck due to enlarged thyroid
Osteoporosis	Calcium	Weakened bones susceptible to fractures
Pellagra	Niacin	Dermatitis, dementia, diarrhea, and death
Rickets	Vitamin D	Bowed legs and other abnormalities of the skeleton
Scurvy	Vitamin C	Mouth sores and bleeding gums

radical, formed upon oxidation, is stable and usually stops the chain reaction. This free radical cannot function as either a vitamin or an antioxidant until it is regenerated. Ascorbic acid (vitamin C) can be converted to dehydroascorbic acid, which in turn regenerates α-tocopherol. Dehydroascorbic acid does have some vitamin C activity, but it can be irreversibly degraded to lose all of its vitamin and regenerating capacity. Bioflavanoids, found in many berries, can regenerate the ascorbic acid as part of a protective cascade. As a result of this protective cascade, consumption of vitamins A, C, E, bioflavanoids, and other compounds with antioxidant activity is increasing in an effort to avoid the diseases of civilization. Although a deficiency of these compounds may increase our chances of getting one of these diseases, clinical tests have not shown clear evidence that larger doses effectively prevent the diseases.

Europeans and Americans have become interested in supplementing the diet with higher levels of vitamins and minerals and other compounds that appear to have health benefits but don't qualify as nutrients by classical definitions. Nutraceuticals and functional foods described in Chapter 2 fit into this category. A list of some food components with possible benefits can be found in Insert 11.9. Much information on supplements is passed from person to person and not verified by thorough scientific study. Many consumers tend to believe that if a little of a compound is good, more is better. They do not understand that compounds can be beneficial at one level and toxic at a higher level. As more detailed scientific studies are conducted, we will be able to separate fact from fiction and design foods that are both healthy and satisfying. At present, some of these dietary supplements can interact with nutrients and drugs, leading to unintended consequences.

Insert 11.9 Potential benefits of functional components in food products. (Adapted from International Food Information Council Foundation Backgrounder on Functional Foods, November 2006, at http://ific.org/ nutrition/functional.)

Potential benefit	Component	Food source
May bolster cellular antioxidant defense system	Anthocyanidins	Berries, red grapes
	Caffeic acid	Apples, pears
	Sulphoraphane	Cauliflower, cabbage
May reduce risk of coronary heart disease	Beta glucan	Oat bran, rolled oats
	Omega-3 fatty acids	Salmon, tuna
	Plant sterols	Fortified table spreads
May benefit immune system	Conjugated linoleic acid	Beef, lamb
	Dithiolthiones	Broccoli, collard greens
	Lignans	Flax seeds, rye

Some naturally occurring substances in foods can lower the nutritional quality of our diets. Earlier in this chapter we learned that excess levels of dietary fiber can bind needed nutrients, particularly minerals. Oxalates in spinach and rhubarb can bind calcium and iron and interfere with their absorption. We also learned in this chapter that certain minerals, when consumed in excess, can interfere with the absorption of other minerals. Although soy is highly recommended, it contains protease inhibitors. Proteases are the catabolic enzymes that convert proteins to individual amino acids. Protease inhibitors can help slow cancer development, but can interfere with absorption of proteins. Rice contains iron-binding compounds, which can interfere with the absorption of iron if the iron is consumed at the same time as the rice. Small amounts of any antinutrient will have little effect on people consuming adequate diets. Antinutrients can have an effect on consumers who are on the borderline of getting an adequate supply of a specific nutrient. As emphasized throughout this book, it is important to eat a balanced diet.

Remember this!

- Nutritional deficiency diseases develop slowly over a long period of time.
- Digestion is a metabolic process that extracts nutrients from food products for use by the body.

- Food preparation steps that require heating can destroy minerals and vitamins.
- By careful mixing of amino acids from plant sources such as grains, beans, and seeds, vegetarians can balance their proteins.
- Milk products are high in calcium, protein, potassium, and vitamin D.
- Like vegetables, fruits contribute vitamins, minerals, and fiber to our diets.
- As the primary electrolyte in intracellular fluids, potassium is a critical nutrient in our diets.
- The usefulness of a vitamin or a mineral in a food depends on its bioavailability.
- Responsible consumption of alcohol can be enjoyable and even health promoting, but too often it is misused and abused.
- Carbohydrates are sugars, starches, and nondigestible cell-wall materials.
- Lipids are critical components of our bodies, and we could not function without them.
- Antioxidants and dietary supplements are thought to contribute to health, but their role is not as clearly defined as calories, vitamins, and minerals.

Looking ahead

The book will conclude with chapters on food microbiology, food engineering, and sensory evaluation.

References

International Food Information Council Foundation. 2006. "Functional Foods." http://ific.org/nutrition/functional. Accessed 6/2/2008.

Tuszynski, J. A., and M. Kurzynski. 2003. *Introduction to molecular biophysics*. Boca Raton, FL: CRC Press, Taylor & Francis Group.

Willett, W. C. 2001. *Eat, drink and be healthy*. New York: Simon & Schuster.

Further reading

Bock, G., and J. Goode, eds. 2007. *Dietary supplements and health*. Chichester, U.K.: John Wiley and Sons, Ltd.

Webb, G. P. 2006. *Dietary supplements & functional foods*. Ames, IA: Blackwell Publishing Limited.

Whitney, E. N., and S. R. Rolfes. 2007. *Understanding nutrition*, 9th ed. Belmont, CA: West/Wadsworth.

Wolinsky, I., and J. A. Driskell. 2008. *Sports nutrition: Energy metabolism and exercise*. Boca Raton, FL: CRC Press, Taylor & Francis Group.

chapter twelve

Food microbiology and biological properties of foods

After her bout with *Salmonella* poisoning from the raw cookie dough she had eaten, Amy decided she wanted to learn more about the microbes that cause illness. She wanted to avoid another late-night bout hugging her toilet and to spare others the same experience. Amy majored in food science and graduated with a good record. She decided to stay on for her master's degree and chose to specialize in food microbiology. As part of her degree, she is conducting original research under a member of the faculty in the Food Science Department. She became interested in how *Salmonella* can get stressed under certain conditions (often when nutrients are unavailable). *Salmonella* cells can be injured and difficult to detect. Thus the food being tested for *Salmonella* may show no signs of its presence, but can still be quite dangerous. Thus, to detect the presence of *Salmonella*, a recovery medium must be used to culture an extract of the food sample to recover any injured cells. Amy's work is to determine at what level a specific nutrient is required to improve recovery.

Brian's passion was molecular genetics. It was a field unheard of back in the 1960s. For his science fair project in high school he extracted DNA from *Arabidopsis*, a common weed. He received his bachelor's degree in genetics with a minor in microbiology, but he became disinterested in learning all the theoretical scientific principles. He wondered how all of this knowledge applied to real life. That's when he discovered food science and learned how important genetics was in working with food materials and also in the organisms that contaminate the materials. Like Amy, Brian went on to pursue a master's degree in food science. He was particularly interested in the genetics of the microbes that are used to ferment milk into yogurt. His research looked specifically at the resistance of these bacteria to the bacteriophages (viruses that attack bacteria) that interfere with the conversion of milk to yogurt. He is now ready to start his Ph.D. research, and he will be looking at the genetics of the yeasts that are used in the production of beer and bread.

Looking back

Previous chapters focused on food issues we encounter in our daily lives, with the types of food products we encounter in the marketplace, and with activities that food scientists perform. In the last chapter we emphasized the importance of chemistry in understanding foods. Another basic science we need to know to work effectively with foods is biology, particularly microbiology. Some key points that were covered in previous chapters that prepare us for a basic understanding of food microbiology and other biological aspects of foods are the following:

- The Hazard Analysis and Critical Control Point (HACCP) system is a means of ensuring microbial safety in a product.
- The main function of a package is to prevent microbial or chemical contamination.
- Food is preserved to make it safer by reducing or eliminating harmful microbes.
- The higher the temperature and shorter the time of a heat process, the less vitamins are lost.
- Irradiation is a potent killer of microbes used to preserve foods with little or no heat.
- Fermentation is the only method of food preservation that encourages multiplication of microbes.
- Designing processing techniques requires an understanding of the microbes and the food.
- Preservatives are food additives that prevent or retard spoilage.
- The last meal consumed is not usually the meal responsible for an outbreak of food poisoning.
- The expiration date represents the food scientist's best guess of how long a food will last before it spoils.
- Spoilage is not a good indicator of a safety risk.
- Fresh foods are more likely to contain harmful microbes than processed foods.

Food microbiology

Food scientists are obsessed with microbes because microbes are responsible for almost all food-associated illness and are also the primary cause of food spoilage. Pathogens are microorganisms that cause disease and death. Not all microbes are harmful; there are beneficial ones that can control harmful ones. Beneficial microbes are also important in making yogurt, bread, beer, wine, sausages, and other fermented foods. Food scientists study microbiology to learn about microbes that contaminate food, the environment that affects their growth and behavior, how their presence affects the food and the humans who consume that food, and the conditions in the food itself. With this knowledge, the food scientist will understand how to control the microbes and will be able to produce better, safer, and more wholesome foods. Studying food microbiology will give us a better understanding of the behavior of microbes and their effect on the safety, spoilage, and preservation of foods.

Microbiology is the study of the physiology, genetics, growth characteristics, survival, and behavior of microbes (or more appropriately microorganisms). Microorganisms are generally life forms that cannot be seen by the naked eye. They are in the air we breathe, the beverages we drink, the food we eat, and everything we touch. They even inhabit our bodies! They represent the dominant life form on earth. Life would be impossible for humans without microorganisms. We depend on them in many aspects of our lives: for the air we breathe and for many of the crops we grow. Microbiology is a rapidly growing field of study. The more we learn about microorganisms, the more we learn how little we really know. It is estimated that 99% of bacterial species remain undiscovered. Much of microbiology focuses on microorganisms of medical or commercial significance. Food microbiologists focus on those that contaminate or grow in foods.

Types of microorganisms in our foods

Among the most commonly studied microorganisms are bacteria, molds, and yeasts. Viruses and parasitic protozoa are gaining interest. Microorganisms are classified in many ways: by genus and species, size and shape (morphology), chemical reactions, nutrient requirements, metabolic products, and nucleic acid composition.

Bacteria are single-celled organisms enclosed by a cell wall. Bacteria are classified as prokaryotes because they are without defined nuclei. Under the microscope, bacteria usually are round (*cocci*) or oblong (*rods*). Another way of distinguishing between bacteria is by using the Gram stain technique. Gram-positive bacteria have a thicker but less complex cell wall and are purple after staining. Gram-negative bacteria are red after staining. Bacteria reproduce asexually by cell division. Some Gram-positive bacteria form dormant, tough, resistant structures called *spores* under adverse conditions, thus ensuring the survival of the bacterium. Certain bacteria are active agents in fermentation, whereas others cause spoilage and food-associated illnesses.

Molds, yeasts, and mushrooms are known as *fungi*. Fungi are classified as eukaryotes because they have clearly defined nuclei. The composition of the fungi cell wall is different from the bacterial cell wall, and fungi can produce reproductive and nonreproductive spores. Yeasts are single-celled organisms, whereas molds form long filaments consisting of many cells. Under the right conditions, however, yeasts can also form filaments. Fungi reproduce sexually and asexually, in ways such as *budding* and *fission*. Although the individual cells of fungi cannot be seen by the naked eye, most of us have seen colonies of mold that have contaminated a food and rendered it unappetizing.

Protozoa are parasites that do not have cell walls. They are eukaryotes that are mobile and digest other microorganisms to get their nutrients. Most students are introduced to the protozoa by way of the *Amoeba* or *Paramecium* in high school or introductory biology classes. Once primarily the cause of intestinal diseases in developing countries, it is now becoming a more common source of contaminated foods around the world.

Viruses are not considered microorganisms. They are single or double strands of RNA or DNA, and they must contaminate a cell to replicate. Although viral contamination is generally associated with a disease, not all contamination causes a disease. Viruses that attack bacteria are called *bacteriophages*.

Microbial genetics

The genetic program of a cell is located in its DNA. For eukaryotes, the DNA is called the *chromosome* and is located in its nucleus. For prokaryotes the DNA is distributed throughout the cell. DNA is composed of two complimentary chains of nucleotides. DNA is organized into units called genes with each gene coding for a specific protein. Each nucleotide contains a five-carbon sugar, a phosphate group and a nitrogenous base. The two chains form a double helix. DNA is replicated by separating the two strands and making a new complimentary strand from each original strand, resulting in a total of four strands and two identical helices. Protein molecules, which affect the cell's growth and survival, are synthesized in a two-step process: *transcription* (synthesizing RNA from DNA) and *translation* (linking amino acids to synthesize proteins from RNA). Each amino acid is coded by a three-base sequence on the RNA. The proteins that have the most immediate effect on cell metabolism are enzymes. Not all genes are expressed (transcribed and translated) in the cell at all times. Expression may be induced by environmental conditions. Indeed, the ability to withstand environmental conditions and respond to specific stresses is embedded in the genes.

Everything that a cell is or does is coded in its genes. The color, size, and shape of a cell are inherent in the code. A *genotype* is the basic genetic makeup of a specific organism, which differentiates it from all other organisms. A mutation modifies the base-pair sequence of the DNA and thus the genotype. Manipulation of the genes within an organism to achieve a change in that organism is called *genetic engine*ering. These changes can promote the expression of genes that encode desirable characteristics or the changes can block the expression of genes that encode undesirable ones.

Cell physiology and reproduction

In contrast to the genotype, the *phenotype* represents the physiological response of the organism to a given environment. Frequently, microorganisms in foods are responding to environmentally stressful conditions. The reason one species can survive when another cannot is because of its ability to adapt to the environment, including the chemical composition of the food. Cell physiology includes all of the biochemical reactions that go on during cell replication, transcription, and translation. It includes all metabolic processes within the cell: the assimilation and synthesis of nutrients and their conversion to useful compounds, the transport of these compounds to the location where they are needed, and the breakdown of nutrients and other compounds. The eukaryotic cell contains organelles, such as mitochondria, chloroplasts, and the endoplasmic reticulum, which have specific functions in the metabolic process. The cell is not the static entity we see in Insert 12.1, rather it is always under reconstruction. Cells undergo the process of turnover, constantly replacing molecules within cell structures while continuing to maintain the same outward appearance because the cell's genetic material encodes these traits. When the organism is placed under stress, a complex signaling mechanism is triggered causing some genes to be expressed in order to increase synthesis of a specific enzyme to help the cell adapt to the new conditions. In addition,

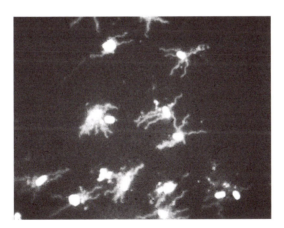

Insert 12.1 Photomicrograph of *Salmonella typhi* cells. (Photo available at http://www.asm.org/division/c/gramneg.htm and reprinted with permission of Dr. Michael J. Miller, Centers for Disease Control [CDC], Atlanta, Georgia.)

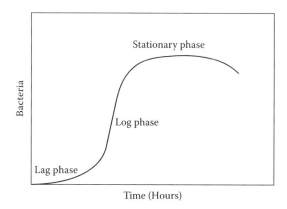

Insert 12.2 Exponential growth curve for microorganims.

other genes may be "turned off" (not expressed) to decrease the level and activity of other enzymes.

Microorganisms reproduce themselves in many ways, but the most common way is by cell division. Basically, a cell grows to the point where it divides into two identical cells. Given optimal conditions, bacterial cells can multiply rapidly, doubling in as little as thirty minutes. Such rapid multiplication can produce a million cells from one cell in a period of less than ten hours! When a cell is transferred into a new environment, it must adjust to the new conditions and will not multiply at an optimal rate until it has time to synthesize necessary proteins. This period of delay is called the *lag phase* (see Insert 12.2). Once the microorganism adapts to its new environment, it begins to multiply rapidly in the *log phase*. At a certain point, however, the level of available nutrients is not sufficient to sustain further division and the organism enters the *stationary phase*. In the stationary phase, the number of cells dividing is balanced by the numbers that are dying. Maintenance cannot be sustained forever unless the cells are transferred to a new environment. Thus, the last phase is known as the *death phase*. In stressed conditions, a microorganism may stop growing at levels way below that of a typical stationary phase. Under adverse environmental conditions, some Gram-positive bacteria will form spores from normal vegetative cells. Spores are more resistant to harsh environments than vegetative cells and much more likely to survive under these conditions. If the spores are provided with rich sources of nutrients and optimal growth temperatures, they will germinate into vegetative cells.

Sources of microbial contamination

As indicated in the first chapter, the primary sources of harmful microorganisms are soil, feces, and people. When food microbiologists like

Amy and Brian need to isolate a particular pathogen, the two places they are most likely to look are soil and feces. Most contamination of fresh fruits, vegetables, and meats is on the surface because the internal tissues of plants and animals are usually free of microorganisms. Animal intestines are colonized by both harmful and beneficial microorganisms, and feces are a rich source of these microorganisms. Our intestinal walls and immune system prevent these harmful microorganisms from getting into our bloodstream where they can do damage. However, the failure to wash our hands after defecation permits them into our stomachs where they can make us very ill. The transfer of microorganisms from contaminated hands to food or water is known as the fecal–oral route. Likewise, plant tissues can become contaminated from the soil, from improperly manufactured organic fertilizers, and from people. Animal tissues can become contaminated from hair, feathers, feces, and intestinal tracts during slaughter, collection, or milking. Adequate pasteurization of milk and juices eliminates harmful microorganisms. As we learned in Chapter 10, proper sanitation is necessary to prevent the spread of microorganisms.

Every step in the handling or processing of food products represents an opportunity for microorganisms to contaminate the food and for food scientists to control them. The goal of proper sanitation is to minimize the contamination in the first place. The goal of proper handling is to minimize the growth and spread of microorganisms. The goal of processing is to kill harmful microorganisms or prevent their growth. The goal of packaging is to prevent the recontamination of a safe product. In the farm environment, plants come in contact with soil and animals come in contact with feces. Contaminated irrigation water can also increase the level of contamination. To reduce cross-contamination, fresh fruits and vegetables should not be shipped in trailers that have transported live animals on the previous haul. Once at a warehouse or processing plant, food storage temperatures are kept low to slow microbial growth. Equipment must be kept clean to prevent the transfer of microorganisms from contaminated equipment to food products. It is particularly critical that products that will receive no heat treatment are not contaminated.

Although it is not a pleasant thought, fresh foods and raw materials are likely to contain filth. Filth in foods includes such disgusting things as fly eggs, insect parts, maggots, rodent hair and feces, and weevils. Most of these contaminants cannot be seen with the naked eye and are just as likely to be in fresh vegetables picked from a backyard garden as those bought at a farmer's market or supermarket. A systems approach to minimizing contaminants in our foods starts with the use of pesticides in the field to reduce insect populations, and carries through to adequate sanitation during harvesting, handling, and processing.

Environmental conditions affecting microbial growth

Contrary to popular belief, very few foods are sterile. The microflora (the microorganisms that are characteristic of or associated with that food) present in a food are affected by the composition of the food and its environment. As we learned in Chapter 4, canned foods are sterile, but most other foods are not. Most foods contain mixed populations of microorganisms: some could be pathogens (lead to illness), others could cause spoilage, but many have little or no effect on the food or on the consumer. Environmental conditions can increase, decrease, or change the types of microorganisms present. Eventually, however, one species tends to dominate. Which species predominates depends on which one is the quickest at adapting to the composition of the food and the environmental conditions. If the predominant species is pathogenic, then consumption of the food could cause serious health problems. If the predominant species causes spoilage, then the food is likely to spoil before it becomes a health hazard. The metabolites of the predominating species may result in self-inhibition (the microorganism may kill itself off) and conditions that favor other species.

The temperature, relative humidity, and gaseous composition of the environment affect microbial growth. The most effective way to slow microbial growth is to control the temperature. Most microorganisms grow best between 50 and 95°F (10 to 35°C). That is why we need to keep hot foods hot (higher than 50°C) and cold foods cold (less than 7°C). Bacteria that can survive and grow at high temperatures are called *thermophiles*; those that can grow at low temperatures are called *psychrophiles*. Bacteria grow at high relative humidity and can outcompete other microbes. As the relative humidity is lowered, however, bacteria lose their competitive advantage and yeast or molds are more likely to predominate. All molds require oxygen to grow and survive, but some bacteria and yeasts can grow and survive without oxygen. Organisms that can only grow in the absence of oxygen are called *anaerobes*; organisms that can grow with very small amounts of oxygen are called *microaerophiles*.

Food compositional factors affecting microbial growth

Microbial growth in foods is affected by the chemical composition of the food. Water activity, pH, oxidation-reduction potential, nutrients, growth factors, and inhibitors affect microbial growth in foods. Every microorganism must have a carbon source and a nitrogen source for basic microbial nutrition. Different species have different growth requirements. There are a wide variety of growth factors that help some species of microorganisms grow at the expense of others. Most of these growth factors are vitamins. Inhibitors are chemical compounds that slow the growth of microorganisms, with some species being more susceptible to specific compounds than others. Inhibitors added directly to foods to prevent microbial growth are called *preservatives*. Water activity (a_W) is the amount of water available to the microorganism, and is related to the relative humidity within the food itself. Bacteria require higher water activity levels than yeasts and molds. As the acidity increases (pH decreases), bacteria are less able to grow or survive. Foods with a pH of 4.5 or higher are low-acid foods; those with a pH lower than 4.5 are high-acid foods. The level of oxygen (related to the oxidation-reduction potential) is another factor that affects the competitive advantages of different species as discussed in the previous section. The observant student will recognize that the conditions affecting microorganisms are very similar to those affecting enzymes. The reason is that microorganisms grow and produce chemical products through chemical reactions that are catalyzed by enzymes, such that conditions that inhibit enzymes are likely to inhibit microbial growth.

Competition among microorganisms determines the relative populations. Some species of microorganisms require amino acids or other complex sources of nitrogen, but others can use a chemical like caffeine as both a carbon and nitrogen source. Sugars and salts bind water and thus reduce water activity. Microorganisms that can tolerate higher levels of salts than other species are called *halophiles*; those that can tolerate higher levels of sugars are called *osmophiles*. Changing the amount of sugar or salt in a product or the gases in its package can affect the competitive advantages of organisms within a food product.

One of the most popular ways of isolating and culturing microorganisms is on agar, usually in a Petri dish (see Insert 12.3 to see some examples of selected cultures). Standard plate count agar can theoretically be used to grow all the different species (microflora) present in a food. Colonies on the agar represent billions of microbial cells but are assumed to come from a single cell. Thus, by counting the colonies, we can gain a rough approximation of how many microorganisms inhabit a particular food (usually expressed as colony-forming units (CFU) per gram of a solid food or per milliliter of a liquid. Since some species can outcompete others in the food and on the plate, we must be careful in the conclusions we draw. If we wish to find a certain genera or species or related groups of microorganisms, special agar is used that has growth factors, carbon sources, nitrogen sources, and inhibitors that select for specific genera or species. Sometimes a color change in the agar provides proof of the presence of a specific organism by relying on a biochemical reaction characteristic of the organism. Physical signs include the size, shape, or color of the cells, clear rings around the colony, or production of slime. Chemical factors include the production of gas or conversion of specific chemicals

Insert 12.3 Gram-stained smear from a *Campylobacter jejuni* culture incubated for 48 hours at 42°C. (Photo available at http://www.asm.org/division/c/gramneg. htm and reprinted with permission of Dr. Michael J. Miller, Centers for Disease Control [CDC], Atlanta, Georgia.)

to reaction products, usually inducing a physical change in the growth medium. Immunological tests, based on molecules called *antibodies*, are more specific for proteins on the outer surfaces of some microorganisms. There are also tests based on the DNA content of the microorganism that are more specific.

Certain species can become injured. In the injured state, they might be present but not be capable of growing under ordinary circumstances. The food microbiologist might believe there are no organisms present, and may be surprised when the food causes an outbreak later. Such an event might occur because the microorganism in question was present at a sufficient level (although not a detectable level) to cause an outbreak or the conditions in the food or environment became more favorable for growth. Genera like *Salmonella* can be injured in food. The bacterium may not be detectable in this state and may cause a false negative result. To decrease the chances of this happening, the cells can be nurtured in special recovery broth to determine if any are present, as Amy does in her research.

Fermenting microorganisms

Although we tend to think that microorganisms in food are undesirable, some of our more popular foods are produced by beneficial microorganisms (see Insert 12.4). As indicated earlier, fermentation is the only primary method of food preservation that encourages multiplication of microorganisms. During fermentation the nutritional quality of foods

Insert 12.4 Fermented products, substrates (starting material), and the responsible microorganisms.

Product	Substrate	Fermenting microorganism
Beer	Barley	*Saccharomyces cerevisae*
Bread	Wheat	*Saccharomyces cerevisue*
Pickles	Cucumber	*Lactobacillus plantarum*
Sauerkraut	Cabbage	*Leuconostoc mesenteroides*
Soy sauce	Soybeans	*Aspergillus oryzae, Zygosaccharomyces rouxii*
Wine	Grapes	*Saccharomyces cerevisae*
Yogurt	Milk	*Streptococcus thermophilus, Lactobacillus delbruekii*

can be improved because vitamins are synthesized. The bioavailability of vitamins and other nutrients can be enhanced during fermentation, and carbohydrates may be broken down into smaller compounds by hydrolysis. *Fermentation* is generally referred to as the biological process in which microorganisms induce a series of chemical reactions leading to food preservation. A stricter chemical definition of fermentation involves carbohydrate degradation in the absence of oxygen. An example of a true fermentation involves the conversion of lactose to lactic acid to produce yogurt, sour cream, and some cheeses. The smooth mouthfeel of yogurt comes from the coagulation of milk proteins. A bacterial culture (usually a mix of *Lactobacillus delbruekki* subsp. *bulgaricus* and *Streptococcus thermophilus*) is used to start the fermentation. The distinct sour taste is due to lactic acid, but the characteristic aroma is from acetaldehyde. Yogurt is considered by many to be a "health food" primarily because it contains active bacterial cultures. The idea is that these active cultures can colonize our intestines, improving our microflora and replacing or outcompeting less desirable species. Scientific research suggests that few, if any, active yogurt cultures can survive the digestion process because stomach acid can be lethal to the culture. Thus, active cultures may not significantly affect the microflora in our intestines. *Lactobacillus acidophilus*, on the other hand, is more likely to survive our stomach's environment and be beneficial. Acidophilus products, although not as popular as yogurt, are usually available in the dairy section of most supermarkets.

Many fermented products are widely available (see Insert 12.5 for responsible microorganisms). Sometimes a single species is responsible, but at other times it takes more than one to produce an acceptable product. In yogurt, the fermentation is started with *Lactobacillus delbruekki* subsp. *bulgaricus*, which produces lactic acid and lowers the pH to about 5.5, a point at which it can no longer grow effectively. Then *Streptococcus thermophilus* starts growing and lowers the pH to the final level of 4.5. Note that the lower pH acts as a "natural" preservative because fewer species

Insert 12.5 Comparison of two fermentation microorganisms.

Microorganism	*Lactobacillus bulgaricus*	*Saccharomyces cerevisiae*
Type	Gram-positive bacterium	Yeast
Food products	Yogurt	Breads & alcoholic beverages
Chemical products	Lactic acid	Carbon dioxide & ethanol
Signature	Proteolysis	Gas production
Favorable conditions	Acid, 40–45°C	Carbohydrate-rich, 30–35°C

can grow in the high-acid environment that the two microorganisms have created in this synergistic effort. Careful control of the fermentation temperature is necessary to maintain an appropriate balance between the strong acid production of the *Lactobacillus* and the greater flavor development by the *Streptococcus*.

Other fermentations are produced by yeasts: the production of carbon dioxide gas to help dough rise in bread baking and the production of alcohol in alcoholic beverages. Remarkably the same species of yeast (*Saccharomyces cerivisiae*) is responsible for most bread and alcohol products, emphasizing the importance of the specific strain and the starting substrate. Other specialized species may be used for specific applications such as *Saccharomyes carlsbergensis* for lager beers.

A successful fermentation is dependent upon the microorganisms, raw materials, type of process, and environmental conditions. The starter culture is added to begin the fermentation with the idea that a single microorganism will predominate to induce the desired chemical changes that will produce a high quality final product. Too little of the starter may cause inadequate growth and change, but too much can be expensive. An incorrect starter culture formulation may produce a product of poor quality. Contamination of the raw material or the starter culture may promote the growth of other species and produce unwanted changes. The temperature, relative humidity, as well as food compositional factors like pH should be controlled to favor the growth of the desired species.

Spoilage microorganisms

By now we should have a clear understanding that the microorganisms that cause food spoilage are not the same ones that cause food poisoning. Food spoilage produces undesirable aromas, color defects, and slime formation. Pathogens, on the other hand, do not usually alter the flavor, color, or texture of foods because the number of pathogens necessary to cause illness is fewer than the number needed to cause spoilage. Thus, aroma is not a reliable guide in determining microbial safety. Spoilage can be

Insert 12.6 Comparison of two spoilage microorganisms.

Microorganism	*Erwinia caratova*	*Rhizopus stolinifer*
Type	Gram-negative bacterium	Mold
Food products	Fruits & vegetables	Breads
Quality defects	Soft rots	Visible mold growth
Signature	Breakdown of cell walls	Black and fuzzy
Favorable conditions	Low oxygen/30°C	High humidity/30–35°C

caused by the growth of microorganisms or by the metabolic products of microbial enzymes. Only a relatively small number of microbial genera are responsible for food spoilage (see Insert 12.6). Although it is a good idea not to eat spoiled food, we are just kidding ourselves if we think we can tell when a food is safe just by using our eyes and nose.

As we learned earlier, fresh foods are more likely to contain harmful microorganisms than processed ones. *Psychrotrophic bacteria* are spoilage organisms that effectively outcompete other microorganisms at refrigerated temperatures. *Thermophilic bacteria* can survive heat treatment and can cause spoilage of pasteurized products. *Aciduric bacteria* can tolerate high acid levels (low pH) that normally inhibit microbial growth. Generally, symptoms of spoilage do not occur until there are 10,000,000 microbial cells per gram. Sometimes eliminating microorganisms from food may provide an advantage to pathogens. Cooking followed by contamination and improper storage can speed up growth of harmful microorganisms because we have killed off their competitors. That's why it is important to refrigerate foods after they have been prepared. Objectionable odors in spoiled foods are usually breakdown products of lipids or proteins. Unsightly colors come from the synthesis of pigments. Slime can be produced by the accumulation of too many cells or the synthesis of dextran.

Some important food-spoilage bacterial genera include *Bacillus*, *Enterobacter, Erwinia, Lactobacillus, Pediococcus,* and *Psuedomonas.* Although considered commercially sterile, canned foods can spoil if *Bacillus stearothermophilus* spores or cells survive the heating process. *Bacillus stearothermophilus* is more heat stable than *Clostridium botulinum.* Various species of *Bacillus* and *Lactobacillus* are responsible for the objectionable flavors encountered in spoiled pasteurized milk. The bright fluorescent sheen on cold cuts left in the refrigerator too long is due to *Pseudomonas fluorescens,* and the strong objectionable odor could be due to various *Enterobacter* or *Serratia* species. Spoilage of fruits, vegetables, and their products is frequently caused by *Bacillus, Erwinia,* or *Pseudomonas* species of bacteria or *Penicillium* and *Alternaria* molds. Lactic acid bacteria and yeasts can cause the spoilage of fruit juices and salad dressings. Moldy bread is caused by *Penicillium* species and *Rhizopus stolinifer* (producing a nice black, fuzzy culture on the surface of the loaf). Remember that the conditions in the food itself (like pH, water activity, and presence of inhibitors) and the environment around the food (temperature and presence or absence of oxygen) determine which spoilage microorganism(s) predominate.

Few things are more frustrating than buying a food product and finding that it has spoiled before we ever have a chance to eat it. One way to help avoid that problem is to look at the expiration date. Expiration dates are usually only on perishable products like juices, milk, cold cuts, and chips, but they are becoming more common on many products. An expiration date is the food engineer's best guess at a product's shelf life. As should be obvious by now, fresh foods are more perishable than processed products. Shelf-life estimations are determined by predicting the increase of an undesirable characteristic or the decrease of a desirable characteristic. Quality losses tend to be either zero-order (linear) or first-order (semilogarithmic) reactions. Stability tests are conducted under defined conditions and mathematical models are developed. Accelerated storage tests can provide quicker but less reliable estimates of shelf life by looking at losses of product quality at temperatures higher than those at which the product would normally be stored. A prediction of shelf life is made based on expected conditions, and an expiration date is placed on the product. Despite some popular beliefs, the food does not change from perfectly good the day before the expiration date to completely spoiled the day after the expiration date. Rather, it is the cumulative result of small changes in quality. Small losses that occur each day add up to spoilage after many weeks or months. The estimate must be such that the product will be acceptable to most consumers at the expiration date, but it should not be such that people will toss out large amounts of perfectly good food just because the date has passed. Shelf life is primarily based on spoilage and not safety. If the product has recently expired, yet shows no indication of spoilage, it should be fine to eat. If the food is well past its expiration date, however, the prudent thing to do would be to toss it.

Pathogenic microorganisms

The main reason food scientists are obsessed with microorganisms is that 95% of reported outbreaks in the United States are due to microorganisms, with 4% due to chemicals and 1% due to parasites. Food-associated outbreaks are most often caused by improper storage temperatures, poor personal hygiene, inadequate cooking temperatures, cross-contamination, and unsanitary or contaminated raw materials.

Most outbreaks can be traced to a single pathogenic microorganism contaminating a single food that was either mishandled or prepared improperly (see Insert 12.7). The most common food-associated disease outbreaks are caused by *Campylobacter jejuni*, *Salmonella* species,

Insert 12.7 Comparison of three pathogenic microorganisms.

Microorganism	*Escherichia coli* O157:H7	*Listeria monocytogenes*	*Staphylococcus aureus*
Type	Gram-negative bacterium	Gram-negative bacterium	Gram-positive bacterium
Illness/ symptoms	Hemorrhagic colitis, hemorrhagic uremic syndrome	Gastroenteritis, fever	Gastroenteritis
Incubation time	2–12 days	1–7 days	1–8 hours
Food products	Meats, bean sprouts	Luncheon meats, cheese	Meats, salads, baked goods
Signature	Bloody stools	Transmission between humans	Heat-stable enterotoxins
Prevention	Pasteurization and cooking	Proper cooking	Refrigeration or heat

and *Staphylococcus aureus*. Among the most dangerous food-associated microorganisms are *Clostridium botulinum* and *Escherichia coli* O157:H7. Formulated and processed products are more susceptible to attack by a single spoilage or pathogenic microorganism for reasons discussed earlier. Most food poisoning outbreaks are caused by one of a small number of harmful microorganisms. Food-associated pathogens affect immuno-compromised individuals most severely. This group includes the very young, whose immune system has not yet fully developed, the elderly, whose immune system may be failing, pregnant women, and people suffering from chronic diseases (HIV, liver cirrhosis, etc.).

Food poisoning outbreaks are caused by either intoxications or infections. Food intoxications are caused by ingesting a toxin that has been produced by the contaminating microorganism. Only the toxin, not the pathogen, causes the illness. As the pathogen grows in the food, it produces the toxin. Some toxins are more heat stable than the microorganism that produces them. Although the toxin may survive a heat treatment, the bacterial cell does not. *Staphylococcus aureus* causes illness by this route. Food infections cause illness because ingested microorganisms that were in the food grow inside the human body. In most cases, the infecting microorganisms grow in the food. In some cases, however, very low levels of contamination are sufficient to induce illness.

Food poisoning outbreaks do not usually occur immediately after the consumption of the contaminated food and may not become evident until as much as 24 hours after consumption or longer. Although many symptoms of food-associated illnesses are the same, each organism seems to have its signature symptom. These illnesses are categorized by their incubation time (time between eating the food and becoming ill), the symptoms, and the duration of the illness. For example, botulism (an intoxication) develops within 12 to 36 hours of consumption of contaminated food. It starts with constipation, fatigue, and muscle weakness, and proceeds to droopy eyelids, blurred or double vision, dry mouth, slurred speech, and swallowing difficulties, ultimately leading to paralysis and respiratory failure in severe cases. *Escherichia coli* O157:H7 can induce hemorrhagic colitis, characterized by bloody stools; hemolytic uremic syndrome, which is the leading cause of kidney failure in children; or thrombotic thrombocytopenic purpura, which combines kidney failure with brain damage. It can be caused by either an infection or intoxication. *Clostridium perfringens,* a common source of food-associated infection, causes profuse diarrhea and acute abdominal pain within 6 to 24 hours of consumption. In most cases of bacterial food-associated illnesses, the symptoms will go away without medical treatment. Use of antibodies to treat food-associated illness is controversial because it is not clear whether they provide a cure for the illness.

Epidemiology

The science of identifying the causes of disease outbreaks is called *epidemiology.* The more we understand the cause of a specific outbreak, the greater the chance that future outbreaks will not occur. Most outbreaks are caused by a combination of mistakes that permit the presence or growth of one of a small number of pathogens in the contaminated food. In the investigation of a food-associated outbreak, the symptoms of those who become sick provide clues as to the responsible organism. Recent food consumption records are then collected for those who became sick and common foods are identified. The length of time between consumption of the offending foods and the beginning of the outbreak is also an important clue. Some patients may develop more severe symptoms than others. Possible explanations include variations in the amount of contaminated food consumed, the level of contamination in different locations in the food, and individual sensitivities to the pathogen. Potential causes of the outbreak may be traced to improper handling or preparation of the food. Fecal samples from the victims are taken, as well as samples of the suspected food, which are then analyzed and compared to see if they are both contaminated with the same pathogen. Possible points of contamination are identified by analyzing the distribution pattern from harvest to consumption. After the collection of all the information, a pattern usually emerges: the responsible species, the initial cause of contamination, and any human errors that allowed growth of the microorganism. Immediate efforts should then be taken to prevent further spreading of the outbreak. The case study can also be used to help set up guidelines to prevent similar occurrences from happening again.

Epidemiology is relatively simple when a large percentage of a group that attended the same social function comes down with an illness 12 to 16 hours after eating the same food. It becomes more difficult when only five or six people in different locations across several states report similar symptoms over a period of two weeks. Epidemiological teams from the Centers of Disease Control and Prevention, in collaboration with state and local health boards, have become very proficient at identifying causes of outbreaks and ways to help reduce their occurrences. Food-associated epidemiology presents a special challenge. Many victims do not seek medical treatment because they are better in a few days. As a result, cases are underreported in an outbreak.

Controlling microorganisms in food

The most effective way of controlling microorganisms in foods is to kill them. Food processes are primarily designed to kill microorganisms or at least to minimize their growth. Unfortunately, the same processes that can limit microbial growth can also affect quality. Since fresh foods are more likely to be unsafe than processed ones, other techniques are needed to decrease the chances for outbreaks caused by fresh foods. *Hurdles* are barriers to microbial growth in fresh and minimally processed food products. Adequate sanitation is important in preventing illness and spoilage by lowering the chances for initial contamination. Processing decreases microbial populations and packaging prevents recontamination. The hurdle concept takes advantage of the combination of several techniques to slow the growth of microorganisms, to lengthen shelf life, and to maintain safety. Using hurdle technology, microorganisms may become injured by one technique and killed by another. For example, adding an organic acid to a product may injure the cell, and then a heat treatment may kill it. A lower heat treatment can be used because the cell was already weakened by the acid. Mathematical models are developed to predict microbial growth and used to estimate the safety and stability of foods as affected by these hurdles under different storage conditions. Examples of potential hurdles include low-heat treatment, low pH, low water activity, modification of atmospheres, preservatives, and ultrahigh hydrostatic pressure.

Preservatives are food additives that prevent or retard spoilage. Sucrose (table sugar) and sodium chloride (table salt) are the two most widely used preservatives. Many foods that are advertised as having "no preservatives" are actually preserved by sucrose or sodium chloride. Both of these preservatives function mainly to reduce water activity, although sodium chloride can also directly inhibit many species of microorganisms. Organic acids, like lactic and citric acid, lower pH, inhibiting the growth of most bacteria. Benzoates and sorbates are salts of organic acids that inhibit microbial growth. Hydrogen peroxide destroys *Salmonella* in egg and dairy products. Nitrates help prevent growth of *Clostridium botulinum* in vacuum-packaged, cured meat products. Sulfur dioxide and sulfites are added to fruit juices, dried fruit, and wines to prevent growth of unwanted species. These are just a few examples of the many applications of the preservatives that are used in the food industry.

The importance of food preparation in maintaining safety must be emphasized because improperly prepared foods can cause disease and death. Sanitation is critical with raw meats, since all raw meats should be considered contaminated with pathogenic microorganisms. Anything raw meat touches, such as a cutting board, knife, or fork can become contaminated. If these are then used to prepare foods that are not going to be heated (like a salad), the cross-contamination can be dangerous. Cooking kills the microorganisms on the surface of the meat, but the microorganisms in the uncooked salad will continue to multiply. In a beef steak that has not been poked with a fork or other implement, the inside of the muscle tissue is considered sterile. That is why a rare beef steak is generally considered safe to eat, but a rare hamburger (where the surface has been ground and has contaminated the rest of the meat) is not considered safe to eat. Poultry and pork needs to be thoroughly cooked to be safe.

Biological properties of foods

As we saw in the previous chapter, food scientists are very interested in chemistry. In this chapter we learned about the importance of microbiology for a basic understanding of foods. Although not emphasized as much in food science, the biology of food materials is also important. Most raw materials of foods were living tissues at one time. Living organisms are governed by physiological processes. Many raw products are distributed from the farm to the supermarket with little or no processing. Fresh fruits and vegetables have been detached from a parent plant. The changes that occur during storage of fresh fruits and vegetables are caused by the physiology of life, as well as the physiology of stress. Fresh meats are derived from animal muscles. The changes that occur during the conversion of muscle to meat are caused by the physiology of death. Most raw items are perishable and require refrigeration or processing to prevent losses. Raw milk is a biological fluid and is kept under sanitary conditions, pasteurized, and then refrigerated. Understanding the basic physiological principles governing food tissues and fluids is important in providing the best quality foods to the consumer.

Plants and animals are eukaryotes. Unlike microorganisms, plant and animal cells become specialized into tissues within the larger organism. These tissues perform specific functions within the plant or animal when alive and convey different sensory sensations related to food quality. Even though cells from different tissues are very different, they all contain the same genetic code that all other cells in that organism contain. Cell differentiation is a complex process that is not fully understood. The genetic code for many plants and animals is being studied to identify the genes that are responsible for specific characteristics. Manipulation of these genes (genetic engineering) can change the code and modify expression. Since almost all quality characteristics of a food tissue are the result of enzyme activity, changes in key enzymes can have a dramatic effect on the final product quality. Examples of foods that have had recent genetic modifications include soybeans with less saturated lipids, tomatoes that ripen with fuller flavor as they soften more slowly, and rice that contains vitamin A.

Postharvest physiology of fresh fruits and vegetables

When we say that fresh fruits and vegetables are alive we mean that they respire, transpire, senesce, and die. Postharvest physiologists study these processes in edible, detached, living plant tissues, which the rest of us call fruits and vegetables. *Respiration* is the plant equivalent of breathing and transpiration is similar to sweating. Respiration involves the intake of oxygen and the production of carbon dioxide. If you learned in biology class that plants reduce the level of carbon dioxide and increase the level of oxygen in the earth's atmosphere, you are probably confused. Green plants are net users of carbon dioxide and producers of oxygen through photosynthesis, which requires light. Although photosynthesis provides energy for the plant, respiration operates continually in both the light and the dark. Since most storage of fresh fruits and vegetables is in the dark, postharvest physiologists tend to ignore photosynthesis and focus on respiration. *Transpiration* involves the transfer of water vapor into the environment. *Senescence* (somewhat like senility) has been associated with a general degradation of plant tissue over time, leading to death. Once thought primarily to be a function of age, it is now clear that senescence is preprogrammed into the genes. Strategies for extending the shelf life of fresh fruits and vegetables centers around slowing respiration, transpiration, and senescence, as well as preventing microbial decay. Many of these strategies place stress on the fruit or vegetable and can cause premature spoilage.

The maturity of fresh fruits and vegetables at harvest affects their quality and storage stability. A fruit is considered mature when it is either fully ripe or can ripen after detachment from the plant. Fruits like apples, bananas, and tomatoes that do ripen off the plant are called *climacteric fruits* (see Insert 12.8). In addition to respiring, they produce ethylene gas, which also functions as the ripening hormone. Generally speaking, as a climacteric fruit becomes more mature or ripe it also becomes more

Insert 12.8 Examples of climacteric fruits (ripen off the plant) and nonclimacteric fruits (do not ripen off the plant). Note that some of these "fruits" are normally considered vegetables, but all items listed below are fruits botanically. True vegetables are nonclimacteric.

Climacteric	Nonclimacteric
Apple	Bell pepper
Avocado	Cherry
Banana	Grape
Mango	Orange
Peach	Pineapple
Pear	Strawberry
Tomato	Watermelon

perishable. If it is harvested too ripe, it is likely to spoil (by either microbial decay or senescence) before it is consumed. On the other hand, if it is harvested too early, a climacteric fruit may not develop its full flavor and won't be as desirable when ripe. Apples and bananas are examples of a climacteric fruit that can be held for long periods in controlled atmospheres (high carbon dioxide, low oxygen, no ethylene) to slow respiration prior to controlled ripening in special chambers. *Nonclimacteric fruit*, like oranges and grapes, should not be harvested until they are fully ripe. They tend to deteriorate more rapidly after detachment than they do on the plant. Unlike fruits, vegetables are more desirable (tender and succulent) and perishable when they are immature and more stable as they mature. Thus, harvesters of fresh fruits and vegetables must carefully match the maturity of their harvested crops to the expected storage and distribution requirements of the market.

More and more fruits and vegetables are being cut, sliced, and packaged to provide more convenience to the consumer. The quality of fresh-cut salad products is dependent on sophisticated packaging systems. Cutting is like wounding and the cut cells initiate a wound response and generate ethylene. Cutting tends to accelerate quality deterioration, so cutting of highly perishable product may be done in processing facilities close to the consumer. Although the generation of ethylene ripens fruits, many vegetables are sensitive to ethylene and will ripen too quickly, which will lower the quality of the vegetable. Gas composition in a package is modified by increasing carbon dioxide and lowering oxygen to slow spoilage of perishable products. If all the oxygen is depleted, however, the tissue can no longer respire and fermentation begins. Although alcohol may be desirable to a beer drinker, it and the other odors produced are not considered desirable qualities in fresh fruits and vegetables. Some vegetables, like lettuce, are sensitive to carbon dioxide, so handlers must be aware of the specific needs of the item in the package.

Physiology of muscle foods

When we eat meat we are eating muscle tissue. In the living animal, muscles respire, taking in oxygen and expelling carbon dioxide. Upon slaughter, the muscle no longer has a fresh supply of oxygen. Respiration ceases and adenosine triphosphate (ATP), the major way cells store energy, is no longer produced. As the muscle uses up the remaining ATP, the two proteins (actin and myosin) in muscle tissue responsible for contraction become irreversibly bound to each other. This binding results in a toughening of the muscle known as *rigor*. Rigor mortis in beef and pork is particularly unpleasant. Rigor can be resolved by hanging the carcasses up in cold storage. Enzymes called *proteases* break down these proteins at places away from the main binding site, thus having a tenderizing effect.

The color of fresh red muscle is due to the pigment myoglobin. Fresh meat is purple and not red as popularly thought. Upon exposure to oxygen, myoglobin is converted to oxymyoglobin and the muscle turns red. However, oxymyoglobin degrades into metmyoglobin, which is brown in color. As meat ages, proteases from bacteria can release odors that we associate with age and spoiled meats.

Remember this!

- Food scientists study microbiology to learn about the types of microbes that can contaminate food and how their presence affects the food and the people who consume the food.
- Among the most commonly studied microorganisms in foods are bacteria, molds, and yeasts.
- Every step in the handling or processing of food products represents an opportunity for microorganisms to contaminate the food and for food scientists to control them.
- Contrary to popular belief, very few foods are sterile.
- Microbial growth in foods is affected by the chemical composition of the food.
- Some primary symptoms of food spoilage include undesirable aromas, color defects, and slime formation.
- Most outbreaks are caused by a combination of mistakes that permit the presence or growth of one of a small number of pathogens in the contaminated food.
- Hurdles are barriers to microbial growth.

- Most raw agricultural crops are perishable and require careful handling or processing to prevent losses.
- Fresh fruits and vegetables are alive.

Looking ahead

This chapter introduced you to food microbiology. Chapter 13 introduces food engineering, including the design of processes to protect us from disease and foods from spoilage due to unwanted microorganisms. The final chapter describes the science associated with the use of humans to evaluate the quality of food products.

Further reading

Doyle, M. P., and L. R. Beuchat. 2007. *Food microbiology: Fundamentals and frontiers*, 3rd ed. Washington, DC: ASM Press.

Jay, J. M., M. J. Loessner, and D. A. Golden. 2005. *Modern food microbiology*, 7th ed. New York: Springer Science.

Montville, T. J., and K. R. Matthews. 2005. *Food microbiology: An introduction*. 2005. Washington, DC: ASM Press.

Ortega, Y. R. 2006. *Foodborne parasites*. New York: Springer Science.

Ray, B. 2004. *Fundamental food microbiology*, 3rd ed. Boca Raton, FL: CRC Press, Taylor & Francis Group.

Riemann, H. P., and D. O. Cliver. 2006. *Foodborne infections and intoxications*. 2006. San Diego, CA: Academic Press.

chapter thirteen

Food engineering

Rakesh K. Singh and Robert L. Shewfelt

Shaundra was concerned about excess water use by food processors and about water pollution. She decided to study food engineering to be able to help lower water use and reduce water pollution by the food industry. She soon learned that she would need to become familiar with many concepts such as mass and energy balance, biological oxygen demand, and the carbon-to-nitrogen ratio. The more she learned, the more she realized that there are many things that can be done to better conserve our resources, but food safety is always the biggest concern of a food engineer.

Frank wondered why some of his protein shakes had a better mouth-feel than others. In one of his classes he was exposed to a science called *rheology*, which uses engineering principles to understand the properties of liquid and semisolid food products. He decided to do an undergraduate research project on the properties of several of the shakes he had tried to see what made some better than others. He measured the viscosity of the products and compared the values with the ingredients listed on the label. Now he understands why it is so difficult to get those protein powders into solution.

Looking back

The two previous chapters provided a perspective on food chemistry and nutrition. To preserve foods and to understand the physical structure of foods, a food scientist must also be familiar with basic engineering principles. Concepts developed in earlier chapters that recognize the importance of food engineering include:

- Every step in the handling or processing of food products represents an opportunity for microorganisms to contaminate the food and for food scientists to control them.
- Molecular mobility encompasses the rate of movement of all molecules within a food product.

- A process developer uses a pilot plant, consisting of miniaturized pieces of equipment that mimic those in the manufacturing plant, to optimize process operations.
- Food processors must be careful to prevent contamination, beginning with the raw materials and ingredients through every unit operation, using sanitation as the key to preventing contamination.
- Food processing and preservation increases the shelf stability of a raw material, usually at the cost of affordability, nutrition, and quality of the product.
- Unit operations are distinct steps common to many food processes.
- Designing processing techniques requires an understanding of the microbes and the food.
- The higher the temperature and shorter the time of a heat process, the less vitamins are lost.
- Technology has transformed our food supply: More formulated foods are produced, leading to the consumption of less whole foods.
- Fresh foods are more likely to contain harmful microbes than processed foods.

Engineering principles

Food engineering is a subdiscipline of food science, dedicated to the engineering processes related to food production, particularly industrialized foods. Food engineers apply engineering principles to help understand food materials undergoing various processes and the optimization of the processing operation. Genetic engineering of plants and animals is not normally the work of a food engineer. A food engineer by training is interested in numbers, shows a natural curiosity about the workings of a system, and has the ability to design working systems.

Mass balance

The food industry extensively uses the concept of *conservation of mass* via mass balance (or material balance). The mass, in the form of raw ingredients or semiprocessed, come into the processing facility and the same mass must either leave the facility as a final product or waste, or be accumulated in the warehouse as inventory for future use. This concept is used in all the processes such as drying, evaporation, concentration, peeling, cutting, canning, etc. An example of a canning retort is shown in Insert 13.1. The form of mass could be changed from liquid to solid or added liquid in solid depending on the final product. The concept of mass balance can be applied on total mass

$$\text{Mass In} = \text{Mass Out} + \text{Accumulation}$$

Insert 13.1 Retort for sterilization of packaged foods used in producing canned foods in flexible pouches, such as those used for Meals Ready to Eat (MREs) for the military and disaster relief. (Photo by Katherine Erickson.)

or on individual components, such as water fraction, solid fraction, minerals, or proteins. Sometimes the mass will also change phase, such as in evaporation of water or condensation of steam to water.

Energy balance

Similar to the conservation of mass concept, the law of *energy conservation* is also used in food engineering. The law of conservation of energy states that the total amount of energy in any isolated system remains constant but cannot be recreated, although it may change forms; for example, friction turns kinetic energy into thermal energy. In short, the law of conservation of energy states that energy cannot be created or destroyed; it can only be changed from one form to another. The unit of energy is the BTU (British thermal unit) or kilo Joules (kJ).

Determining an energy balance for a system is very similar to determining the mass balance, but there are a few differences to remember:

- a specific system might be closed in a mass balance sense, but open as far as the energy balance is concerned, and
- while it is possible to have more than one mass balance for a system, there can be only one energy balance.

If a balance is determined for total energy, the energy balance becomes

Energy In = Energy Out + Accumulation.

Notice that there is no production term since energy cannot be produced, only converted. If instead, some kinds of energy are ignored, for example, if a heat balance is made, the energy balance becomes

Energy In + Production = Energy Out + Accumulation.

If heat is consumed, the production term is negative. Energy may enter or leave a system but is always conserved. Therefore, one should always be able to account for all the energy or energy flows associated with a process.

There are only two basic forms of energy: *kinetic energy*, which is associated with motion, and *potential energy*, which is associated with the position or proximity to other systems. Energy is normally computed or measured relative to some reference state. The basic measure of energy is the work required to bring the system to its designated state or condition relative to some initial reference state in an adiabatic (no heat) process. The total energy of a system is the sum of its bulk kinetic and potential energies, plus all the microscopic kinetic and potential energies, or internal energy.

Heat transfer

Heating and cooling are extensively used in food processing and storage. The cooling of fruits and vegetables, freezing of foods, sterilization of canned foods, pasteurization of milk, and drying of foods are some examples of processes that employ heat transfer.

Heat transfer is the transmission of energy from a higher temperature object to a lower temperature object. There are three modes of heat transfer: (1) conduction, (2) convection, and (3) radiation. *Conduction* occurs within a solid object or from one body to another by interchange of kinetic energy between molecules being heated or cooled. *Convection* transfers heat by the combined action of conduction, energy storage, and mixing. Thermal *radiation* describes the electromagnetic radiation that has been observed to be emitted at the surface of a body that has been thermally excited. The emitted radiation may be reflected, transmitted, or absorbed by another body. The rate at which a body reflects or absorbs radiation is termed *emissivity*. Freezing equipment is shown in Insert 13.2.

All materials involved in heat transfer are either conductors or insulators. If a material transmits heat readily, it is called a *conductor*; if it resists heat transfer, it is an *insulator*. The ability to transfer heat is dependent on the chemical composition, molecular structure, and spatial homogeneity of the material.

Insert 13.2 Cabinet for quick freezing of meat with liquid nitrogen. (Photo by Katherine Erickson.)

Thermal conductivity is a physical property of materials used for predicting the heating of products. It basically denotes the ease with which heat is transmitted through a homogeneous material. It incorporates heat flux, area for heat transfer, and thickness of the material. Impingement ovens heat the object from both sides simultaneously (Insert 13.3). The opposite of thermal conductivity is called *thermal resistivity*, which reflects the insulation value of a material. Many insulating materials are specified by their insulation value or R-value.

Thermal conductivity is dependent on the characteristics of materials, their moisture content, and temperature. High-moisture foods have very little airspace and their thermal conductivity is very close to that of water. Also, foods show higher thermal conductivity at higher temperatures. Thermal conductivity can be directly measured and used for prediction

Insert 13.3 Impingement oven. An impingement oven is frequently used in restaurants. It heats the item, such as a hamburger, on each side, allowing heat penetration in half the time. (Photo by Katherine Erickson.)

of heating and cooling times, as well as for developing models that can predict thermal conductivity based on liquid and solid fractions in the food. However, there are some limitations in predicting thermal conductivity of fibrous materials, such as beef, due to the fact that the value parallel to fibers is approximately 15 to 20% higher than that perpendicular to fibers.

Mass transfer

Mass transfer is commonly used in engineering for physical processes that involve transport of atoms and molecules within physical systems. Mass transfer includes both fluid flow and separation unit operations. Some common examples of mass transfer processes are:

- Evaporation of water during dehydration or concentration
- Diffusion of salt or nutrients in food products
- Separation of components in a distillation column or during membrane filtration
- Migration of flavors or moisture through a package

Mass transfer may take place at a surface or inside materials. The driving force is a difference in concentration as the random motion of molecules causes a net transfer of mass from an area of high concentration to an area of low concentration. The amount of mass transfer can be quantified through the calculation and application of mass transfer coefficients. Mass transfer finds extensive application in food process engineering, where material balance on components is performed.

For example, water evaporates from a food surface that is exposed to an unsaturated air stream at the same temperature as the water. In this case the water vapor pressure at the food surface is higher than that in the air, and thus the evaporation of water occurs from the food to the air. The required heat of evaporation is taken from the water itself, and the temperature of water drops, causing the heat to flow from air to water at a rate proportional to the temperature difference. This action results in a dynamic equilibrium in which the water temperature does not decrease any further and the heat supplied by the air is just sufficient to balance the heat required for steady evaporation. The temperature of water under this condition is called the *wet bulb temperature*, which is particularly useful in calculating the drying rate and drying times. See Insert 13.4 for a picture of an air dryer.

Rheology

Engineering can be rather theoretical and boring, but it can also be very useful. For example, food engineers can improve the taste and mouthfeel of a product. Studies have shown that increasing viscosity reduces flavor perception. The ability to perceive flavors relies upon the flavoring

Insert 13.4 Air dehydration of sweet potato slices.

compound's ability to diffuse from the food onto the tongue. Diffusion is an engineering process because it deals with rate of mass transfer from the product to the taste buds. The level of gelling or thickening agent increases the product's viscosity, thereby slowing the flavor release from food to taste and aroma receptors. As a result, increasing gel strength decreases the perceived flavor intensity. The low melting point temperature of gelatin facilitates the release of flavoring compounds at mouth temperatures of 30 to 35°C. The same amount of flavoring in a texturizer like carrageenan under similar texture conditions will yield lower perceived flavor. However, stronger gels will take longer to chew and increase the flavor perception time. Thus, a food engineer can tailor the viscosity of a product to control the diffusion of flavor or its perception time.

A food engineer can also help make the mouthfeel or viscosity of a low-carbohydrate beverage similar to that of a full-carbohydrate beverage so that it doesn't seem as "watery." In a low-sugar (low-carbohydrate) beverage, 15% of sugar is replaced with water and a very small amount of sweetener, which makes the viscosity so low that the product tastes thin and has very poor mouthfeel. This problem can be corrected by adding 0.10% high-methoxy-pectin, which increases the viscosity and improves the mouthfeel.

Engineers can also help stabilize the protein beverages Frank lives on. Whey and soy proteins are becoming common for producing healthy drinks, but the sedimentation is a major problem with whey proteins. Manufacturers can modify the whey ingredient to improve heat stability. Hydrolyzing some of the whey protein is one way to increase the heat stability for whey ingredients used in retort or UHT (ultra-high temperature) products. Another way to improve protein stability is by reducing the particle size or adding stabilizers that keep insoluble components in suspension. Again, viscosity and flow characteristics of liquid foods are extremely important in developing stable and flavorful products.

Fluid flow is an abstract concept, but an understanding of fluid flow can lead to better products. Many food products exist in liquid form and undergo manufacturing processes to make them palatable and safe. The handling of fluid products (fruit juices, sauce, pastes, etc.) is done by pumping them through pipes and processing equipment. Fluid motion requires energy input to compensate for flow losses. The manner in which a fluid flows depends on its viscosity, density, velocity, and size, shape, and the roughness of the pipe or duct through which it flows.

Viscosity is a measure of resistance to flow of a fluid. The molecules of a fluid are constantly in a random motion, but the net velocity in any direction is zero unless an external force is applied to cause the fluid to flow. The amount of force required to cause flow at a certain velocity is related to the viscosity of that fluid. Flow occurs when fluid molecules slip past one another in a particular direction, which means that there is a difference in velocity among adjacent molecules. The resistance of the fluid to flow in a defined plane is called *shear stress*, and the velocity difference between adjacent molecules is called *shear rate*. If a plot of shear stress (Y-axis) against shear rate (X-axis) shows a linear relationship and passes through the origin (0, 0), the fluid is a *Newtonian fluid*, and if the relationship is nonlinear, it is a *non-Newtonian fluid*. The slope of the linear plot is called *viscosity*. Water, milk, and dilute sugar solutions are examples of Newtonian fluids. The viscosity of water is 1 centi Poise (cP) and that of milk is about 2 cP. Thick pastes, purees, and concentrated juices are examples of non-Newtonian fluids. The non-Newtonian fluids are either shear thinning (pseudoplastic) or thickening (dilatant). Apple sauce and mayonnaise are psuedoplastic products, and some types of honey are dilatant.

Some fluids exhibit a threshold shear stress, called *yield stress*, which must be overcome before the fluid starts to flow. If the relationship between the shear stress and shear rate is linear and the fluid exhibits yield stress, it is a *Bingham plastic*; if a nonlinear relationship is observed with yield stress, it is called a *Bingham pseudoplastic* or *Herschel-Bulkley plastic*. Tomato paste and fat-free cream cheese are examples of Bingham plastic products, while a full-fat cream cheese is an example of a Herschel-Bulkley plastic. The effect of lipids on textural properties of cream cheese is an example of the importance of the functional properties of chemical components in a product on sensory quality.

Fluid flow may be broadly characterized as *laminar* or *turbulent*. In laminar flow, the path of each molecule of fluid follows a well-defined streamline, which can be demonstrated by injecting a fine stream of dye in the flow stream when a clear fluid is pumped through a straight transparent pipe. Laminar flow happens when the flow rate or velocity of fluid is relatively low. Turbulent flow occurs when the flow velocity increases, leading to more frequent collision of fluid molecules and mixing of the streamlines (see Insert 13.5). Eddy currents develop and cause a certain amount of mixing in the fluid. This can be easily demonstrated by increasing the flow rate of a colored dye during experiments for laminar flow until the mixing of dye is noticed.

The distinction between laminar and turbulent flow is made on the basis of the Reynolds number (Re), which is a dimensionless quantity. The Re is calculated by multiplying tube diameter, average velocity, and density of the fluid, and dividing that number by the viscosity. If the value of Re is below 2,100, the fluid flow is laminar, and if it is higher than 4,000, the flow is turbulent. However, the flow is considered transitional from laminar to turbulent for Re between 2,100 and 4,000.

Food engineers measure viscosity using instruments called *viscometers* (an example of a viscometer is shown in Insert 13.6). Viscosity of Newtonian fluids, such as water and carbonated beverages, can be easily measured by inducing measurable flow and measuring the applied

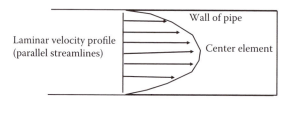

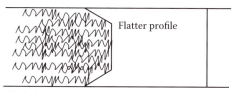

Turbulent velocity profile (mixing of stream lines)

Insert 13.5 Diagram of the differences between laminar and turbulent flow in a pipe. (From R. K. Singh, "Aseptic Processing," in *Advances in Thermal and Non-Thermal Food Preservation*, ed. G. Tewari and V. K. Juneja, Ames, IA: Blackwell Publishing Professional, 2007. With permission.)

force. The geometry in which the flow is induced must be simple so that the force and flow can be translated into shear stress and shear rate. Most common viscometers are either capillary tube or rotational types. Many formulated foods require a tight control on viscosity, which can be monitored on the processing line during the manufacturing of the product. The viscosity of ketchup is monitored during the blending of its ingredients, tomato paste with sugar, vinegar, and flavorings.

Another important concept in understanding fluid dynamics is *pressure drop*. The flow of fluid foods through a tube causes pressure drop. The drop in pressure is equivalent to stress, also called *frictional resistance*, which must be overcome by external force to induce the flow. The pressure drop is a function of fluid properties, like density and viscosity, and the dimensions of the tube. Thus the pressure drop must be calculated for the fluid under the specified flow condition. The pressure drop is used for calculation of the power required to pump that product. The pumping requirement is calculated using a mechanical energy balance on the flow system.

Water management

Water is used for cleaning equipment, cooling containers, and as a component of some foods. In each case, only potable (drinkable) water

Insert 13.6 Viscometers measure viscosity (resistance to flow) of liquid and semisolid products. (Photo by George Cavender.)

should be used, and it may therefore be necessary to treat water before it is used. There are two types of treatments: removal of suspended soils and removal or destruction of microorganisms. Suspended soils can be removed by allowing them to settle out in settling tanks, and/or filtering the water through specially designed water filters. Both processes are relatively slow, and large storage tanks are necessary if water is needed for washing or incorporation into the product. Water for cooling can be recirculated. A microfiltration unit is shown in Insert 13.7.

Although some types of water filters also remove microorganisms, the easiest way of destroying them is to add chlorine solution (5 to 8 ppm final concentration of chlorine obtained by diluting bleach to 0.02 to 0.04%). Lower chlorine levels (e.g., 0.5 ppm) are needed if the water is to be used in a product to prevent off-flavors. Chlorination of water supplies can be simply arranged by allowing bleach to drip, at a fixed rate, into storage tanks or pipelines. The rate of bleach addition is found by experiment, using simple chlorine paper or more sophisticated probes to check the

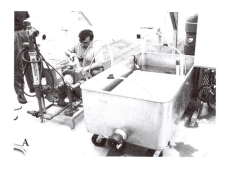

Insert 13.7 Purification of water in a processing plant using a module micro-filtration system. (A) Permeate is drawn on suction side of pump. Low trans-membrane pressure (< 1 Bar); One pump can handle 4 modules with total membrane area of 64 m². (B) Air sparging outside membrane prevents foulant buildup on surface.

chlorine concentration. A less suitable alternative is to boil water to steril-ize it. Water should be heated to boiling and then boiled vigorously for at least ten minutes. This of course has a high fuel requirement and will therefore increase processing costs. Some relatively newer techniques to sanitize water for food processing applications are ultraviolet light (UV) or ozone treatments.

Water can be a medium for heat transfer from a heating unit to food. Heat is transferred throughout water through conduction and convection or by phase change if steam is used. Boiling, steaming, and simmering are popular cooking methods that often require immersing food in water or its gaseous state, steam.

Water plays many critical roles within the field of food science. It is important for a food scientist to understand the roles that water plays within food processing to ensure the success of their products. Water hardness is also a critical factor in food processing. The hardness of the water relates to its mineral content. Hard water has much higher levels of minerals than soft water. Mineral content of water can dramatically affect the quality of a product as well as playing a role in sanitation. Water hardness is classified based on the amounts of removable calcium car-bonate salt it contains per gallon. Water hardness is measured in grains; 0.064 g calcium carbonate is equivalent to one grain of hardness. Water is classified as soft if it contains 1 to 4 grains, medium if it contains 5 to 10 grains, and hard if it contains 11 to 20 grains. The hardness of water may be altered or treated by using a chemical ion exchange system. The hardness of water also affects its pH balance, which plays a critical role in food processing. For example, hard water prevents successful production of clear beverages. Water hardness also affects sanitation; with increasing hardness, there is a loss of effectiveness for its use as a sanitizer.

Shaundra was pleased to learn that saving water also saves wastewater treatment costs by reducing the volume of wastewater. Food processing plants must monitor their water use in specific unit operations and use various techniques to reduce usage. One of the techniques to reduce the volume of wash water is to use timers on wash nozzles and use high pressure air with low volumes of water to clean the floors and walls. Another common strategy used is to recycle and reuse the water wherever possible without compromising food safety.

Energy use in food processing plants

The food industry, just like any manufacturing industry, uses energy in the form of electricity and fossil fuel. Energy use in the U.S. food processing system has grown rapidly in the last 70 years. Energy use in the food industry increased at an annual percentage rate of 3.8% from 1940 to 1970 as indicated by Unklesbay and Unklesbay (1982). This energy use growth rate was double the population growth rate over the same period. Industry's increasing use of energy was further impacted by higher fuel costs in the 1970s, but prices returned to lower levels in the 1980s until 2004. An increase in energy consumption between now and 2020 is projected to be at an annual rate of 1.6% compared to a projected increase of 2.8% in U.S. industry based on a report by Unruh (2002). Increased fuel costs may decrease consumption as companies in the food and other industries seek to reduce manufacturing costs. Today's new food products and packages include microwaveable individual servings and shelf-stable foods—all energy-consuming items. Also, there is a trend to more convenience and away-from-home food consumption, which requires even more energy for further processing and refrigerated storage.

The food industry uses energy for food preservation, safe and convenient packaging, and storage. Food preservation is dependent on strict temperature controls. Safe and convenient packaging is extremely important in food manufacturing and is also energy intensive. The newest packaging techniques require aseptic techniques and electrochemical changes. Proper storage is also energy dependent. Freezing and drying are the most crucial methods of food storage. Freezing operations require a large portion of electricity used by industries. Drying procedures usually depend on fossil fuels. Older dehydration systems were designed to operate with maximum throughput, disregarding energy efficiency. Newer systems are designed with recirculating dampers and thermal energy recovery equipment to cut energy use by 40%.

Approximately half of all energy end-use consumption is used for converting raw materials into products (process use). Process uses include process heating and cooling, refrigeration, machine drive (mechanical energy), and electrochemical processes. Less than 8% of the energy consumed by manufacturing is for nonprocess uses, including facility heating, ventilation, refrigeration, lighting, facility support, on-site transportation, and conventional electricity generation. Boiler fuel represents nearly one-third of end-use consumption. This energy was transformed into another energy source. For example, boiler fuel can be used to produce steam, which can have end uses. Since several food processing operations require steam, a large amount of energy in terms of enthalpy of vaporization (latent heat) is needed to convert water into steam. Steam tables help food engineers calculate the enthalpy or energy associated with steam or water used in processing plants.

Major energy savings in the food processing industry can be accomplished by energy conservation, waste utilization, and heat recovery from one unit operation to be used in another. Many food processing plants reuse the steam condensate from heat exchangers to heat the boiler as boiler feed water. Recovery of vapors from juice or milk evaporation plants is used for energy conservation in mechanical vapor recompression units. Hot water usage in terms of amount and duration is also curtailed to save energy. Such initiatives are ongoing efforts by the food industry to minimize energy usage. Some food processing operations, such as sugar refineries, use their waste to generate energy via either direct combustion or fermentation of sugars into alcohol. The economics of energy generation are dependent upon the characteristics of food processing waste and the process used for energy generation.

Handling processing wastes

Every food processing operation generates waste in solid and liquid forms. All food processing plants, including canneries, meat and fresh produce packinghouses, cheese manufacturers, and wineries generate solid and liquid wastes, which contain significant quantities of organic matter, nutrients, and salts. Therefore, both types of wastes can be either reprocessed to make by-products or disposed of properly. Examples of wastes include fruit and vegetable peels, seeds, pits, cheese whey, blood, bone, tofu whey, process water, and wastewater treatment sludge (see Insert 13.8).

Processing facilities must follow the U.S. Food and Drug Administration (FDA) Good Manufacturing Practices (GMP) to maintain the facility and grounds in an acceptable manner (21 CFR [Code of Federal Regulations] Part 110 Food Regulation, 110.20 Plant and grounds):

The grounds about a food plant under the control of the operator shall be kept in a condition that will protect against the contamination of food. The methods for adequate maintenance of grounds include, but are not limited to:

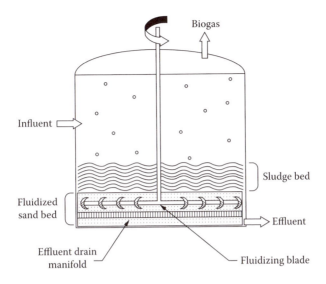

Insert 13.8 Diagram of a NewBio hybrid bioreactor for purifying process-ing wastes. (From R. A. Korus, "Anaerobic Processes for the Treatment of Food Processing Wastes," in *Food Biotechnology*, ed. K. Shetty, G. Paliyath, A. Pometto, and R. E. Levin, Boca Raton, FL: CRC Press, Taylor & Francis Group, 2006, chap. 3.21. With permission.)

1. Properly storing equipment, removing litter and waste, and cutting weeds or grass within the immediate vicinity of the plant buildings or structures that may constitute an attractant, breeding place, or harborage for pests.
2. Maintaining roads, yards, and parking lots so that they do not con-stitute a source of contamination in areas where food is exposed.
3. Adequately draining areas that may contribute contamination to food by seepage, foot-borne filth, or providing a breeding place for pests.
4. Operating systems for waste treatment and disposal in an adequate manner so that they do not constitute a source of contamination in areas where food is exposed.

Most food processing residuals (solid wastes) are used for animal feed. They may be fed fresh, as silage, or as dried waste solids. Citrus, pineapple, and potato processing residuals are almost entirely fed to ani-mals. Other fruit and vegetable residuals, however, are not widely used in animal feed rations because they are available only during the short processing session. The food processing waste characteristics of interest for animal uses include energy and fat, fiber, minerals, protein, and water. Solid waste that cannot be used as animal feed due to their nutrient or microbial characteristics can be used as soil conditioner or compost. Some characteristics of food processing waste that are important for its use as soil conditioner include biological oxygen demand (BOD), calcium carbonate balance, carbon-to-nitrogen ratio, fats and oils, odor, heavy metals, pathogens, pH, soluble salts, and toxicity.

Some liquid wastes such as cheese whey can easily be processed to make lactose and whey proteins, dried whey powder, or whey concentrate. Similarly, some of the wastewater generated by washing machinery and floors can still be recycled if treated properly according to the legal codes designated for that particular operation. However, the majority of liquid wastes must be disposed of quickly to prevent environmental contamination. If the wastewater is to be disposed of on land, it must be screened through 10- to 20-mesh screens to remove silt and other insoluble solids. The pH of the wastewater is usually adjusted from 6.4 to 8.4 because the pH outside this range will render some nutrients in the wastewater inaccessible to plants. The most commonly used methods for land disposal of food processing residuals are sprinkler irrigation and surface irrigation. The wastewater can also be ultrafiltered to recover clean water and/or feed-grade ingredients, such as starch and protein from wet milling operations.

Food processing waste must be properly treated before stream disposal. Biological secondary treatment is used to remove both the BOD and suspended solids from the wastewater. It may be carried out aerobically or in the absence of oxygen. The activated sludge process is most widely used for waste treatment. The treatment unit consists of a bioreactor, which provides an environment for converting soluble waste solids into insoluble microbial cells under aerobic conditions, and a clarifier where microbial cells are allowed to settle. The settled cells or sludge may be returned to the bioreactor or wasted. Removal of BOD by this method ranges from 80 to 99% depending on the waste characteristics, loading rate, and other operating conditions. Food processing wastes usually have a nutrient deficiency and require the addition of both nitrogen and phosphorus. For optimum biological stabilization of the wastewater, it is necessary to have a BOD-to-nitrogen-to-phosphorous ratio of 100:5:1.

Many opportunities exist for recovery of residuals for human and animal uses. A variety of processes have been developed for converting these residuals into fuels, food ingredients, and other valuable products. A number of waste treatment processes are available to make the wastewater suitable for discharge. Most solid wastes are disposed of by returning them to the land. The key to minimizing the disposal costs is to remove excessive moisture from the waste materials.

Remember this!

- Every food processing operation generates waste in solid and liquid forms.
- The food industry uses energy for food preservation, safe and convenient packaging, and storage.
- Water is used for cleaning equipment, cooling containers, and as a component of some foods.
- Viscosity is a measure of resistance to flow of a fluid.
- Mass transfer is used for physical processes that involve transport of atoms and molecules within physical systems.
- Heat transfer is the transmission of energy from a higher-temperature object to a lower-temperature object.
- There are only two basic forms of energy: kinetic energy, which is associated with motion, and potential energy, which is associated with the position or proximity to other systems.
- The mass in the form of raw ingredients or semiprocessed come into the processing facility, and the same mass must either leave the facility as a final product or waste, or be accumulated in the warehouse as inventory for future use.

Looking ahead

The final chapter in the book describes how we evaluate the quality of foods using humans instead of instruments.

References

Korus, R. A. 2006. Anaerobic processes for the treatment of food processing wastes. In *Food biotechnology*, ed. K. Shetty, G. Paliyath, A. Pometto, and R. E. Levin. Boca Raton, FL: CRC Press, Taylor & Francis Group, chap. 3.21.

Singh, R. K. 2007. Aseptic processing. In *Advances in thermal and non-thermal food preservation*, ed. G. Tewari and V. K. Juneja. Ames, IA: Blackwell Publishing Professional.

Unklesbay, N., and K. Unklesbay. 1982. *Energy management in food service*. Westport, CT: AVI Publishing Co.

Unruh, B. 2002. Delivered energy consumption projections by the food industry. In *Annual energy outlook*, Energy Information Administration, U.S. Department of Energy, http://tonto.eia.doe.gov/FTPROOT/forecasting/consumption.pdf. Accessed December 10,2008.

Further reading

Hang, Y. D. 2004. Management and utilization of food processing wastes, *J. Food Science* 69: CRH104.

Klemeš, J., R. Smith, and J. K. Kim. 2008. *Handbook of water and energy management in food processing.* Cambridge, U.K.: Woodhead Publishing Limited.

Liu, S. X. 2007. *Food and agricultural wastewater utilization and treatment.* Ames, IA: Blackwell Publishing.

Masanet, E., E. Worrell, W. Graus, and C. Galitsky. 2007. *Energy efficiency improvement and cost saving opportunities for the fruit and vegetable processing industry— An ENERGY STAR® guide for energy and plant managers.* Berkeley, CA: Ernest Orlando Lawrence Berkeley National Laboratory.

Rao, M. A., S. S. H. Rizvi, and A. K. Datta. 2005. *Engineering properties of foods*, 3rd ed. Boca Raton, FL: CRC Press, Taylor & Francis Group.

Singh, R. P., and D. R. Heldman. 2001. *Introduction to food engineering*, 3rd ed. New York: Academic Press.

Tewari, G., and V. K. Juneja. 2007. *Advances in thermal and non-thermal food preservation.* Ames, IA: Blackwell Publishing Professional.

Toledo, R. T. 2007. *Fundamentals of food process engineering*, 3rd ed. New York: Springer Science.

U.S. Food and Drug Administration, 2008. Titled 21-Code of Federal Regulations, Part 110 "Current good manufacturing practice in manufacturing, packing, or holding human food." http://www.accessdata.Fda.gov/scripts/cdrh/cfdocs/cFcfr/CFRSearch.cfm?Fr=110.20. Accessed December 10, 2008.

Vaclavik ,V. A., and E. W. Christian. 2008. *Essentials of food science*, 3rd ed. New York: Springer Scientific.

Wilhelm, L. R., D. A. Suter, and G. H. Brusewitz. 2004. Energy use in food processing. In *Food process engineering technology*. St. Joseph, MI: American Society of Agricultural Engineers, chap. 11.

chapter fourteen

Sensory evaluation

Paula is skeptical of genetically modified foods, yet she is intrigued with the idea of golden rice. Now here is one GMO (genetically modified organism) that would really help many people around the world. Then her Asian friends told her that they didn't think most Asians would eat it no matter how hungry they were. For her graduate project she wants to test consumer acceptability of golden rice to see if the color really turns off consumers.

Jennifer is disgusted with the flavor of the tomatoes she buys in the supermarket. She remembers how her dad grew great tomatoes in their backyard. This memory has sparked her interest, and she wants to know why his tomatoes taste so much better than the store-bought variety. She also wants to know the chemistry behind the difference.

Rob likes steak more than any other food. Although he understands the importance of steak flavor, he realizes that a steak must be tender but not too tender. He wants to know just what it is that makes steaks and other meats tender.

Paula, Jennifer, and Rob are food scientists who need to learn sensory techniques to answer the questions they have posed. Although technology is advancing rapidly, there are still some measurements that are best made by humans rather than by instruments.

Looking back

Previous chapters focused on food issues we deal with in our daily lives, with the types of food products we encounter in the marketplace, with activities that food scientists perform, and with the basic sciences that provide the foundation for this discipline. Although food scientists use many instruments and other physicochemical techniques to analyze foods, some research techniques require human subjects. Sensory evaluation of food products is one of these techniques. The following key points covered in previous chapters will help prepare us for sensory studies:

- Primary symptoms of food spoilage include undesirable aromas, color defects, and slime formation.
- Flavor can be narrowly defined as the combination of the senses of taste and smell, or broadly defined to encompass all sensations encountered in the eating experience.

- During market testing, image characteristics are more closely evaluated than sensory characteristics.
- Product quality is defined by the properties of the food that can be measured by food scientists.
- Quality includes sensory characteristics, which can be readily detected by the five senses, and hidden characteristics, which are not readily detectable by consumers.
- For a quality measurement to be useful, it must be accurate, precise, sensitive, and relevant.
- The ultimate goal in ensuring product quality is to satisfy the consumer.
- While quality focuses on the characteristics of the product, acceptability must consider the attitudes of the consumer.
- The conditions that affect nutritional value also affect sensory quality.
- Sensory properties include color, flavor, and texture.

We perceive quality through the five senses (sight, hearing, smell, taste, and touch). Color and appearance are judged by sight, flavor by the senses of smell and taste, and texture by touch. When food scientists measure quality with human subjects, it is called *sensory evaluation*. Sensory evaluation is important because it provides us with insight into how consumers react to the quality of a product. Sensory evaluation is performed using human panelists to make judgments on the differences in samples or to describe sensory attributes.

Sensory quality of foods

Color is our first indicator of whether a food is acceptable. We expect certain colors with certain foods and tend to reject those that do not meet these expectations. We are also likely to inspect a product for blemishes or other visual defects, even if they are not a good indicator of a product's flavor. For example, most American consumers reject bananas with brown spots on them, even though a banana with brown spots is probably sweeter than one without spots. Flavor is the combination of taste and aroma. When we say that a food tastes great, we generally mean that it has good flavor. Taste is generally limited to the sensations of sweet, sour, salty, and bitter. Flavor is what we usually associate with quality, but texture also plays a role. We reject foods if they are too hard, too mushy, or too gummy. Finally, the snap, crackle, pop of a popular cereal or the sizzle of a grilled steak add to the quality of a product.

Sensory perception and physiological response

Our sensory experience usually begins by looking at the food. We see the color or colors of a particular food, shininess, bruises, and visual texture.

We may then feel the texture of the food as we handle it and smell the aroma in the air. The aroma of the food as perceived by the nose before it enters the mouth is called *orthonasal perception.*

Once the product is in the mouth, a different sensory experience occurs. Taste is perceived by receptors on our tongues and classified as sweet, sour, bitter, salty, or umami (a meaty taste). Aroma is the perception of volatile compounds by receptors in the nose both before and during the eating process. Aromas are not only perceived orthonasally before consuming the food but also retronasally through the back of the throat during chewing. Flavor perception is a result of an interaction between the palate and a mixture of organic and inorganic compounds, which changes with time and mouth action. Textural properties such as crispness, crunchiness, stickiness, and sliminess also become evident during chewing.

To separate taste sensations from aroma sensations, we can block our nasal passages with a nose clip or chew with our mouth closed. These two simple techniques prevent aroma perception while permitting full taste perception. Some individuals are super tasters in that they have a very low threshold for taste compounds, particularly bitter. They tend to be much more sensitive to small changes in formulations that affect taste. Regardless of how we interact with the sensory properties of the food, signals are sent from the receptor site (nerve, tastebud, etc.) to the brain for interpretation. These messages are communicated through a series of chemical impulses that can be recorded.

Color and appearance

The colors we see in a food product are the result of natural pigments in the food or the addition of natural or artificial colorants. Among the natural colorants are those produced in plants and animals or by microbes. The green in vegetables is due to the presence of the pigment chlorophyll. As a peach ripens, the chlorophyll disappears to reveal red anthocyanins and yellow carotenoids and flavanoids. The yellow is actually present throughout the peach during ripening, but it is masked by other pigments. Most peaches are only red on the side that faces the sun when it is on the tree, as sunlight is needed to synthesize the anthocyanin pigment. Anthocyanins are responsible for the reds, blues, and purples of many fruits and vegetables, such as strawberries, blueberries, grapes, and eggplant. The red color of tomatoes and watermelon flesh is due to the carotenoid, lycopene, whereas the red of peppers is capsanthin. Other carotenoids such as β-carotene provide the orange of carrots, cantaloupes, and salmon. Oxymyoglobin provides the bright red of fresh hamburger, whereas nitrosomyoglobin is the pigment responsible for the characteristic color of luncheon meats like bologna. Browning of fresh fruits and vegetables is due to the formation of quinine polymers from the action of the enzyme polyphenoloxidase. The brown of chocolate and bread crust is due to melanoidins formed by Maillard browning reactions. The microbe, *Colletotrichum musae*, causes anthracnose on banana peel in the late stages of ripening to form the brown spots.

Plant and animal pigments tend to be unstable, frequently deteriorating more quickly than flavor, thus causing us to reject food that is otherwise safe and flavorful. Food processes, particularly those involving heat, cause colors to fade due to the degradation of the pigments. Artificial colorants are brighter and more stable than many natural pigments. They are also usually more potent and can be added in much smaller quantities to achieve the same effect, thus they are the ingredients of choice for most formulated foods. Food labels that have ingredients such as Yellow 5, Blue 1, or Red 40 contain artificial colorants.

Color perception is affected by more than food chemistry. Individuals who are color-blind have difficulty perceiving certain colors, such as red and green. The color of foods may not be as important to consumers who are color-blind as they are to those who are not. The amount and type of light can also affect our color perception. The appearance of a fresh apple under fluorescent lighting may not be as appealing as it is in bright sunlight. Certain colors clash with other colors, whereas some colors are complimentary. This knowledge is important to designers of food packages who face the challenge to attract the attention of food purchasers and motivate them to buy the product.

According to the color opponent theory, red and green are opposite colors as are yellow and blue. Colorimeters, instruments that measure the color of an object, take advantage of the color opponent theory to provide the hue, value, and chroma of a food. The colorimeter provides three readings: L_*, a_*, and b_*. L_* is the lightness or value of the product with a reading of 100 pure white and 0 pure black as shown in the color space in Insert 14.1. Red and green character is measured by a_* with a positive number indicating that it is more red than green, and a negative number indicating that it is more green than red. Likewise, a positive reading for b_* indicates that the product is more yellow than blue and a negative b_* reading indicates that it is more blue than yellow. Using a trigonometric transformation ($\tan^{-1} b_*/a_*$), we can calculate hue angle, which relates to the color name such that red has a hue angle of 0°; orange, 45°; yellow, 90°; green, 180°; blue, 270°; and purple, 315°.

Chroma ($\sqrt{a_*^2 + b_*^2}$) is an indication of saturation or brightness: the higher the number the brighter the color. Colorimeters are very useful with products that have a homogeneous color. With samples that have different colors, such as red and white stripes, they will only provide an average color.

Since color perception can influence flavor perception, it may become necessary, when conducting sensory panels, to screen out the color by

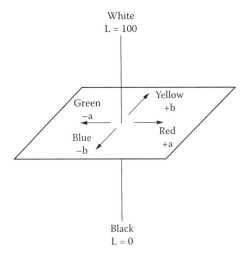

Insert 14.1 HunterLab colorspace used to quantify color measurements of foods. (Line drawing reprinted with permission from HunterLab; previously published in *Applications Note* 8, no. 9 (1996): 1.)

using colored lights. The use of red lights, for example, can prevent a panelist from seeing the color differences of food samples.

Flavor

Taste is the perception of specific compounds on the tongue. Sugars like sucrose and fructose, as well as artificial compounds, interact with receptors on the tongue to elicit a sweet taste. Organic acids, like citric acid in oranges and acetic acid in vinegar, are responsible for the taste of sourness. In some sweet fruits, the perception of sweetness is related to the ratio of sugars to acids. Sodium chloride registers as salty on the tongue, whereas compounds like caffeine, quinine, and tannins are bitter. Monosodium glutamate is the compound most closely associated with umami.

Aroma is much more complex than taste. Thousands of chemical compounds interact with receptors in our nose, either orthonasally or retronasally, to convey characteristic aromas. Some foods have character-impact compounds in which a single compound can convey the aroma of that food. Raspberries, for example, contain 4-(4-hydroxyphenyl)-butan-2-one, a character-impact compound known as the raspberry ketone. Benzaldehyde is the character-impact compound for cherries and almonds. Some character-impact compounds are shown in Insert 14.2. The characteristic odor of most foods is, however, due to an interaction of many compounds. The aromatic compounds of fresh fruits develop during ripening. When these compounds accumulate to too high a level, the fruit is

4-(4-hydroxyphenyl)-butan-2-one

Benzaldehyde

Iso-amy l acetate

Nootkatone

Insert 14.2 Character-impact compounds of fresh fruits. Can you guess the fruit associated with each structure? Answers are at the end of the chapter.

considered overripe. The homegrown tomatoes Jennifer likes so much are picked and consumed close to their peak flavor. Supermarket tomatoes, however, are picked before they reach their peak flavor to allow for shipment to the store and storage in the home. Early picking is done so that the tomatoes will not be overripe by the time they reach the supermarket.

The physiology of fresh meats during storage can produce both desirable and undesirable aromas. Likewise, cheese aroma develops over time in a process also know as ripening. Heating during cooking and other food processes transforms flavor compounds into those we associate with toasted bread, grilled chicken, pasteurized milk, hot chocolate, vegetable casseroles, and numerous other foods.

Flavor perception involves a complex series of reactions of the food with our nose, tongue, and other parts of our mouth. Flavor perception starts with the peeling of an orange, brewing of coffee, or baking of fresh bread, which sends volatile aromatic compounds into the air. The volatile compounds travel to the olfactory receptors in our nose and start our saliva and digestive juices flowing. When the food enters the mouth, chewing breaks the food down into smaller components and releases nonvolatile taste and volatile aromatic compounds. These compounds interact to produce a unique flavor for each food item. Flavor perception changes as we continue to chew. Most consumers integrate these sensations into a single response that is generally considered pleasant, neutral, or unpleasant. Expert tasters, however, can detect specific flavor notes and evaluate the changes using time-intensity techniques.

Flavor profiles can be developed for each food product. Flavor compounds contribute to the initial impact of the product, referred to as the *top note*. Enhancers and masks may have little direct impact on the flavor, but they can affect the perception of other compounds in the food.

Still other compounds provide a product with background, which adds richness to the flavor. Also, some compounds contribute to an aftertaste, which can be either pleasant or unpleasant.

Analytical chemists have developed sophisticated techniques to determine the chemical components of food products that contribute to taste and aroma. High-pressure (or high-performance) liquid chromatography (HPLC) is used to separate individual sugars and organic acids. Gas chromatography (GC) is used to separate volatile aromatic compounds. By splitting the signal, the individual compounds are sent to a nose port, where an analyst can sniff them and describe the specific aroma associated with that compound. When a gas chromatograph is coupled with a mass spectrometer, the volatile compounds can be identified.

Another instrumental approach to flavor evaluation is the use of electronic noses and tongues. These instruments use a series of sensors to detect different patterns of response to the chemicals present, either the volatile components in the headspace above the sample or the components in solution. Advanced statistical analysis can then be used to compare the patterns with each other and cluster similar patterns. Although these techniques provide no information on the chemical composition of the products, they can be useful in quality control to identify items that do not provide a close enough match to the standard. Thorough testing must be conducted to calibrate these instruments to ensure that items classified as similar to the standard are acceptable and that those classified as not similar are not acceptable. These instruments also have a tendency to "drift," meaning that after continued use they may provide different responses to the same samples. Periodic recalibration is necessary.

Texture

The texture of a food relates to the sense of touch. Although just looking at a product can give us an idea of whether it is rough or smooth, firm or mushy, many times we will touch the product with our hands or with a utensil to decide whether we wish to consume it. A hard peach or tomato is not likely to be eaten. Many items that cannot be cut by the side of a fork are not considered acceptable for eating. The tender steaks Rob likes are the result of many factors including the resolution of rigor and perhaps the addition of meat tenderizers.

Once we get a product into our mouth, the texture of the product becomes even more important in determining if it is acceptable. For example, most consumers like their chips and celery to be crisp, their shakes to be thick and creamy, their meat and vegetables to be tender, and their bread to be light and fluffy. Conversely most consumers do not like their creamed potatoes to be lumpy, their okra or luncheon meats to be slimy, their peaches or cherries to be crunchy, or their soft drinks to be watery.

There are instruments that measure textural properties of food products. Individual instruments have been designed to measure the firmness and toughness of fruits and vegetables, the toughness of wieners and other meat products, or the consistency and viscosity of semisolid and liquid products. One instrument incorporates a set of dentures to help determine the effect of chewing on the breakdown of foods. An Instron Universal Testing Machine comes with attachments that allow an analyst to perform different types of tests on many different types of foods. The texture profile technique is accomplished using an Instron to measure such characteristics as gumminess, chewiness, and adhesiveness.

Sensory tests

Although there are advantages to using instruments to measure sensory characteristics, there are some attributes that are very difficult to determine accurately with instruments. Sensory panels are often used to measure quality because the results can be more relevant to consumer acceptability than those obtained from instrumental tests. Some food scientists dislike sensory tests because they consider them to be "subjective" and not "objective." Sensory scientists argue that a sensory test that follows a given protocol closely can be as accurate and precise as an instrumental test, and that a well-conducted sensory test that measures an attribute relevant to consumer perception of quality is better than an "objective" test that has no relevance to consumer perception.

All quality tests must be designed carefully to obtain meaningful information, but attention to detail is even more critical in sensory testing because of the difficulties associated with using human subjects. Human subjects are more likely to be rushed, less likely to work nights and weekends, more likely to complain, less concerned about repeating an experiment, and more likely to get sick than instruments. These reasons are why sensory testing requires careful planning and more complex logistics than most experiments. Any sensory test must have a clear-cut objective. A specific test is then chosen to match the objective of the study. In universities, any test involving human subjects must be approved by an institutional review board to ensure that such subjects are not at risk. Samples must be prepared using proper sanitary standards and served at the appropriate temperature for consumption. Samples must be presented to the panelists in random order to prevent panelist bias. The panel room must be held at a comfortable temperature for the panelists and free from distractions such as noise and interfering odors (see a panel room in Insert 14.3). Panelists must be separated so they are unable to compare responses.

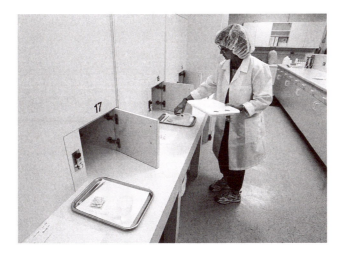

Insert 14.3 Analytical sensory tests are conducted in controlled booths. (Photo by John Amis.)

Sensory tests can take many forms based on the objective. For example, if a company wants to tell if there is a difference between the leading brand and its copycat version, or between the current formulation and a new one with less-expensive ingredients, there are difference tests that can provide the answer. If there are differences between two products, a sensory descriptive panel can determine the specific differences and help quantify those differences. Consumer tests can help establish what consumers like and what they don't like.

Difference tests

Many times we just want to know whether two samples are different from each other. Too often experimenters design elaborate sensory tests when a simple difference test is what is really needed. Difference tests are very useful when we want to know if a product is comparable to the competition, what effect a new ingredient has on the acceptability of a current or new product, or if the quality of a product is deteriorating during storage. Several types of tests are used to determine differences; most of them involve three samples. In a triangle test, a panelist is asked to choose the different one. In a duo-trio test, two samples are identified as different and the panelist is asked to match the third sample to one of the first two. Another variation is to match a standard with one of the other two samples. Difference test ballots are shown in Insert 14.4.

These tests are very good at identifying a sample that is different, but they have limitations. Sometimes a test is designed to determine

a. Triangle Test

Project Name:						Date:			
Age range									
<20	15-20	21-25	26-30	31-35	36-40	41-45	46-50	51-55	56-60	>60

Gender:	F/M		Smoker?	Y/N

Instructions:
1. Taste the samples from left to right.
2. Two of the samples are identical. Check the box that corresponds to the odd sample.
   If no difference is present, you must guess.

☐ 965   ☐   043   ☐   180

Comments:

_____

_____

_____

b. Duo-Trio Test

Project Name:						Date:			
Age range									
<20	15-20	21-25	26-30	31-35	36-40	41-45	46-50	51-55	56-60	>60

Gender :	F/M		Smoker?	Y/N

Instructions:
1. Taste the samples from left to right.
2. One of the samples is identical to the reference sample. Check the box that
   corresponds to the reference sample. If no difference is present, you must guess.

☐ reference   ☐ 970   ☐ 799

Comments:

_____

_____

_____

Insert 14.4 Ballots for sensory difference tests.

differences in flavor, but panelists are able to distinguish differences based on color or texture. To solve this problem, colored lights can be used to screen out differences in color. Texture, however, is more difficult to hide. Also, because of the statistical tests used, negative results don't necessarily mean there is no difference. It just means that any differences are not detectable using that test. In most cases the lack of a statistically significant difference is good enough to conclude that the samples are similar, but occasionally such conclusions can prove costly.

For example, a beverage company reformulated its product using a less expensive ingredient, performed a difference test, found no difference, and changed their formulation. Over the next two years they made more changes in ingredients and performed a series of difference tests on the subsequent formulations. When a marked drop in sales of their product was reported, a sensory scientist compared their current product with the original one and found that the two were quite different. Although each change in itself was not enough to produce a noticeable difference, the sum of the changes was large enough to produce consumer dissatisfaction and decreased sales.

Another limitation of difference tests is that they don't tell us which sample is better or why. One way to get more information is to use a difference-preference test, which asks the panelist which sample they prefer and why they prefer it. The problem with such questions is that they can take a panelist's mind off the task at hand and provide erroneous results. Also, a preference test usually requires a larger sample size than a difference test to determine if there is a true preference. This problem can be solved by performing a difference test first. If it is determined that there is a difference, then a preference test can be performed to find out which sample is best. A descriptive test can be conducted to help understand why one sample is preferred to another sample.

Thresholds

Threshold testing is performed to determine the level at which an ingredient or chemical compound can be detected or recognized. This type of testing is done when it is important to know if an ingredient or a compound is making a difference in a product. There are two types of thresholds, *detection* and *recognition*. A detection threshold is the concentration of a substance at which a panel can consistently detect a difference. A recognition threshold of a substance is the level at which the panel can determine the identity of the substance. For example, if peppermint was added to a chewing gum, the detection threshold would be the amount of peppermint that would need to be added to make a difference in the aroma or flavor of the gum, but it would not be recognized as peppermint. More peppermint would need to be added to clearly identify the flavor as peppermint. The threshold for a given substance is calculated as the geometric mean of panelist responses.

Thresholds can be very useful, but the tests are very labor intensive and time consuming. Only one substance can be tested at a time, and it must be sampled at many dilutions. Frequently they are diluted with water, which may or may not be related to their role in a food product with many other ingredients.

Sensory descriptive analysis

Many times we need to know more about a product or sample than just the difference or more about an ingredient than its threshold. Sensory descriptive analysis is performed to provide a description of the components of sensory quality that can be detected. These components are called *notes*. The notes can be quantified to better understand the relative contribution of each note. Differences between samples can be evaluated to help understand why they are different and what can be done to improve food quality.

Sensory descriptive tests are very complex and require extensive preparation. Panelists are screened for the ability to detect small differences. They are then trained to ensure that the panel obtains consistent results. Descriptive panels are more sensitive to small differences in products than the general public. It is not a good idea to include trained descriptive panelists in other types of panels. For many products there are published lexicons, a list of specific descriptive notes for that product. If there is no lexicon for a product, the panel director works with the panel to develop a lexicon. Typical notes include sweet, acid or sour, salty, bitter, fruity, and earthy. There are many different types of descriptive analysis. In many of them a set of standards is developed for each note, such as different concentrations of sugar for the sweet note and of caffeine for the bitter note. Descriptive tests can also be used to partition taste and aroma. Taste and aroma are partitioned by using a swimmer's nose plug to block the nasal passage of panelists so that taste can be detected but not aroma. The taste notes can be evaluated first followed by the aromatic notes. A ballot is shown in Insert 14.5.

Sensory descriptive tests provide a profile of the notes that contribute to the flavor or texture of a sample. Differences between samples can be demonstrated by spider-web graphs as shown in Insert 14.6. Like other types of sensory tests, descriptive analysis has limitations. Although a trained panel may find pronounced differences in different products, an untrained panel may not be able to detect overall differences in the samples. Sensory descriptive analysis also does not determine which product is best. Preference testing is needed to determine which products are best.

Integrating sensory and physicochemical tests

Sensory tests are difficult to conduct. They require hours of sample preparation and the participation of sensory panelists. Sensory panelists usually have other duties and responsibilities and are rarely fully compensated for their time and effort. Sensory tests are necessary to make certain determinations, but most food scientists would prefer to develop a physical test using an instrument or direct chemical test to make comparisons between

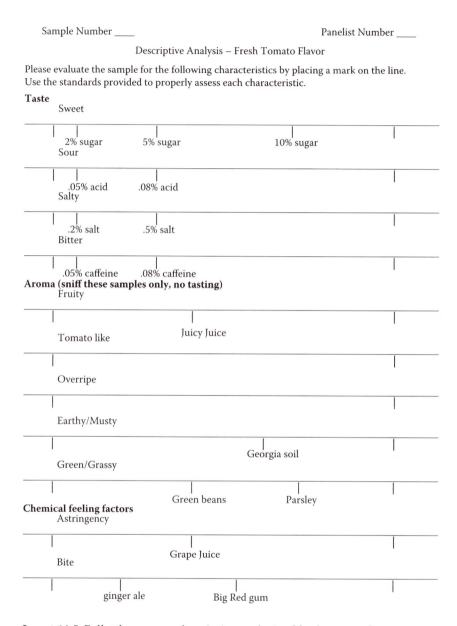

Sample Number ____   Panelist Number ____

Descriptive Analysis – Fresh Tomato Flavor

Please evaluate the sample for the following characteristics by placing a mark on the line.
Use the standards provided to properly assess each characteristic.

Taste
    Sweet

    2% sugar   5% sugar   10% sugar
    Sour

    .05% acid   .08% acid
    Salty

    .2% salt   .5% salt
    Bitter

    .05% caffeine   .08% caffeine
Aroma (sniff these samples only, no tasting)
    Fruity

                          Juicy Juice
    Tomato like

    Overripe

    Earthy/Musty

                              Georgia soil
    Green/Grassy

                          Green beans   Parsley
Chemical feeling factors
    Astringency

                          Grape Juice
    Bite

            ginger ale   Big Red gum

Insert 14.5 Ballot for sensory descriptive analysis of fresh tomato flavor.

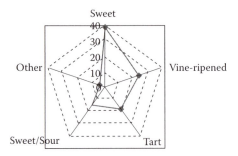

Insert 14.6 Significant supermarket segments: self-identified segment preference of sweet tomato samples. (From J. R. West, "Segmentation of Tomato Consumers by Preferences in Flavor Acceptability," M.S. thesis, University of Georgia, 2000.)

products or to trace quality changes of that product during storage. Frequently, when conducting a sensory test, food scientists will also conduct physical and chemical tests (known as physicochemical tests) to see if one of these tests accurately relates to the sensory response. Statistical correlation is generally used to provide a rough idea of what tests may be useful. Mathematical models are developed to establish predictive relationships. If a single physicochemical test or combination of tests helps predict the results of the sensory panel, the food scientist can use that mathematical model as an indicator of sensory quality. The indicator can then be used for routine testing in quality control or new product development instead of gathering together sensory panels. It is generally a good idea to test the model occasionally with sensory testing to make sure that the model is still valid.

Consumer testing

Sensory testing is divided into two types—analytical and affective. Difference, threshold, and descriptive tests are *analytical*. Preference or consumer tests are *affective*. Consumer testing is less controlled than other types of sensory testing, and consumers are less reliable as panelists than those typically used in analytical tests. The only way to find out what consumers think, however, is to ask the consumer. Two types of consumer testing involve the use of focus groups and actual large consumer tasting panels. Focus groups are used by marketing specialists to test new product concepts on consumers before a product is developed. They may also be used later in the development process. Large sensory panels are usually used late in the development process.

Focus groups usually contain eight to twelve people who have an interest in the product. They are led by a group leader who attempts to elicit a wide range of responses from the group. Focus groups present ideas that can be developed further. Through careful probing of the participants, the group leader can uncover valuable information about why certain consumers like certain food quality characteristics and why they don't like others. They can also be useful in sorting out market segments, such as mild, medium, sharp, and extra sharp cheese. Large consumer panels are usually conducted to determine preference. A series of samples are presented and the consumers are asked to pick the best sample or rank them in order from best to worst. Sometimes large consumer panels are used to determine differences between two samples as well.

In conducting a consumer test it is critical to separate typical consumers from potential consumers and nonconsumers of the product or product type. Each group may have a very different reaction to the product being tested. It is also important to look at market segments. For example, Paula is interested in GMO rice, but she fails to understand that most Asians are much more sensitive to subtle differences in rice flavor than Americans. The carotenoids that contribute to the color of the golden rice can also contribute to the flavor. Many of the flavor compounds in tomatoes are derived from carotenoids. These compounds might be below the flavor thresholds of a typical American, but they could be objectionable to Asians who eat rice as a staple.

Focus groups can be very useful in helping sensory scientists decide which product concepts to reject and which concepts to develop further. There are not enough participants in a focus group to measure preference or help detect differences. Large consumer panels can tell us what consumers like and what they don't like, but they can't tell us why. If the consumer tests are conducted without an understanding of potential market segments, the results may not be clear or useful.

Integrating sensory and consumer tests

Fundamental differences exist in traditional sensory (analytical) and consumer (affective) tests. Consumer tests tell us what consumers like and dislike, whereas sensory tests help determine if products are different from each other (difference), and how they differ and by how much (descriptive analysis). None of these tests when done separately give us a complete picture of a product. When they are conducted sequentially or simultaneously, however, we can get a much better understanding of what factors affect the acceptability of a food product, what characteristics contribute to acceptability, and what we can do to improve acceptability. A mathematical model integrating sensory and consumer tests is shown in Insert 14.7.

Insert 14.7 Mathematical model to explain tomatolike aroma in terms of chemical composition. (Adapted from K. S. Tandon, E. A. Baldwin, J. W. Scott, and R. L. Shewfelt. "Linking Sensory Descriptors to Volatile and Non-volatile Components of Fresh Tomato Flavor." *Journal of Food Science* 68 (2003): 2366.)

Tomatolike aroma = 82.3 + 1.1(Ethanol) + 2.2(Hexanal) − 1.4(*cis*-3-Hexenal) − 18.4(Methylbutanol) + 3.8(*trans*-2-Hexenal) − 43(6-Methyl-5-hepten-2-one) + 192.8(Isobutylthiazole) − 10.7(Geranylacetone) + 419.1(β-Ionone) − 47.7(Penten–3-one) + 13.3(% Soluble solids) − 106.4(Titratable acidity) + 90.2(Total sugars) − 59.8(Sucrose equivalents) ($R^2 = 0.95$)

Remember this!

- Flavor is the combination of taste and aroma.
- Flavor perception results from interactions within a complex mixture of organic and inorganic compounds.
- The color of a food product is the result of natural pigments in the food or the addition of natural or artificial colorants.
- The texture of a food is related to the sense of touch.
- Sensory panels are used to help produce measurements that are more relevant to consumer acceptability than those obtained from instrumental tests.
- Difference tests are very useful when we want to know if a product is comparable to the competition, what effect a new ingredient has on the acceptability of a current or new product, or if the quality of a product is deteriorating during storage.
- Threshold testing is performed to determine the level at which an ingredient or chemical compound can be detected or recognized.
- Sensory descriptive analysis is performed to provide a description of the components of sensory quality that can be detected.
- The only way to find out what consumers think is to ask the consumer.

Answers to chapter questions

Insert 14.2

Iso-amy l acetate—bananas
Nootkatone—grapefruit

Find the answers for the other two compounds in the text.

References

Tandon, K. S., E. A. Baldwin, J. W. Scott, and R. L. Shewfelt. 2003. Linking sensory descriptors to volatile and non-volatile components of fresh tomato flavor. *Journal of Food Science* 68: 2366.

West, J. R. 2000. Segmentation of tomato consumers by preferences in flavor accept-ability, M.S. thesis, University of Georgia.

Further reading

Krueger, R. A., and M. A. Casey. 2000. *Focus groups: A practical guide for applied research*, 3rd ed. Thousand Oaks, CA: Sage Publications, Inc.

Lawless, H. T., and H. Heymann. 1997. *Sensory evaluation of food: Principles and practices*. Gaithersburg, MD: Aspen Publishers, Inc.

MacFie, H. 2007. *Consumer-led food product development*. Cambridge, U.K.: CRC Press, Woodhead Publishing Limited.

Meilgaard, M., G. V. Civille, and B. T. Carr. 2006. *Sensory evaluation techniques*, 4th ed. Boca Raton, FL: CRC Press.

Stone, H., and J. L. Sidell. 2004. *Sensory evaluation practices*, 3rd ed. Boston: Elsevier Academic Press.

Index

A

Accelerated storage test, 298
Acesulfame K, 130
Acrylamide, 229, 232
Advertising, 27, 67–8, 76–7, 178, 208–9, 221, 301
Agroterrorism, 225, 232
Alcohol
    beverages, 11, 46, 60–1, 63–4, 86, 95, 107, 116–7, 126, 218, 296
    calories, 194, 273
    consumption, 38, 57, 281, 284
    diet recommendations, 61
    fortification, 118
    manufacture of, 118, 296, 320
    myths about, 57
    regulation of, 218
    religious significance, 64
Allergies, 6, 18–9, 23, 206, 220, 229, 232
Alternaria, 297
Amino acid, 41, 123, 125, 255–8, 262, 267–8, 271, 276–7, 279–80, 283–4, 288, 293
    commercial products, 41
    deficiencies, 281
    essential, 267–8
Anabolism, 280–1
Anaphylactic shock, 18, 229
Antibiotics, 13, 35, 94, 113, 146, 225, 239, 245
Antibodies, 294
Anticaking agents, 239, 245
Antimicrobial compounds, 40, 93, 225
Antinutrient, 100, 103, 175, 281, 283
Antioxidants, 40, 133, 226, 239, 245, 255, 267, 273, 281–4
    activity, 43, 282
    structures, 246
synthetic, 133, 245, 255

vitamins, 38, 123, 133, 249, 255, 270, 281
Anorexia, 42–4, 280–1
Appetite suppressants, 60
Arabidopsis, 285
Artificial
    antioxidant, 246
    casings, 120
    chemicals, 17–8, 21, 56, 84, 238
    flavors, 117, 122–3, 131, 248, 329
    food colors, 18, 43, 132, 165, 183, 248, 327–8, 340
    ingredients, 17
    sweeteners, 129–30, 213, 246–7
Aseptic
    packaging, 99, 319, 323
    processing, 87, 89, 98, 100, 104, 106, 146, 316
Astringency, 166, 248, 337
Atherosclerosis, 39, 265, 270
Athletes, 27, 30, 40–1, 62, 73, 256, 268

B

Bacillus, 88, 202, 297
Bacteriophage, 95, 285, 288
BATF, 218, 221
Beer, 26, 46, 66, 110, 118, 183, 186, 194, 218, 247, 273–4, 285–6, 295–6, 304
    as a diuretic, 116
Bioactivity, 123, 227, 230
Bioavailability, 17, 168, 272, 276–7, 284, 295
Biofilm, 225
Bioflavanoids, 282
Bioinformatics, 226, 230
Biological control, 17
Biological oxygen demand (BOD), 307, 322
Bioreactor, 321–2

unsafe, 19
Insects, 7, 14, 16–7, 23, 64, 97, 105, 132, 175, 177, 185, 204, 215, 217, 251, 291
Institutional Review Board, 332
Instron Universal Testing Machine, 168–9, 332
Insulators, 310–1
Integrated Pest Management, *see* IPM
Intermediate-moisture foods, 90, 101, 149, 190–2, 243
International Organization for Standardization, *see* ISO
Intoxication
  alcohol, 64, 117
  food, 7, 9, 24, 299–300, 306
  water, 46
IPM, 16
IQF, 89–90
Iron, 37, 39–40, 47–52, 58–9, 122–4, 152–3, 163, 198, 206, 218, 239, 242, 250, 272, 276, 278, 282–3
Irradiation, 87, 96–7, 100–2, 104–6, 286
ISO, 173

J

Juice, 298
  adulteration, 21
  blending, 117, 197
  cartons, 89
  clarification, 100, 117, 127
  classification, 34
  concentrated, 90, 92, 117, 127, 315
  fast, 41
  fermented, 118
  fresh, 6, 22, 94, 117, 163, 165, 197, 240
  fruit, 6, 22, 32, 39, 75, 94, 116–7, 127, 162, 166, 217, 240, 245, 275, 297, 302, 314, 327
  meat, 15
  pasteurization, 291
  powder, 51
  processing, 94–5, 116–7, 156, 320
  vegetable, 32, 94, 117, 127, 152

K

Ketosis, 30, 42, 60
Kosher laws, 64–5, 78
Kwashiorkor, 37, 281

L

Labels, 48, 118, 188, 205–6, 214, 216, 219–20
  design, 200, 204–5
  ingredients, 19, 48, 188–9, 196, 206–7, 210, 216, 307, 328
  instructions, 20, 205
  nutritional, 50, 53, 56–7, 110, 123–6, 138, 147, 162, 165, 167, 183–4, 197, 214, 238
    guidelines, 279
  reading, 31, 47–8, 55, 77, 84, 94, 109, 110–1, 117, 125, 138, 278
  regulation of, 218, 231, 266
  warnings, 12
Lactic acid, 95, 96, 107, 123, 295–7, 301
Lactose intolerant, 19, 39, 129, 146–7, 195, 252, 276, 295, 322
Lactobacillus, 295–7
Laminar flow, 315–6
LDL, 270, 277
Leavening, 64, 65, 67, 113, 119, 123
Lecithin, 122–3, 133, 245, 253, 261
Leftovers, 14–5, 70, 111, 145–6, 191
Legumes, 18, 233
Line extension, 185, 188, 197, 203, 210
Lipid oxidation, 100–1, 255, 261–2
Listeria, 4, 7, 10, 22–3, 121, 143–5, 202, 299
Loose stools, 31, 270
Low Density Lipoproteins, *see* LDL
Low-acid foods, 88, 293
Lycopene, 40, 43, 240, 246, 255, 327

M

Macronutrients, 47, 51, 60, 251, 255
Malnutrition, 37–8, 42–3, 51, 281
Marasmus, 37, 281
Market
  demand, 209
  segment, 178, 339
  test, 207–10, 326
Mass transfer, 312–4, 323
Meals, 25, 42, 61, 63, 65, 69, 83, 109, 121, 162, 164, 281
  away from home, 69, 165
  cost, 71
  home-cooked, 83–4, 161
  on Wheels, 155
  patterns, 64, 73–8, 84
  preparation, 14–5, 72, 148, 154, 165, 192
  Ready to Eat, 309